Neural Systems for Control

Neural Systems for Control

Edited by

Omid Omidvar
University of the District of Columbia

David L. Elliott
University of Maryland

ACADEMIC PRESS
San Diego   London   Boston   New York   Sydney   Tokyo   Toronto

This book is printed on acid-free paper. ∞

Copyright © 1997 by Academic Press

All rights reserved.
No part of this publication may be reproduced or transmitted in any form or by any means, electronic or mechanical, including photocopy, recording, or any information storage and retrieval system, without permission in writing from the publisher.
Chapter 2 is reprinted with permission from A. Barto, "Reinforcement Learning," *Handbook of Brain Theory and Neural Networks*, M.A. Arbib, ed., The MIT Press, Cambridge, MA, pp. 804–809, 1995.
Chapter 4, Figures 4–5 and 7–9 and Tables 2–5, are reprinted with permission, from S. Cho, "Map Formation in Proprioceptive Cortex," *International Journal of Neural Systems*, 5 (2): 87–101.

ACADEMIC PRESS
525 B Street, Suite 1900, San Diego, CA 92101-4495, USA
1300 Boylston Street, Chestnut Hill, MA 02167, USA
http://www.apnet.com

ACADEMIC PRESS LIMITED
24–28 Oval Road, London NW1 7DX, UK
http://www.hbuk.co.uk/ap/

Library of Congress Cataloging-in-Publication Data

Neural systems for control / edited by Omid Omidvar, David L. Elliott.
    p. cm.
    Includes bibliographical references and index.
    ISBN 0-12-526430-5
    1. Neural networks (Computer science)   2. Automatic control.
  I. Omidvar, Omid. II. Elliott, David L.
QA76.87.N4925 1997
629.8'9--dc21   96-29556
                                                                  CIP

Printed in the United States of America
96 97 98 99 00 EB 9 8 7 6 5 4 3 2 1

Contents

Contributors		ix
Preface		xiii
1	**Introduction: Neural Networks and Automatic Control** *David L. Elliott*	**1**
	1 Control Systems	1
	2 What is a Neural Network?	3
2	**Reinforcement Learning** *Andrew G. Barto*	**7**
	1 Introduction	7
	2 Nonassociative Reinforcement Learning	8
	3 Associative Reinforcement Learning	12
	4 Sequential Reinforcement Learning	20
	5 Conclusion	26
	6 References	27
3	**Neurocontrol in Sequence Recognition** *William J. Byrne and Shihab A. Shamma*	**31**
	1 Introduction	31
	2 HMM Source Models	32
	3 Recognition: Finding the Best Hidden Sequence	33
	4 Controlled Sequence Recognition	34
	5 A Sequential Event Dynamic Neural Network	42
	6 Neurocontrol in Sequence Recognition	49
	7 Observations and Speculations	52
	8 References	56
4	**A Learning Sensorimotor Map of Arm Movements: a Step Toward Biological Arm Control** *Sungzoon Cho, James A. Reggia and Min Jang*	**61**
	1 Introduction	61
	2 Methods	63
	3 Simulation Results	70
	4 Discussion	82
	5 References	83

5 Neuronal Modeling of the Baroreceptor Reflex with Applications in Process Modeling and Control   87
Francis J. Doyle III, Michael A. Henson, Babatunde A. Ogunnaike, James S. Schwaber, and Ilya Rybak

1	Motivation	87
2	The Baroreceptor Vagal Reflex	89
3	A Neuronal Model of the Baroreflex	93
4	Parallel Control Structures in the Baroreflex	102
5	Neural Computational Mechanisms for Process Modeling	116
6	Conclusions and Future Work	121
7	References	122

6 Identification of Nonlinear Dynamical Systems Using Neural Networks   129
A. U. Levin and K. S. Narendra

1	Introduction	129
2	Mathematical Preliminaries	131
3	State space models for identification	139
4	Identification Using Input–Output Models	142
5	Conclusion	154
6	Appendix: Proof of Lemma 1	156
7	References	158

7 Neural Network Control of Robot Arms and Nonlinear Systems   161
F. L. Lewis, S. Jagannathan, and A. Yeşildirek

1	Introduction	161
2	Background in Neural Networks, Stability, and Passivity	163
3	Dynamics of Rigid Robot Arms	167
4	NN Controller for Robot Arms	169
5	Passivity and Structure Properties of the NN	183
6	Neural Networks for Control of Nonlinear Systems	187
7	Neural Network Control with Discrete-Time Tuning	193
8	Conclusion	207
9	References	207

8 Neural Networks for Intelligent Sensors and Control — Practical Issues and Some Solutions   213
S. Joe Qin

1	Introduction	213
2	Characteristics of Process Data	215
3	Data Preprocessing	217
4	Variable Selection	220
5	Effect of Collinearity on Neural Network Training	222
6	Integrating Neural Nets with Statistical Approaches	225

	7	Application to a Refinery Process	230
	8	Conclusions and Recommendations	230
	9	References	231

9 Approximation of Time-Optimal Control for an Industrial Production Plant with General Regression Neural Network 235
Clemens Schäffner and Dierk Schröder

1	Introduction	235
2	Description of the Plant	236
3	Model of the Induction Motor Drive	238
4	General Regression Neural Network	239
5	Control Concept	242
6	Conclusion	257
7	References	257

10 Neuro-Control Design: Optimization Aspects 259
H. Ted Su and Tariq Samad

1	Introduction	259
2	Neuro-Control Systems	260
3	Optimization Aspects	273
4	PNC Design and Evolutionary Algorithm	279
5	Conclusions	281
6	References	283

11 Reconfigurable Neural Control in Precision Space Structural Platforms 289
Gary G. Yen

1	Connectionist Learning System	289
2	Reconfigurable Control	293
3	Adaptive Time-Delay Radial Basis Function Network	295
4	Eigenstructure Bidirectional Associative Memory	297
5	Fault Detection and Identification	302
6	Simulation Studies	304
7	Conclusion	309
8	References	312

12 Neural Approximations for Finite- and Infinite-Horizon Optimal Control 317
Riccardo Zoppoli and Thomas Parisini

1	Introduction	317
2	Statement of the Finite-Horizon Optimal Control Problem	320
3	Reduction of Problem 1 to a Nonlinear Programming Problem	321
4	Approximating Properties of the Neural Control Law	323

5	Solution of Problem 2 by the Gradient Method	327
6	Simulation Results	330
7	The Infinite-Horizon Optimal Control Problem and Its Receding-Horizon Approximation	335
8	Stabilizing Properties of the Receding-Horizon Regulator	337
9	Neural Approximation for the Receding-Horizon Regulator	340
10	Gradient Algorithm for Deriving the RH Neural Regulator; Simulation Results	344
11	Conclusions	348
12	References	348

Index   **353**

Contributors

- Andrew G. Barto*
  Department of Computer Science
  University of Massachusetts
  Amherst, MA 01003, USA
  E-mail: barto@cs.umass.edu

- William J. Byrne*
  Center for Language and Speech Processing, Barton Hall
  Johns Hopkins University
  Baltimore, MD 21218, USA
  E-mail: byrne@cspjhu.ece.jhu.edu

- Sungzoon Cho*
  Department of Computer Science and Engineering
  POSTECH Information Research Laboratories
  Pohang University of Science and Technology
  San 31 Hyojadong
  Pohang, Kyungbook 790-784, South Korea
  E-mail: zoon@zoon.postech.ac.kr

- Francis J. Doyle III*
  School of Chemical Engineering
  Purdue University
  West Lafayette, IN 47907-1283, USA
  E-mail: fdoyle@ecn.purdue.edu

- David L. Elliott*
  Institute for Systems Research
  University of Maryland
  College Park, MD 20742, USA
  E-mail: delliott@isr.umd.edu

- Michael A. Henson
  Department of Chemical Engineering
  Louisiana State University
  Baton Rouge, LA 70803-7303, USA
  E-mail: henson@nlc.che.lsu.edu

- S. Jagannathan
  Controls Research, Caterpillar, Inc.

Tech. Ctr. Bldg. "E", M/S 855
14009 Old Galena Rd.
Mossville, IL 61552, USA
E-mail: saranj@cat.com

- Min Jang
Department of Computer Science and Engineering
POSTECH Information Research Laboratories
Pohang University of Science and Technology
San 31 Hyojadong
Pohang, Kyungbook 790-784, South Korea
E-mail: jmin@zoon.postech.ac.kr

- Asriel U. Levin*
Wells Fargo Nikko Investment Advisors, Advanced Strategies and Research Group
45 Fremont Street
San Francisco, CA 94105, USA
E-mail: asriel.levin@bglobal.com

- Frank L. Lewis*
Automation and Robotics Research Institute
University of Texas at Arlington
7300 Jack Newell Blvd. S
Fort Worth, TX 76118, USA
E-mail: flewis@arrirs04.uta.edu

- Kumpati S. Narendra
Center for Systems Science
Department of Electrical Engineering
Yale University
New Haven, CT 06520, USA
E-mail: Narendra@koshy.eng.yale.edu

- Babatunde A. Ogunnaike
Neural Computation Program, Strategic Process Technology Group
E. I. Dupont de Nemours and Company
Wilmington, DE 19880-0101, USA
E-mail: ogunnaike@esspt0.dnet.dupont.com

- Omid M. Omidvar
Computer Science Department
University of the District of Columbia
Washington, DC 20008, USA
E-mail: oomidvar@udcvax.bitnet

- Thomas Parisini*
Department of Electrical, Electronic and Computer Engineering

DEEI–University of Trieste, Via Valerio 10, 34175 Trieste, Italy
E-mail: thomas@dist.dist.unige.it

- S. Joe Qin*
  Department of Chemical Engineering, Campus Mail Code C0400
  University of Texas
  Austin, TX 78712, USA
  E-mail: qin@che.utexas.edu

- James A. Reggia*
  Department of Computer Science, Department of Neurology, and
  Institute for Advanced Computer Studies
  University of Maryland
  College Park, MD 20742, USA
  E-mail: reggia@avion.cs.umd.edu

- Ilya Rybak
  Neural Computation Program, Strategic Process Technology Group
  E. I. Dupont de Nemours and Company
  Wilmington, DE 19880-0101, USA
  E-mail: rybaki@eplrx7.es.dupont.com

- Tariq Samad
  Honeywell Technology Center
  Honeywell Inc.
  3660 Technology Drive, MN65-2600
  Minneapolis, MN 55418, USA
  E-mail: samad@htc.honeywell.com

- Clemens Schäffner*
  Siemens AG
  Corporate Research and Development, ZFE T SN 4
  Otto–Hahn–Ring 6
  D – 81730 Munich, Germany
  E-mail: Clemens.Schaeffner@zfe.siemens.de

- Dierk Schröder
  Institute for Electrical Drives
  Technical University of Munich
  Arcisstrasse 21, D – 80333 Munich, Germany
  E-mail: eat@e-technik.tu-muenchen.de

- James A. Schwaber
  Neural Computation Program, Strategic Process Technology Group
  E. I. Dupont de Nemours and Company
  Wilmington, DE 19880-0101, USA
  E-mail: schwaber@eplrx7.es.dupont.com

- Shihab A. Shamma
  Electrical Engineering Department and the Institute for Systems Research
  University of Maryland
  College Park, MD 20742, USA
  E-mail: sas@isr.umd.edu

- H. Ted Su*
  Honeywell Technology Center
  Honeywell Inc.
  3660 Technology Drive, MN65-2600
  Minneapolis, MN 55418, USA
  E-mail: tedsu@htc.honeywell.com

- Gary G. Yen*
  USAF Phillips Laboratory, Structures and Controls Division
  3550 Aberdeen Avenue, S.E.
  Kirtland AFB, NM 87117, USA
  E-mail: yeng@plk.af.mil

- Aydin Yeşildirek
  Measurement and Control Engineering Research Center
  College of Engineering
  Idaho State University
  Pocatello, ID 83209-8060, USA
  E-mail: yesiaydi@fs.isu.edu

- Riccardo Zoppoli
  Department of Communications, Computer and System Sciences
  University of Genoa, Via Opera Pia 11A
  16145 Genova, Italy
  E-mail: rzop@dist.unige.it

* Corresponding Author

Preface

If you are acquainted with neural networks, you will find that automatic control problems provide applications — industrially useful — of your knowledge, and that they have a dynamic or evolutionary nature lacking in static pattern-recognition. Control ideas are also prevalent in the study of the natural neural networks found in animals and human beings.

If you are interested in the practice and theory of control, you will find that artificial neural networks offer a way to synthesize nonlinear controllers, filters, state observers and system identifiers using a parallel method of computation.

The purpose of this book is to acquaint those in either field with current research involving both. The book project originated with O. M. Omidvar. Chapters were obtained by an open call for papers and by invitation. The topics requested included mathematical foundations; biological control architectures; applications of neural network control methods (neurocontrol) in high technology, process control, and manufacturing; reinforcement learning; and neural network approximations to optimal control. The responses included leading edge research, exciting applications, surveys and tutorials to guide the reader who needs pointers for research or application. The authors' addresses are given in the Contributors list; their work represents both academic and industrial thinking.

This book is intended for a wide audience — those professionally involved in neural network research, such as lecturers and primary investigators in neural computing, neural modeling, neural learning, neural memory, and neurocomputers. *Neural Systems for Control* focuses on research in natural and artificial neural systems directly applicable to control or making use of modern control theory.

Each of the chapters was refereed; we are grateful to those anonymous referees for their careful work.

Omid M. Omidvar, University of the District of Columbia

David L. Elliott, University of Maryland, College Park

Neural Systems for Control

Chapter 1

Introduction: Neural Networks and Automatic Control

David L. Elliott

1 Control Systems

Through the years artificial neural networks (Frank Rosenblatt's *perceptrons*, Bernard Widrow's *adalines*, Albus' CMAC) have been invented with both biological ideas and control applications in mind, and the theories of the brain and nervous system have used ideas from control system theory (e.g. Norbert Wiener's *cybernetics*). This book attempts to show how the control system and neural network researchers of the present day are cooperating. Since members of both communities like signal flow charts, I will use a few of these schematic diagrams to introduce some basic ideas.

Figure 1 is a stereotypical control system. (The dashed lines with arrows indicate the flow of signals; Σ is a summing junction where the feedback is subtracted from the command to obtain an error signal.)

One box in the diagram is usually called the plant, or the object of control. It might be a manufactured object like the engine in your automobile, or it might be your heart–lung system. The arrow labeled *command* then might be the accelerator pedal of the car, or a chemical message from your brain to your glands when you perceive danger — in either case the command being to increase the speed of some chemical or mechanical process. The *output* is the controlled quantity. It could be the engine revolutions-per-minute, which shows on the tachometer; or it could be the blood flow

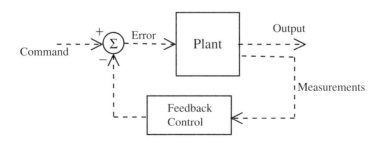

FIGURE 1. Control system.

to your tissues. The measurements of the internal state of the plant might include the output plus other engine variables (manifold pressure for instance) or physiological variables (blood pressure, heart rate, blood carbon dioxide). As the plant responds to the command, somewhere under the car's hood or in your body's neurochemistry, a local *feedback* control may use these measurements to regulate the response.

Automobile design engineers may try, perhaps using electronic fuel injection, to give you fuel economy and keep the emissions of unburnt fuel low at the same time; such a design uses modern control principles, and the automobile industry is beginning to implement these ideas with neural networks.

To be able to use mathematical or computational methods to improve the control system's response to its input command, the plant and the feedback controller are modeled mathematically by differential equations, difference equations, or, as will be seen, by a neural network with internal time lags as in Chapter 6.

Some of the models in this book are industrial rolling mills (Chapter 9), a small space robot (Chapter 12), robot arms (Chapter 7), and in Chapter 11 aerospace vehicles that must adapt or reconfigure their controls after the system has changed, perhaps from damage. Industrial control is often a matter of adjusting one or more simple controllers capable of supplying feedback proportional to error, accumulated error ("integral"), and rate of change of error ("derivative") — a so-called PID controller. Methods of replacing these familiar controllers with a neural network–based device are shown in Chapter 10.

The motivation for control system design is often to optimize a cost, such as the energy used or the time taken for a control action. Control designed for minimum cost is called *optimal control*.

The problem of approximating optimal control in a practical way can be attacked with neural network methods, as in Chapter 12; its authors, control theorists, use the new "receding-horizon" approach of Mayne and Michalska. Chapter 7 also is concerned with control optimization by neural network methods. One type of optimization (achieving a goal as fast as possible under constraints) is applied by such methods to the real industrial problem of Chapter 9.

The control systems in our bodies, such as sensory, pulmonary and circulatory systems, have evolved well enough to keep us alive and running in a dangerous world. Control aspects of the human nervous system are addressed in Chapters 3, 4, and 5. Chapter 3 is from a team using neural networks in signal processing; it shows some ways that speech processing may be simulated and sequences of phonemes recognized using *hidden Markov* methods. Chapter 4, whose authors work in neurology and computer science, uses a neural network with inputs from a model of the human arm to see how the arm's motions may map to the cerebral cortex in a computational way. Chapter 5, which was written by a team representing control engineering, chemical engineering, and human physiology, examines the workings of blood pressure control (the vagal baroreceptor reflex) and shows how to mimic this control system for chemical process applications.

2 What is a Neural Network?

The "neural networks" referred to in this book are *artificial neural networks*, a technique for using physical hardware or computer software to model computational properties analogous to some that have been postulated for real networks of nerves, such as the ability to learn and store relationships. A neural network can efficiently approximate and interpolate multivariate data that might otherwise require huge databases; such techniques are now well accepted for nonlinear statistical fitting and prediction ("ridge regression").

A commonly used artificial neuron, shown in Figure 2, is a simple structure, having just one nonlinear function of a weighted sum of several data inputs $x_1, \ldots, x_n$; this version, often called a *perceptron*, computes what statisticians call a ridge function (as in "ridge regression"),

$$y = \sigma(w_0 + \sum_{i=1}^{n} w_i x_i),$$

and for the discussion below assume that the function σ is a smooth, increasing, bounded function.

Examples of sigmoid functions (so called from their "S" shape) in common use are

$$\begin{aligned}
\sigma_1(u) &= \tanh(u), \\
\sigma_2(u) &= 1/(1+\exp(-u)), \\
\sigma_3(u) &= u/(1+|u|).
\end{aligned}$$

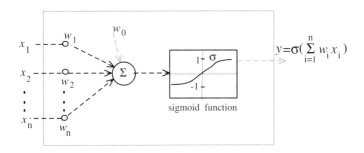

FIGURE 2. Feedforward neuron.

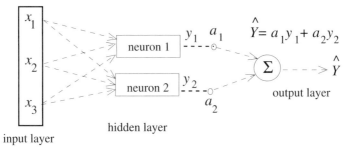

FIGURE 3. A small feedforward network.

The weight-adjustment algorithm will use the derivatives of these sigmoid functions, which are easily evaluated for the examples we have listed by using the differential equations they satisfy:

$$\begin{aligned} \sigma_1' &= 1 - (\sigma_1)^2, \\ \sigma_2' &= \sigma_2(1 - \sigma_2), \\ \sigma_3' &= (1 - |\sigma_3|)^2. \end{aligned}$$

Statisticians use many other such functions, including sinusoids. In proofs of the adequacy of neural networks to represent quite general smooth functions of many variables, the sinusoids are an important tool.

The weights w_i are to be selected or adjusted to make this ridge function approximate some function which may or may not be known in advance. The basic principles of weight adjustment were originally motivated by ideas from the psychology of learning (see Chapter 2).

In order to learn functions more complex than ridge functions, one must use networks of perceptrons. The simple example of Figure 3 shows a *feedforward perceptron network*, the kind you will find most often in the following chapters.[1] Thus the general idea of feedforward networks is that they allow us to realize functions of many variables by adjusting the network weights. Here is a typical scenario corresponding to Figure 2:

- From experiment, obtain numerical data samples of each of three different "input" variables, which we arrange as an array $X = (x_1, x_2, x_3)$, and an "output" variable Y that has a functional relation to the inputs, $Y = F(X)$.

- X is used as input to two perceptrons with adjustable weight arrays $[w_{1j}, w_{2j} : j = 1, 2, 3]$; their outputs are y_1, y_2.

- This network's single output is $\hat{Y} = a_1 y_1 + a_2 y_2$, where a_1, a_2 can

---

[1] There are several other kinds of neural network in the book, such as CMAC and radial basis function networks.

also be adjusted; the set of all the adjustable weights is

$$W = \{w_{10}, w_{11}, \cdots, w_{23}, a_1, a_2\}.$$

- The network's input–output relationship is now

$$\hat{Y} \triangleq \hat{F}(X; W) = \sum_{i=1}^{2} \left[a_i \sigma \left(w_{0i} + \sum_{j=1}^{3} w_{ij} x_j \right) \right].$$

- Systematically search for values of the numbers in W that give the best approximation for Y by minimizing a suitable cost. Often, this cost is the sum of the squared errors taken over all available inputs; that is, the weights should achieve

$$\min_{W} \sum_{X} (F(X) - \hat{F}(X; W))^2.$$

The purpose of doing this is that now we can rapidly estimate Y using the optimized network, with good interpolation properties (called *generalization* in the neural network literature). In the technique just described, *supervised training*, the functional relationship $Y = F(X)$ is available to us from many experiments, and the weights are adjusted to make the squared error (over all data) between the network's output $\hat{Y}$ and the desired output Y as small as possible. Control engineers will find this notion natural, and to some extent neural adaptation as an organism learns may resemble weight adjustment. In biology the method by which the adjustment occurs is not yet understood; but in artificial neural networks of the kind just described, and for the quadratic cost described above, one may use a convenient weight-adjustment method, based on the "chain rule" from advanced calculus, called *backpropagation*; see the Index for examples.

The kind of weight adjustment (learning) that has been discussed so far is called *supervised learning*, because at each step of adjustment, target values are available. In building model-free control systems one may also consider more general frameworks in which a control is evolved by minimizing a cost, such as the time-to-target or energy-to-target. Chapter 2 is a scholarly survey of a type of unsupervised learning known as *reinforcement learning*, a concept that originated in psychology and has been of great interest in applications to robotics, dynamic games, and the process industries. Stabilizing certain control systems, such as the robot arms and similar nonlinear systems considered in Chapter 7, can be achieved with on-line learning.

One of the most promising current applications of neural network technology is to "intelligent sensors," or "virtual instruments," as described in Chapter 8 by a chemical process control specialist; the important variables in an industrial process may not be available during the production run, but with some nonlinear statistics it may be possible to associate them with the available measurements, such as time-temperature histories. (Plasma-etching of silicon wafers is one such application.) That chapter considers practical statistical issues including the effects of missing data, outliers, and data that are highly correlated. Other techniques of intelligent control, such as fuzzy logic, can be combined with neural networks as in the reconfigurable control of Chapter 11.

If the input variables x_t are samples of a time-series, and a future value Y is to be predicted, the neural network becomes dynamic. The samples $x_1, \ldots, x_n$ can be stored in a delay-line, which serves as the input layer to a feedforward network of the type illustrated in Figure 3. (Electrical engineers know the linear version of this computational architecture as an *adaptive filter*.) Chapter 6 uses fundamental ideas of nonlinear dynamical systems and control system theory to show how dynamic neural networks can identify (replicate the behavior of) nonlinear systems. The techniques used are similar to those introduced by F. Takens in studying turbulence and chaos.

Most control applications of neural networks currently use high-speed microcomputers, often with coprocessor boards that provide single-instruction, multiple-data parallel computing well suited to the rapid functional evaluations needed to provide control action. The weight adjustment is often performed off-line, with historical data; provision for on-line adjustment or even for on-line learning, as some of the chapters describe, can permit the controller to adapt to a changing plant and environment. As cheaper and faster neural hardware develops, it becomes important for the control engineer to anticipate where it may be intelligently applied.

Acknowledgments: I am grateful to the contributors, whose addresses are listed in the preceding pages. They have been patient with the process of revision, providing LaTeX and PostScript^{TM} files where it was possible and other media when it was not; errors introduced during translation, scanning, and redrawing may be laid at my door.

The Institute for Systems Research at the University of Maryland has kindly provided an academic home during this work; employer NeuroDyne, Inc. has provided practical applications of neural networks and collaboration with experts; and my wife Pauline Tang has my thanks for her encouragement and help.

Chapter 2

Reinforcement Learning

Andrew G. Barto

ABSTRACT Reinforcement learning refers to ways of improving performance through trial-and-error experience. Despite recent progress in developing artificial learning systems, including new learning methods for artificial neural networks, most of these systems learn under the tutelage of a knowledgeable "teacher" able to tell them how to respond to a set of training stimuli. But systems restricted to learning under these conditions are not adequate when it is costly, or even impossible, to obtain the required training examples. Reinforcement learning allows autonomous systems to learn from their experiences instead of exclusively from knowledgeable teachers. Although its roots are in experimental psychology, this chapter provides an overview of modern reinforcement learning research directed toward developing capable artificial learning systems.

1 Introduction

The term *reinforcement* comes from studies of animal learning in experimental psychology, where it refers to the occurrence of an event, in the proper relation to a response, that tends to increase the probability that the response will occur again in the same situation [Kim61]. Although the specific term "reinforcement learning" is not used by psychologists, it has been widely adopted by theorists in engineering and artificial intelligence to refer to a class of learning tasks and algorithms based on this principle of reinforcement. Mendel and McLaren, for example, used the term "reinforcement learning control" in their 1970 paper describing how this principle can be applied to control problems [MM70]. The simplest reinforcement learning methods are based on the commonsense idea that if an action is followed by a satisfactory state of affairs or an improvement in the state of affairs, then the tendency to produce that action is strengthened, i.e., reinforced. This basic idea follows Thorndike's [Tho11] classic 1911 "Law of Effect":

> Of several responses made to the same situation, those which are accompanied or closely followed by satisfaction to the animal will, other things being equal, be more firmly connected with the situation, so that, when it recurs, they will be more likely to recur; those which are accompanied or closely followed

by discomfort to the animal will, other things being equal, have their connections with that situation weakened, so that, when it recurs, they will be less likely to occur. The greater the satisfaction or discomfort, the greater the strengthening or weakening of the bond.

Although this principle has generated controversy over the years, it remains influential because its general idea is supported by many experiments and it makes such good intuitive sense.

Reinforcement learning is usually formulated mathematically as an optimization problem with the objective of finding an action, or a strategy for producing actions, that is optimal in some well-defined way. Although in practice it is more important that a reinforcement learning system continue to improve than that it actually achieve optimal behavior, optimality objectives provide a useful categorization of reinforcement learning into three basic types, in order of increasing complexity: *nonassociative*, *associative*, and *sequential*. Nonassociative reinforcement learning involves determining which of a set of actions is best in bringing about a satisfactory state of affairs. In associative reinforcement learning, different actions are best in different situations. The objective is to form an optimal *associative mapping* between a set of stimuli and the actions having the best immediate consequences when executed in the situations signaled by those stimuli. Thorndike's Law of Effect refers to this kind of reinforcement learning. Sequential reinforcement learning retains the objective of forming an optimal associative mapping but is concerned with more complex problems in which the relevant consequences of an action are not available immediately after the action is taken. In these cases, the associative mapping represents a strategy, or policy, for acting over time. All of these types of reinforcement learning differ from the more commonly studied paradigm of supervised learning, or "learning with a teacher," in significant ways that I discuss in the course of this chapter.

This chapter is organized into three main sections, each addressing one of these three categories of reinforcement learning. For more detailed treatments, the reader should consult references [Bar92, BBS95, Sut92, Wer92, Kae96].

2   Nonassociative Reinforcement Learning

Figure 1 shows the basic components of a nonassociative reinforcement learning problem. The learning system's actions influence the behavior of some process, which might also be influenced by random or unknown factors (labeled "disturbances" in Figure 1). A *critic* sends the learning system a *reinforcement signal* whose value at any time is a measure of the "goodness" of the current process behavior. Using this information, the learning

2. Reinforcement Learning

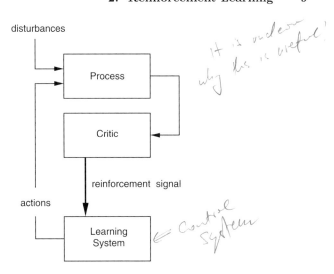

FIGURE 1. Nonassociative reinforcement learning. The learning system's actions influence the behavior of a process, which might also be influenced by random or unknown "disturbances." The critic evaluates the actions' immediate consequences on the process and sends the learning system a reinforcement signal.

system updates its action-generation rule, generates another action, and the process repeats.

An example of this type of problem has been extensively studied by theorists studying *learning automata* [NT89]. Suppose the learning system has m actions $a_1, a_2, \ldots, a_m$ and that the reinforcement signal simply indicates "success" or "failure." Further, assume that the influence of the learning system's actions on the reinforcement signal can be modeled as a collection of success probabilities $d_1, d_2, \ldots, d_m$, where d_i is the probability of success given that the learning system has generated a_i (so that $1 - d_i$ is the probability that the critic signals failure). Each d_i can be any number between 0 and 1 (the d_i's do not have to sum to one), and the learning system has no initial knowledge of these values. The learning system's objective is to asymptotically maximize the probability of receiving "success," which is accomplished when it always performs the action a_j such that $d_j = \max\{d_i | i = 1, \ldots, m\}$. There are many variants of this task, some of which are better known as *m-armed bandit* problems [BF85].

One class of learning systems for this problem consists of *stochastic learning automata* [NT89]. Suppose that on each trial, or time step t, the learning system selects an action $a(t)$ from its set of m actions according to a probability vector $(p_1(t), \ldots, p_n(t))$, where $p_i(t) = Pr\{a(t) = a_i\}$. A stochastic learning automaton implements a commonsense notion of reinforcement learning: if action a_i is chosen on trial t and the critic's feedback is "success," then $p_i(t)$ is increased and the probabilities of the other actions are decreased; whereas if the critic indicates "failure," then $p_i(t)$ is decreased and the probabilities of the other actions are appropriately adjusted. Many methods that have been studied are similar to the following *linear reward–penalty* (L_{R-P}) method:

If $a(t) = a_i$ and the critic says "success," then

$$p_i(t+1) = p_i(t) + \alpha(1 - p_i(t)),$$
$$p_j(t+1) = (1 - \alpha)p_j(t), \quad j \neq i.$$

If $a(t) = a_i$ and the critic says "failure," then

$$p_i(t+1) = (1 - \beta)p_i(t),$$
$$p_j(t+1) = \frac{\beta}{m-1} + (1 - \beta)p_j(t), \quad j \neq i,$$

where $0 < \alpha < 1$, $0 \leq \beta < 1$.

The performance of a stochastic learning automaton is measured in terms of how the critic's signal tends to change over trials. The probability that the critic signals success on trial t is $M(t) = \sum_{i=1}^{m} p_i(t) d_i$. An algorithm is *optimal* if for all sets of success probabilities $\{d_i\}$,

$$\lim_{t \to \infty} E[M(t)] = d_j,$$

where $d_j = \max\{d_i | i = 1, \ldots, m\}$ and E is the expectation over all possible sequences of trials. An algorithm is said to be ϵ-*optimal* if for all sets of success probabilities and any $\epsilon > 0$ there exist algorithm parameters such that

$$\lim_{t \to \infty} E[M(t)] = d_j - \epsilon.$$

Although no stochastic learning automaton algorithm has been proved to be optimal, the L_{R-P} algorithm given above with $\beta = 0$ is ϵ-*optimal*, where α has to decrease as ϵ decreases. Additional results exist about the behavior of groups of stochastic learning automata forming *teams* (a single critic broadcasts its signal to all the team members) or playing *games* (there is a different critic for each automaton) [NT89].

Following are key observations about nonassociative reinforcement learning:

1. *Uncertainty* plays a key role in nonassociative reinforcement learning, as it does in reinforcement learning in general. For example, if the critic in the example above evaluated actions deterministically (i.e., $d_i = 1$ or 0 for each i), then the problem would be a much simpler optimization problem.

2. The critic is an abstract model of any process that evaluates the learning system's actions. The critic does not need to have direct access

to the actions or have any knowledge about the interior workings of
the process influenced by those actions. In motor control, for example, judging the success of a reach or a grasp does not require access
to the actions of all the internal components of the motor control
system.

3. The reinforcement signal can be any signal evaluating the learning
   system's actions, and not just the success/failure signal described
   above. Often it takes on real values, and the objective of learning is
   to maximize its expected value. Moreover, the critic can use a variety of criteria in evaluating actions, which it can combine in various
   ways to form the reinforcement signal. Any value taken on by the
   reinforcement signal is often simply called a *reinforcement* (although
   this is at variance with traditional use of the term in psychology).

4. The critic's signal does not directly tell the learning system what action is best; it only evaluates the action taken. The critic also does not
   directly tell the learning system how to change its actions. These are
   key features distinguishing reinforcement learning from supervised
   learning, and we discuss them further below. Although the critic's
   signal is less informative than a training signal in supervised learning, reinforcement learning is not the same as the learning paradigm
   called *unsupervised learning* because unlike that form of learning, it
   is guided by external feedback.

5. Reinforcement learning algorithms are *selectional processes*. There
   must be *variety* in the action–generation process so that the consequences of alternative actions can be compared to select the best.
   Behavioral variety is called *exploration*; it is often generated through
   randomness (as in stochastic learning automata), but it need not be.
   Because it involves selection, nonassociative reinforcement learning is
   similar to natural selection in evolution. In fact, reinforcement learning in general has much in common with genetic approaches to search
   and problem solving [Gol89, Hol75].

6. Due to this selectional aspect, reinforcement learning is traditionally
   described as learning through "trial and error." However, one must
   take care to distinguish this meaning of "error" from the type of
   error signal used in supervised learning. The latter, usually a vector, tells the learning system the direction in which it should change
   each of its action components. A reinforcement signal is less informative. It would be better to describe reinforcement learning as learning
   through "trial and evaluation."

7. Nonassociative reinforcement learning is the simplest form of learning that involves the conflict between *exploitation* and *exploration*.
   In deciding which action to take, the learning system has to balance

two conflicting objectives: it has to use what it has already learned to obtain success (or, more generally, to obtain high evaluations), and it has to behave in new ways to learn more. The first is the need to *exploit* current knowledge; the second is the need to *explore* to acquire more knowledge. Because these needs ordinarily conflict, reinforcement learning systems have to somehow balance them. In control engineering, this is known as the conflict between control and identification. This conflict is absent from supervised and unsupervised learning, unless the learning system is also engaged in influencing which training examples it sees.

3 Associative Reinforcement Learning

Because its only input is the reinforcement signal, the learning system in Figure 1 cannot discriminate between different situations, such as different states of the process influenced by its actions. In an associative reinforcement learning problem, in contrast, the learning system receives stimulus patterns as input in addition to the reinforcement signal (Figure 2). The optimal action on any trial depends on the stimulus pattern present on that trial. To give a specific example, consider this generalization of the non-associative task described above. Suppose that on trial t the learning system

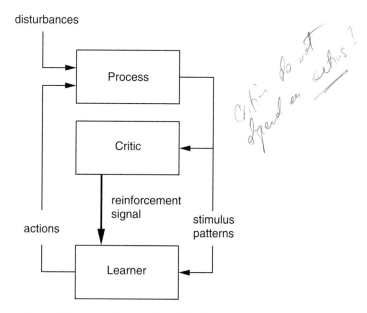

FIGURE 2. Associative reinforcement learning. The learning system receives stimulus patterns in addition to a reinforcement signal. Different actions can be optimal depending on the stimulus patterns.

senses stimulus pattern $x(t)$ and selects an action $a(t) = a_i$ through a process that can depend on $x(t)$. After this action is executed, the critic signals success with probability $d_i(x(t))$ and failure with probability $1 - d_i(x(t))$. The objective of learning is to maximize success probability, achieved when on each trial t the learning system executes the action $a(t) = a_j$, where a_j is the action such that $d_j(x(t)) = \max\{d_i(x(t)) | i = 1, \ldots, m\}$.

The learning system's objective is thus to learn an optimal associative mapping from stimulus patterns to actions. Unlike supervised learning, examples of optimal actions are not provided during training; they have to be *discovered* through exploration by the learning system. Learning tasks like this are related to instrumental, or cued operant, tasks studied by animal learning theorists, and the stimulus patterns correspond to discriminative stimuli.

Several associative reinforcement learning rules for neuron-like units have been studied. Figure 3 shows a neuron-like unit receiving a stimulus pattern as input in addition to the critic's reinforcement signal. Let $x(t)$, $w(t)$, $a(t)$, and $r(t)$ respectively denote the stimulus vector, weight vector, action, and the resultant value of the reinforcement signal for trial t. Let $s(t)$ denote the weighted sum of the stimulus components at trial t:

$$s(t) = \sum_{i=1}^{n} w_i(t) x_i(t),$$

where $w_i(t)$ and $x_i(t)$ are respectively the ith components of the weight and stimulus vectors.

Associative Search Unit — One simple associative reinforcement learning rule is an extension of the Hebbian correlation learning rule. This

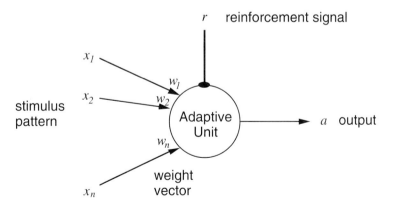

FIGURE 3. A neuron-like adaptive unit. Input pathways labeled x_1 through x_n carry nonreinforcing input signals, each of which has an associated weight w_i, $1 \leq i \leq n$; the pathway labeled r is a specialized input for delivering reinforcement; the unit's output pathway is labeled a.

rule was called the *associative search rule* by Barto, Sutton, and Brouwer [BSB81, BS81, BAS82] and was motivated by Klopf's [Klo72, Klo82] theory of the self-interested neuron. To exhibit variety in its behavior, the unit's output is a random variable depending on the activation level. One way to do this is as follows:

$$a(t) = \begin{cases} 1 & \text{with probability } p(t), \\ 0 & \text{with probability } 1 - p(t), \end{cases} \quad (1)$$

[handwritten: $p(t) = F(s(t))$]

where $p(t)$, which must be between 0 and 1, is an increasing function (such as the logistic function) of $s(t)$. Thus, as the weighted sum increases (decreases), the unit becomes more (less) likely to fire (i.e., to produce an output of 1). The weights are updated according to the following rule:

$$\Delta w(t) = \eta \, r(t) a(t) x(t),$$

where $r(t)$ is $+1$ (success) or -1 (failure).

This is just the Hebbian correlation rule with the reinforcement signal acting as an additional modulatory factor. It is understood that $r(t)$ is the critic's evaluation of the action $a(t)$. In a more real-time version of the learning rule, there must necessarily be a time delay between an action and the resulting reinforcement. In this case, if the critic takes time τ to evaluate an action, the rule appears as follows, with t now acting as a time index instead of a trial number:

$$\Delta w(t) = \eta \, r(t) a(t - \tau) x(t - \tau), \quad (2)$$

where $\eta > 0$ is the learning rate parameter. Thus, if the unit fires in the presence of an input x, possibly just by chance, and this is followed by "success," the weights change so that the unit will be more likely to fire in the presence of x, and inputs similar to x, in the future. A failure signal makes it less likely to fire under these conditions. This rule, which implements the Law of Effect at the neuronal level, makes clear the three factors minimally required for associative reinforcement learning: a stimulus signal, x; the action produced in its presence, a; and the consequent evaluation, r.

Selective Bootstrap and Associative Reward–Penalty Units — Widrow, Gupta, and Maitra [WGM73] extended the Widrow/Hoff, or LMS, learning rule [WS85] so that it could be used in associative reinforcement learning problems. Since the LMS rule is a well-known rule for supervised learning, its extension to reinforcement learning helps illuminate one of the differences between supervised learning and associative reinforcement learning, which Widrow et al. [WGM73] called "learning with a critic." They called their extension of LMS the *selective bootstrap* rule. Unlike the associative search unit described above, a selective bootstrap unit's output is the usual deterministic threshold of the weighted sum:

$$a(t) = \begin{cases} 1 & \text{if } s(t) > 0, \\ 0 & \text{otherwise.} \end{cases}$$

In supervised learning, an LMS unit receives a training signal, $z(t)$, that directly specifies the desired action at trial t and updates its weights as follows:

$$\Delta w(t) = \eta[z(t) - s(t)]x(t). \tag{3}$$

In contrast, a selective bootstrap unit receives a reinforcement signal, $r(t)$, and updates its weights according to this rule:

$$\Delta w(t) = \begin{cases} \eta[a(t) - s(t)]x(t) & \text{if } r(t) = \text{``success''} \\ \eta[1 - a(t) - s(t)]x(t) & \text{if } r(t) = \text{``failure,''} \end{cases}$$

where it is understood that $r(t)$ evaluates $a(t)$. Thus, if $a(t)$ produces "success," the LMS rule is applied with $a(t)$ playing the role of the desired action. Widrow et al. [WGM73] called this "positive bootstrap adaptation": weights are updated as if the output actually produced was in fact the desired action. On the other hand, if $a(t)$ leads to "failure," the desired action is $1 - a(t)$, i.e., the action that was *not* produced. This is "negative bootstrap adaptation." The reinforcement signal switches the unit between positive and negative bootstrap adaptation, motivating the term "selective bootstrap adaptation." Widrow et al. [WGM73] showed how this unit was capable of learning a strategy for playing blackjack, where wins were successes and losses were failures. However, the learning ability of this unit is limited because it lacks variety in its behavior.

A closely related unit is the *associative reward–penalty* (A_{R-P}) unit of Barto and Anandan [BA85]. It differs from the selective bootstrap algorithm in two ways. First, the unit's output is a random variable like that of the associative search unit (Equation 1). Second, its weight–update rule is an *asymmetric* version of the selective bootstrap rule:

$$\Delta w(t) = \begin{cases} \eta[a(t) - s(t)]x(t) & \text{if } r(t) = \text{``success''} \\ \lambda\eta[1 - a(t) - s(t)]x(t) & \text{if } r(t) = \text{``failure,''} \end{cases}$$

where $0 \leq \lambda \leq 1$ and $\eta > 0$. This is a special case of a class of A_{R-P} rules for which Barto and Anandan [BA85] proved a convergence theorem giving conditions under which it asymptotically maximizes the probability of success in associative reinforcement learning tasks like those described above. The rule's asymmetry is important because its asymptotic performance improves as λ approaches zero.

One can see from the selective bootstrap and A_{R-P} units that a reinforcement signal is less informative than a signal specifying a desired action. It is also less informative than the error $z(t) - a(t)$ used by the LMS rule. Because this error is a signed quantity, it tells the unit *how*, i.e., in what direction, it should change its action. A reinforcement signal — by itself — does not convey this information. If the learner has only two actions, as in a selective bootstrap unit, it is easy to deduce, or at least estimate, the desired action from the reinforcement signal and the actual action. However, if there are more than two actions, the situation is more difficult because the reinforcement signal does not provide information about actions that were not taken.

Stochastic Real-Valued Unit — One approach to associative reinforcement learning when there are more than two actions is illustrated by the *stochastic real-valued* (SRV) unit of Gullapalli [Gul90]. On any trial t, an SRV unit's output is a real number, $a(t)$, produced by applying a function f, such as the logistic function, to the weighted sum, $s(t)$, plus a random number $\texttt{noise}(t)$:

$$a(t) = f[s(t) + \texttt{noise}(t)].$$

The random number $\texttt{noise}(t)$ is selected according to a mean-zero Gaussian distribution with standard deviation $\sigma(t)$. Thus, $f[s(t)]$ gives the *expected* output on trial t, and the actual output varies about this value, with $\sigma(t)$ determining the amount of exploration the unit exhibits on trial t.

Before describing how the SRV unit determines $\sigma(t)$, we describe how it updates the weight vector $w(t)$. The weight-update rule requires an estimate of the amount of reinforcement expected for acting in the presence of stimulus $x(t)$. This is provided by a supervised-learning process that uses the LMS rule to adjust another weight vector, v, used to determine the reinforcement estimate $\hat{r}$:

$$\hat{r}(t) = \sum_{i=1}^{m} v_i(t) x_i(t),$$

with

$$\Delta v(t) = \eta [r(t) - \hat{r}(t)] x(t).$$

Given this $\hat{r}(t)$, $w(t)$ is updated as follows:

$$\Delta w(t) = \eta [r(t) - \hat{r}(t)] \left[\frac{\texttt{noise}(t)}{\sigma(t)} \right] x(t),$$

where $\eta > 0$ is a learning-rate parameter. Thus, if $\texttt{noise}(t)$ is positive, meaning that the unit's output is larger than expected, and the unit receives more than the expected reinforcement, the weights change to increase the expected output in the presence of $x(t)$; if it receives less than the expected reinforcement, the weights change to decrease the expected output. The reverse happens if $\texttt{noise}(t)$ is negative. Dividing by $\sigma(t)$ normalizes the weight change. Changing σ during learning changes the amount of exploratory behavior the unit exhibits.

Gullapalli [Gul90] suggests computing $\sigma(t)$ as a monotonically decreasing function of $\hat{r}(t)$. This implies that the amount of exploration for any stimulus vector decreases as the amount of reinforcement expected for acting in the presence of that stimulus vector increases. As learning proceeds, the SRV unit tends to act with increasing determinism in the presence of stimulus vectors for which it has learned to achieve large reinforcement signals. This is somewhat like simulated annealing [KGV83] except that it is stimulus-dependent and is controlled by the progress of learning. SRV units have been used as output units of reinforcement learning networks in a number of applications (e.g., references [GGB92, GBG94]).

Weight Perturbation — For the units described above (except the selective bootstrap unit), behavioral variability is achieved by including random variation in the unit's output. Another approach is to randomly vary the weights. Following Alspector et al. [AMY$^+$93], let δw be a vector of small perturbations, one for each weight, that are independently selected from some probability distribution. Letting J denote the function evaluating the system's behavior, the weights are updated as follows:

$$\Delta w = -\eta \left[\frac{J(w + \delta w) - J(w)}{\delta w} \right], \tag{4}$$

where $\eta > 0$ is a learning-rate parameter. This is a gradient descent learning rule that changes weights according to an estimate of the gradient of $\mathcal{E}$ with respect to the weights. Alspector et al. [AMY$^+$93] say that the method *measures* the gradient instead of *calculating* it as the LMS and error backpropagation [RHW86] algorithms do. This approach has been proposed by several researchers for updating the weights of a unit, or of a network, during supervised learning, where J gives the error over the training examples. However, J can be any function evaluating the unit's behavior, including a reinforcement function (in which case, the sign of the learning rule would be changed to make it a gradient *ascent* rule).

Another weight perturbation method for neuron-like units is provided by Unnikrishnan and Venugopal's [KPU94] use of the *Alopex* algorithm, originally proposed by Harth and Tzanakou [HT74], for adjusting a unit's (or a network's) weights. A somewhat simplified version of the weight-update rule is the following:

$$\Delta w(t) = \eta d(t), \tag{5}$$

where η is the learning-rate parameter and $d(t)$ is a vector whose components, $d_i(t)$, are equal to either $+1$ or -1. After the first two iterations, in which they are assigned randomly, successive values are determined by

$$d_i(t) = \begin{cases} d_i(t-1) & \text{with probability } p(t), \\ -d_i(t-1) & \text{with probability } 1 - p(t). \end{cases}$$

Thus, $p(t)$ is the probability that the direction of the change in weight w_i from iteration t to iteration $t+1$ will be the same as the direction it changed from iteration $t-2$ to $t-1$, whereas $1 - p(t)$ is the probability that the weight will move in the opposite direction. The probability $p(t)$ is a function of the change in the value of the objective function from iteration $t-1$ to t; specifically, $p(t)$ is a positive increasing function of $J(t) - J(t-1)$, where $J(t)$ and $J(t-1)$ are respectively the values of the function evaluating the behavior of the unit at iterations t and $t-1$. Consequently, if the unit's behavior has moved uphill by a large amount, as measured by J, from iteration $t-1$ to iteration t, then $p(t)$ will be large, so that the probability of the next step in weight space being in the same direction as the preceding step will be high. On the other hand, if the unit's behavior moved downhill, then the probability will be high that some of the weights will move in the opposite direction, i.e., that the step in weight space will be in some new direction.

Although weight perturbation methods are of interest as alternatives to error backpropagation for adjusting network weights in supervised learning problems, they utilize reinforcement learning principles by estimating performance through active exploration, in this case achieved by adding random perturbations to the weights. In contrast, the other methods described above — at least to a first approximation — use active exploration to estimate the gradient of the reinforcement function with respect to a unit's *output* instead of its weights. The gradient with respect to the weights can then be estimated by differentiating the known function by which the weights influence the unit's output. Both approaches — weight perturbation and unit-output perturbation — lead to learning methods for networks to which we now turn our attention.

Reinforcement Learning Networks — The neuron-like units described above can be readily used to form networks. The weight perturbation approach carries over directly to networks by simply letting w in Equations 4 and 5 be the vector consisting of all the network's weights. A number of researchers have achieved success using this approach in supervised learning problems. In these cases, one can think of each weight as facing a reinforcement learning task (which is in fact nonassociative), even though the network as a whole faces a supervised learning task. A significant advantage of this approach is that it applies to networks with arbitrary connection patterns, not just to feedforward networks.

Networks of A_{R-P} units have been used successfully in both supervised and associative reinforcement learning tasks ([Bar85, BJ87]), although only with feedforward connection patterns. For supervised learning, the output units learn just as they do in error backpropagation, but the hidden units learn according to the A_{R-P} rule. The reinforcement signal, which is defined to increase as the output error decreases, is simply *broadcast* to all the hidden units, which learn simultaneously. If the network as a whole faces an associative reinforcement learning task, all the units are A_{R-P} units, to which the reinforcement signal is uniformly broadcast (Figure 4). The units exhibit a kind of *statistical cooperation* in trying to increase their common reinforcement signal (or the probability of success if it is a success/failure

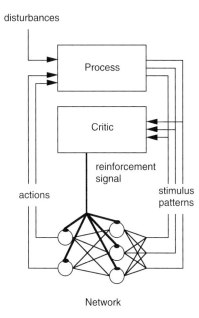

FIGURE 4. A network of associative reinforcement units. The reinforcement signal is broadcast to all the units.

signal) [Bar85]. Networks of associative search units and SRV units can be similarly trained, but these units do not perform well as hidden units in multilayer networks.

Methods for updating network weights fall within a spectrum of possibilities ranging from weight perturbation methods that do not take advantage of *any* of a network's structure to algorithms like error backpropagation, which take full advantage of network structure to compute gradients. Unit-output perturbation methods fall between these extremes by taking advantage of the structure of individual units but not of the network as a whole. Computational studies provide ample evidence that all of these methods can be effective, and each method has its own advantages, with perturbation methods usually sacrificing learning speed for generality and ease of implementation. Perturbation methods are also of interest due to their relative biological plausibility compared to error backpropagation.

Another way to use reinforcement learning units in networks is to use them only as output units, with hidden units being trained via error backpropagation. Weight changes of the output units determine the quantities that are backpropagated. This approach allows the function approximation success of the error backpropagation algorithm to be enlisted in associative reinforcement learning tasks (e.g. reference [GGB92]).

The error backpropagation algorithm can be used in another way in associative reinforcement learning problems. It is possible to train a multi-

layer network to form a model of the process by which the critic evaluates
actions. The network's input consists of the stimulus pattern $x(t)$ as well
as the current action vector $a(t)$, which is generated by another component
of the system. The desired output is the critic's reinforcement signal, and
training is accomplished by backpropagating the error

$$r(t) - \hat{r}(t),$$

where $\hat{r}(t)$ is the network's output at time t. After this model is trained
sufficiently, it is possible to estimate the gradient of the reinforcement signal with respect to each component of the action vector by analytically
differentiating the model's output with respect to its action inputs (which
can be done efficiently by backpropagation). This gradient estimate is then
used to update the parameters of the action-generation component. Jordan
and Jacobs [JJ90] illustrate this approach. Note that the exploration required in reinforcement learning is conducted in the model-learning phase
of this approach instead in the action-learning phase.

It should be clear from this discussion of reinforcement learning networks
that there are many different approaches to solving reinforcement learning problems. Furthermore, although reinforcement learning *tasks* can be
clearly distinguished from supervised and unsupervised learning tasks, it
is more difficult to precisely define a class of reinforcement learning *algorithms*.

4 Sequential Reinforcement Learning

Sequential reinforcement requires improving the long-term consequences of
an action, or of a strategy for performing actions, in addition to short-term
consequences. In these problems, it can make sense to forgo short-term
performance in order to achieve better performance over the long term.
Tasks having these properties are examples of *optimal control problems*,
sometimes called *sequential decision problems* when formulated in discrete
time.

Figure 2, which shows the components of an associative reinforcement
learning system, also applies to sequential reinforcement learning, where
the box labeled "process" is a system being controlled. A sequential reinforcement learning system tries to influence the behavior of the process
in order to maximize a measure of the total amount of reinforcement that
will be received over time. In the simplest case, this measure is the sum of
the future reinforcement values, and the objective is to learn an associative
mapping that at time step t selects, as a function of the stimulus pattern
$x(t)$, an action $a(t)$ that maximizes

$$\sum_{k=0}^{\infty} r(t+k),$$

where $r(t+k)$ is the reinforcement signal at step $t+k$. Such an associative mapping is called a *policy*.

Because this sum might be infinite in some problems, and because the learning system usually has control only over its expected value, researchers often consider the following *discounted sum* instead:

$$E\{r(t) + \gamma r(t+1) + \gamma^2 r(t+2) + \cdots\} = E\{\sum_{k=0}^{\infty} \gamma^k r(t+k)\}, \qquad (6)$$

where E is the expectation over all possible future behavior patterns of the process. The discount factor determines the present value of future reinforcement: a reinforcement value received k time steps in the future is worth γ^k times what it would be worth if it were received now. If $0 \leq \gamma < 1$, this infinite discounted sum is finite as long as the reinforcement values are bounded. If $\gamma = 0$, the robot is "myopic" in being only concerned with maximizing immediate reinforcement; this is the associative reinforcement learning problem discussed above. As γ approaches one, the objective explicitly takes future reinforcement into account: the robot becomes more farsighted.

An important special case of this problem occurs when there is no immediate reinforcement until a goal state is reached. This is a *delayed reward* problem in which the learning system has to learn how to make the process enter a goal state. Sometimes the objective is to make it enter a goal state as quickly as possible. A key difficulty in these problems has been called the *temporal credit-assignment problem*: When a goal state is finally reached, which of the decisions made earlier deserve credit for the resulting reinforcement? A widely studied approach to this problem is to learn an *internal evaluation function* that is more informative than the evaluation function implemented by the external critic. An *adaptive critic* is a system that learns such an internal evaluation function.

Samuel's Checker Player — Samuel's [Sam59] checkers playing program has been a major influence on adaptive critic methods. The checkers player selects moves by using an evaluation function to compare the board configurations expected to result from various moves. The evaluation function assigns a score to each board configuration, and the system make the move expected to lead to the configuration with the highest score. Samuel used a method to improve the evaluation function through a process that compared the score of the current board position with the score of a board position likely to arise later in the game:

> We are attempting to make the score, calculated for the current board position, look like that calculated for the terminal board position of the chain of moves which most probably occur during actual play [Sam59].

As a result of this process of "backing up" board evaluations, the evaluation function should improve in its ability to evaluate long-term consequences of moves. In one version of Samuel's system, the evaluation function was represented as a weighted sum of numerical features, and the weights were adjusted based on an error derived by comparing evaluations of current and predicted board positions.

If the evaluation function can be made to score each board configuration according to its true promise of eventually leading to a win, then the best strategy for playing is to myopically select each move so that the next board configuration is the most highly scored. If the evaluation function is optimal in this sense, then it already takes into account all the possible future courses of play. Methods such as Samuel's that attempt to adjust the evaluation function toward this ideal optimal evaluation function are of great utility.

Adaptive Critic Unit and Temporal Difference Methods — An adaptive critic unit is a neuron-like unit that implements a method similar to Samuel's. The unit is as in Figure 3 except that its output at time step t is $P(t) = \sum_{i=1}^{n} w_i(t) x_i(t)$, so denoted because it is a prediction of the discounted sum of future reinforcement given in Equation 6. The adaptive critic learning rule rests on noting that correct predictions must satisfy a consistency condition, which is a special case of the Bellman optimality equation, relating predictions at adjacent time steps. Suppose that the predictions at any two successive time steps, say steps t and $t+1$, are correct. This means that

$$P(t) = E\{r(t) + \gamma r(t+1) + \gamma^2 r(t+2) + \cdots\},$$
$$P(t+1) = E\{r(t+1) + \gamma r(t+2) + \gamma^2 r(t+3) + \cdots\}.$$

Now notice that we can rewrite $P(t)$ as follows:

$$P(t) = E\{r(t) + \gamma[r(t+1) + \gamma r(t+2) + \cdots]\}.$$

But this is exactly the same as

$$P(t) = E\{r(t)\} + \gamma P(t+1).$$

An estimate of the error by which any two adjacent predictions fail to satisfy this consistency condition is called the *temporal difference (TD) error* [Sut88]:

$$r(t) + \gamma P(t+1) - P(t), \qquad (7)$$

where $r(t)$ is used as an unbiased estimate of $E\{r(t)\}$. The term *temporal difference* comes from the fact that this error essentially depends on the difference between the critic's predictions at successive time steps.

The adaptive critic unit adjusts its weights according to the following learning rule:

$$\Delta w(t) = \eta[r(t) + \gamma P(t+1) - P(t)] x(t). \qquad (8)$$

A subtlety here is that $P(t+1)$ should be computed using the weight vector $w(t)$, not $w(t+1)$. This rule changes the weights to decrease the magnitude of the TD error. Note that if $\gamma = 0$, Equation 8 is equivalent to the LMS learning rule (Equation 3). In analogy with the LMS rule, we can think of $r(t) + \gamma P(t+1)$ as the prediction target: it is the quantity that each $P(t)$ should match. The adaptive critic is therefore trying to predict the next reinforcement, $r(t)$, *plus its own next prediction* (discounted), $\gamma P(t+1)$. The adaptive critic is similar to Samuel's learning method in adjusting weights to make current predictions closer to later predictions.

Although this method is very simple computationally, it actually converges to the correct predictions of the discounted sum of future reinforcement if these correct predictions can be computed by a linear unit. This is shown by Sutton [Sut88], who discusses a more general class of methods, called *TD methods*, that include Equation 8 as a special case. It is also possible to learn nonlinear predictions using, for example, multilayer networks trained by back propagating the TD error. Using this approach, Tesauro [Tes92] produced a system that learned how to play expert-level backgammon.

Actor–Critic Architectures — In an actor–critic architecture, the predictions formed by an adaptive critic act as reinforcement for an associative reinforcement learning component, called the *actor* (Figure 5). To distinguish the adaptive critic's signal from the reinforcement signal sup-

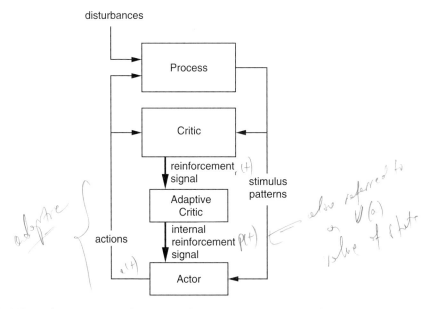

FIGURE 5. Actor–critic architecture. An adaptive critic provides an internal reinforcement signal to an *actor*, which learns a policy for controlling the process.

plied by the original, nonadaptive critic, we call it the *internal reinforcement signal*. The actor tries to maximize the *immediate* internal reinforcement signal, while the adaptive critic tries to predict total future reinforcement. To the extent that the adaptive critic's predictions of total future reinforcement are correct given the actor's current policy, the actor actually learns to increase the total amount of future reinforcement (as measured, for example, by Equation 6).

Barto, Sutton, and Anderson [BSA83] used this architecture for learning to balance a simulated pole mounted on a cart. The actor had two actions: application of a force of a fixed magnitude to the cart in the plus or minus direction. The nonadaptive critic only provided a signal of failure when the pole fell past a certain angle or the cart hit the end of the track. The stimulus patterns were vectors representing the state of the cart–pole system. The actor was an associative search unit as described above except that it used an *eligibility trace* [Klo82] in its weight-update rule:

$$\Delta w(t) = \eta\, \hat{r}(t) a(t) \bar{x}(t),$$

where $\hat{r}(t)$ is the internal reinforcement signal and $\bar{x}(t)$ is an exponentially-decaying trace of past input patterns. When a component of this trace is nonzero, the corresponding synapse is *eligible* for modification. This technique is used instead of the delayed stimulus pattern in Equation 2 to improve the rate of learning. It is assumed that $\hat{r}(t)$ evaluates the action $a(t)$. The internal reinforcement is the TD error used by the adaptive critic:

$$\hat{r}(t) = r(t) + \gamma P(t+1) - P(t).$$

This makes the original reinforcement signal, $r(t)$, available to the actor, as well as changes in the adaptive critic's predictions of future reinforcement, $\gamma P(t+1) - P(t)$.

Action-Dependent Adaptive Critics—Another approach to sequential reinforcement learning combines the actor and adaptive critic into a single component that learns separate predictions for each action. At each time step the action with the largest prediction is selected, except for a random exploration factor that causes other actions to be selected occasionally. An algorithm for learning action-dependent predictions of future reinforcement, called the *Q-learning* algorithm, was proposed by Watkins in 1989, who proved that it converges to the correct predictions under certain conditions [WD92]. The term *action-dependent adaptive critic* was first used by Lukes, Thompson, and Werbos [LTW90], who presented a similar idea. A little-known forerunner of this approach was presented by Bozinovski [Boz82].

For each pair (x, a) consisting of a process state, x, and and a possible action, a, let $Q(x, a)$ denote the total amount of reinforcement that will be produced over the future if action a is executed when the process is in state x and optimal actions are selected thereafter. Q-learning is a simple on-line algorithm for estimating this function Q of state-action pairs. Let Q_t denote the estimate of Q at time step t. This is stored in a look-up table with an entry for each state-action pair. Suppose the learning system observes the process state $x(t)$, executes action $a(t)$, and receives the resulting immediate reinforcement $r(t)$. Then

$$\Delta Q_t(x, a) = \begin{cases} \eta(t)[r(t) + \gamma P(t+1) - Q_t(x, a)] & \text{if } x = x(t) \text{ and } a = a(t), \\ 0 & \text{otherwise,} \end{cases}$$

where $\eta(t)$ is a positive learning-rate parameter that depends on t and

$$P(t+1) = \max_{a \in A(t+1)} Q_t(x(t+1), a),$$

with $A(t+1)$ denoting the set of all actions available at $t+1$. If this set consists of a single action for all t, Q-learning reduces to a look-up-table version of the adaptive critic learning rule (Equation 8). Although the Q-learning convergence theorem requires look-up-table storage (and therefore finite state and action sets), many researchers have heuristically adapted Q-learning to more general forms of storage, including multilayer neural networks trained by backpropagation of the Q-learning error.

Dynamic Programming — Sequential reinforcement learning problems (in fact, all reinforcement learning problems) are examples of stochastic optimal control problems. Among the traditional methods for solving these problems are dynamic programming (DP) algorithms. As applied to optimal control, DP consists of methods for successively approximating optimal evaluation functions and optimal decision rules for both deterministic and stochastic problems. Bertsekas [Ber87] provides a good treatment of these methods. A basic operation in all DP algorithms is "backing up" evaluations in a manner similar to the operation used in Samuel's method and in the adaptive critic and Q-learning algorithms.

Recent reinforcement learning theory exploits connections with DP algorithms while emphasizing important differences. For an overview and guide to the literature, see [Bar92, BBS95, Sut92, Wer92, Kae96]. Following is a summary of key observations.

1. Because conventional dynamic programming algorithms require multiple exhaustive "sweeps" of the process state set (or a discretized approximation of it), they are not practical for problems with very large finite-state sets or high-dimensional continuous state spaces. Sequential reinforcement learning algorithms *approximate* DP algorithms in ways designed to reduce this computational complexity.

2. Instead of requiring exhaustive sweeps, sequential reinforcement learning algorithms operate on states as they occur in actual or simulated

experiences in controlling the process. It is appropriate to view them as *Monte Carlo* DP algorithms.

3. Whereas conventional DP algorithms require a complete and accurate model of the process to be controlled, sequential reinforcement learning algorithms do not require such a model. Instead of computing the required quantities (such as state evaluations) from a model, they estimate these quantities from experience. However, reinforcement learning methods can also take advantage of models to improve their efficiency.

4. Conventional DP algorithms require look-up-table storage of evaluations or actions for all states, which is impractical for large problems. Although this is also required to guarantee convergence of reinforcement learning algorithms, such as Q-learning, these algorithms can be adapted for use with more compact storage means, such as neural networks.

It is therefore accurate to view sequential reinforcement learning as a collection of heuristic methods providing computationally feasible approximations of DP solutions to stochastic optimal control problems. Emphasizing this view, Werbos [Wer92] uses the term *heuristic dynamic programming* for this class of methods.

5 Conclusion

The increasing interest in reinforcement learning is due to its applicability to learning by autonomous robotic agents. Although both supervised and unsupervised learning can play essential roles in reinforcement learning systems, these paradigms by themselves are not general enough for learning while acting in a dynamic and uncertain environment. Among the topics being addressed by current reinforcement learning research are extending the theory of sequential reinforcement learning to include generalizing function approximation methods; understanding how exploratory behavior is best introduced and controlled; sequential reinforcement learning when the process state cannot be observed; how problem-specific knowledge can be effectively incorporated into reinforcement learning systems; the design of modular and hierarchical architectures; and the relationship to brain reward mechanisms.

Acknowledgments: This chapter is an expanded and revised version of "Reinforcement Learning" by Andrew G. Barto, which appeared in the *Handbook of Brain Theory and Neural Networks*, M. A. Arbib, editor, pp. 804-809. MIT Press: Cambridge, Massachusetts, 1995.

6 REFERENCES

[AMY+93] J. Alspector, R. Meir, B. Yuhas, A. Jayakumar, and D. Lippe. A parallel gradient descent method for learning in analog VLSI neural networks. In S. J. Hanson, J. D. Cohen, and C. L. Giles, editors, *Advances in Neural Information Processing Systems 5*, pages 836–844, Morgan Kaufmann, San Mateo, California, 1993.

[BA85] A. G. Barto and P. Anandan. Pattern recognizing stochastic learning automata. *IEEE Transactions on Systems, Man, and Cybernetics*, 15:360–375, 1985.

[Bar85] A. G. Barto. Learning by statistical cooperation of self-interested neuron-like computing elements. *Human Neurobiology*, 4:229–256, 1985.

[Bar92] A. G. Barto. Reinforcement learning and adaptive critic methods. In D. A. White and D. A. Sofge, editors, *Handbook of Intelligent Control: Neural, Fuzzy, and Adaptive Approaches*, pages 469–491. Van Nostrand Reinhold, New York, 1992.

[BAS82] A. G. Barto, C. W. Anderson, and R. S. Sutton. Synthesis of nonlinear control surfaces by a layered associative search network. *Biological Cybernetics*, 43:175–185, 1982.

[BBS95] A. G. Barto, S. J. Bradtke, and S. P. Singh. Learning to act using real-time dynamic programming. *Artificial Intelligence*, 72:81–138, 1995.

[Ber87] D. P. Bertsekas. *Dynamic Programming: Deterministic and Stochastic Models*. Prentice-Hall, Englewood Cliffs, New Jersey, 1987.

[BF85] D. A. Berry and B. Fristedt. *Bandit Problems*. Chapman and Hall, London, 1985.

[BJ87] A. G. Barto and M. I. Jordan. Gradient following without backpropagation in layered networks. In M. Caudill and C. Butler, editors, *Proceedings of the IEEE First Annual Conference on Neural Networks*, II-629–II-636, San Diego, 1987.

[Boz82] S. Bozinovski. A self-learning system using secondary reinforcement. In R. Trappl, editor, *Cybernetics and Systems*. North-Holland, Amsterdam, 1982.

[BS81] A. G. Barto and R. S. Sutton. Landmark learning: An illustration of associative search. *Biological Cybernetics*, 42:1–8, 1981.

[BSA83] A. G. Barto, R. S. Sutton, and C. W. Anderson. Neuronlike elements that can solve difficult learning control problems. *IEEE Transactions on Systems, Man, and Cybernetics*, 13:835–846, 1983. Reprinted in J. A. Anderson and E. Rosenfeld, *Neurocomputing: Foundations of Research*, MIT Press, Cambridge, Massachusetts, 1988.

[BSB81] A. G. Barto, R. S. Sutton, and P. S. Brouwer. Associative search network: A reinforcement learning associative memory. *IEEE Transactions on Systems, Man, and Cybernetics*, 40:201–211, 1981.

[GBG94] V. Gullapalli, A. G. Barto, and R. A. Grupen. Learning admittance mappings for force-guided assembly. In *Proceedings of the 1994 International Conference on Robotics and Automation*, pages 2633–2638, 1994.

[GGB92] V. Gullapalli, R. A. Grupen, and A. G. Barto. Learning reactive admittance control. In *Proceedings of the 1992 IEEE Conference on Robotics and Automation*, pages 1475–1480. IEEE, Piscataway, New Jersey, 1992.

[Gol89] D. E. Goldberg. *Genetic Algorithms in Search, Optimization, and Machine Learning*. Addison-Wesley, Reading, Massachusetts, 1989.

[Gul90] V. Gullapalli. A stochastic reinforcement algorithm for learning real-valued functions. *Neural Networks*, 3:671–692, 1990.

[Hol75] J. H. Holland. *Adaptation in Natural and Artificial Systems*. Univ. of Michigan Press, Ann Arbor, 1975.

[HT74] E. Harth and E. Tzanakou. Alopex: A stochastic method for determining visual receptive fields. *Vision Research*, 14:1475–1482, 1974.

[JJ90] M. I. Jordan and R. A. Jacobs. Learning to control an unstable system with forward modeling. In D. S. Touretzky, editor, *Advances in Neural Information Processing Systems 2*, Morgan Kaufmann, San Mateo, California, 1990.

[Kae96] L. P. Kaelbling, editor. *Special Issue on Reinforcement Learning*, volume 22. *Machine Learning*, 1996.

[KGV83] S. Kirkpatrick, C. D. Gelatt, and M. P. Vecchi. Optimization by simulated annealing. *Science*, 220:671–680, 1983.

[Kim61] G. A. Kimble. *Hilgard and Marquis' Conditioning and Learning*. Appleton-Century-Crofts, New York, 1961.

[Klo72]   A. H. Klopf. Brain function and adaptive systems—A heterostatic theory. Technical Report AFCRL-72-0164, Air Force Cambridge Research Laboratories, Bedford, MA, 1972. A summary appears in *Proceedings of the International Conference on Systems, Man, and Cybernetics*, IEEE Systems, Man, and Cybernetics Society, Dallas, Texas, 1974.

[Klo82]   A. H. Klopf. *The Hedonistic Neuron: A Theory of Memory, Learning, and Intelligence.* Hemisphere, Washington, D.C., 1982.

[KPU94]   K. P. Venugopal K. P. Unnikrishnan. Alopex: A correlation-based learning algorithm for feed-forward and recurrent neural networks. *Neural Computation*, 6:469–490, 1994.

[LTW90]   G. Lukes, B. Thompson, and P. Werbos. Expectation driven learning with an associative memory. In *Proceedings of the International Joint Conference on Neural Networks*, pages I-521 to I-524. Lawrence Erlbaum, Hillsdale, New Jersey, 1990.

[MM70]   J. M. Mendel and R. W. McLaren. Reinforcement learning control and pattern recognition systems. In J. M. Mendel and K. S. Fu, editors, *Adaptive, Learning and Pattern Recognition Systems: Theory and Applications*, pages 287–318. Academic Press, New York, 1970.

[NT89]   K. Narendra and M. A. L. Thathachar. *Learning Automata: An Introduction.* Prentice-Hall, Englewood Cliffs, New Jersey, 1989.

[RHW86]   D. E. Rumelhart, G. E. Hinton, and R. J. Williams. Learning internal representations by error propagation. In D. E. Rumelhart and J. L. McClelland, editors, *Parallel Distributed Processing: Explorations in the Microstructure of Cognition, volume 1: Foundations.* Bradford Books/MIT Press, Cambridge, Massachusetts, 1986.

[Sam59]   A. L. Samuel. Some studies in machine learning using the game of checkers. *IBM Journal on Research and Development*, pages 210–229, 1959. Reprinted in E. A. Feigenbaum and J. Feldman, editors, *Computers and Thought*, pages 71–105. McGraw-Hill, New York, 1963.

[Sut88]   R. S. Sutton. Learning to predict by the method of temporal differences. *Machine Learning*, 3:9–44, 1988.

[Sut92]   R. S. Sutton, editor. *A Special Issue of Machine Learning on Reinforcement Learning*, volume 8. *Machine Learning*, 1992.

Also published as *Reinforcement Learning*, Kluwer Academic Press, Boston, Massachusetts, 1992.

[Tes92] G. J. Tesauro. Practical issues in temporal difference learning. *Machine Learning*, 8:257–277, 1992.

[Tho11] E. L. Thorndike. *Animal Intelligence*. Hafner, Darien, Connecticut, 1911.

[WD92] C. J. C. H. Watkins and P. Dayan. Q-learning. *Machine Learning*, 8:279–292, 1992.

[Wer92] P. J. Werbos. Approximate dynamic programming for real-time control and neural modeling. In D. A. White and D. A. Sofge, editors, *Handbook of Intelligent Control: Neural, Fuzzy, and Adaptive Approaches*, pages 493–525. Van Nostrand Reinhold, New York, 1992.

[WGM73] B. Widrow, N. K. Gupta, and S. Maitra. Punish/reward: Learning with a critic in adaptive threshold systems. *IEEE Transactions on Systems, Man, and Cybernetics*, 5:455–465, 1973.

[WS85] B. Widrow and S. D. Stearns. *Adaptive Signal Processing*. Prentice-Hall, Englewood Cliffs, New Jersey, 1985.

Chapter 3

Neurocontrol in Sequence Recognition

William J. Byrne
Shihab A. Shamma

ABSTRACT An artificial neural network intended for sequence modeling and recognition is described. The network is based on a lateral inhibitory network with controlled, oscillatory behavior so that it naturally models sequence generation. Dynamic programming algorithms can be used to transform the network into a sequence recognizer (e.g., for speech recognition). Markov decision theory is used to propose alternative, more neural recognition control strategies as alternatives to dynamic programming.

1 Introduction

Central to many formulations of sequence recognition are problems in sequential decision-making. Typically, a sequence of events is observed through a transformation that introduces uncertainty into the observations, and based on these observations, the recognition process produces a hypothesis of the underlying events. The events in the underlying process are constrained to follow a certain loose order, for example by a grammar, so that decisions made early in the recognition process restrict or narrow the choices that can be made later. This problem is well known and leads to the use of *dynamic programming* (DP) algorithms [Bel57] so that unalterable decisions can be avoided until all available information has been processed.

DP strategies are central to *hidden Markov model* (HMM) recognizers [LMS84, Lev85, Rab89, RBH86] and have also been widely used in systems based on neural networks (e.g., [SIY+89, Bur88, BW89, SL92, BM90, FLW90]) to transform static pattern classifiers into sequence recognizers. The similarities between HMMs and neural network recognizers are a topic of current interest [NS90, WHH+89]. The neural network recognizers considered here will be those that fit within an HMM formulation. This covers many networks that incorporate sequential decisions about the observations, although some architectures of interest are not covered by this formulation (e.g., [TH87, UHT91, Elm90]).

The use of dynamic programming in neural network–based recognition systems is somewhat contradictory to the motivating principles of neurocomputing. DP algorithms first require precise propagation of probabilities, which can be implemented in a neural fashion [Bri90]. However, the component events that make up the recognition hypothesis are then found by backtracking, which requires processing a linked list in a very nonneural fashion.

The root of this anomaly is that the recognition process is not restricted to be local in time. In the same way that neural computing emphasizes that the behavior of processing units should depend only on physically neighboring units, the sequential decision process used in recognition ideally should use only temporally local information. Dynamic programming algorithms that employ backtracking to determine a sequence of events are clearly not temporally local.

This problem has also been addressed in HMMs. In many applications, it is undesirable to wait until an entire sequence of observations is available before beginning the recognition process. A related problem is that the state space required by the DP algorithms becomes unmanageably large in processing long observation sequences. As solutions to these problems, approximations to the globally optimal DP algorithms are used. For example, the growth of the state space is restricted through pruning, and real-time sequence hypotheses are generated through partial-traceback algorithms.

Suboptimal approximations to the globally optimal DP search strategies are therefore of interest in both HMM and neural network sequence recognition. One approach to describing these suboptimal strategies is to consider them as *Markov decision problems* (MDPs) [Ros83]. In this work the theoretical framework for such a description is presented. The observation sequence is assumed to be generated by an HMM source model, which allows the observation and recognition process to be described jointly as a first-order controlled Markov process. Using this joint formulation, the recognition problem can formulated as an MDP, and recognition strategies can be found using stochastic dynamic programming.

The relationship of this formulation to neural network –based sequence recognition will be discussed. A stochastic neural network architecture will be presented that is particularly suited to use in both sequence generation and recognition. This novel architecture will be employed to illustrate this MDP description of sequence recognition. The intended application is to speech recognition.

2 HMM Source Models

Computational models that describe temporal sequences must necessarily balance accuracy against computational complexity. This problem is addressed in HMMs by assuming that there is an underlying process that controls the production of the observed process. The underlying, or hidden, process is assumed to be Markov, and the observations are generated independently as a function of the current hidden state. The hidden state process models event order and duration. Observation variability or uncertainty is described by the state-dependent observation distributions. The value of this formulation is that statistics required for training and recognition can be computed efficiently. Brief definitions of the HMMs considered in this chapter are presented here.

The observation sequences are assumed to be generated by a discrete time, discrete observation HMM source with hidden process S and observations I. The source is assumed to have N states, and the model parameters are $\Lambda = (a, b)$, with transition probabilities a and state-dependent observation probabilities b.

The hidden process is a first-order Markov process that produces a state sequence $S = \{S_t\}_{t=1}^{T}$, where the process state takes values in $\{1, \ldots, N\}$ and T is random. For convenience, it will be assumed that this process is "left-to-right," so that the sequence begins with the value 1, ends with the value N, and has intermediate values satisfying $S_t \leq S_{t+1}$.

The state transition probabilities are

$$Pr(S_{t+1} \mid S_t) = \begin{cases} 1 - a_n, & S_{t+1} = n, \quad S_t = n, \\ a_n, & S_{t+1} = n+1, \quad S_t = n, \\ 0, & \text{otherwise,} \end{cases}$$

where a_n is the probability of a transition from state n to state $n+1$. $Pr(S_{t+1} \mid S_t)$ is denoted by $a_{S_t, S_{t+1}}$.

At each time instant, the source generates an observation I_t according to the distribution

$$Pr(I_t | S_t) = b_{S_t}(I_t). \tag{1}$$

Given a hidden state sequence, the observations are independently generated. When the process leaves state N, the sequence ends; an end-of-string symbol is generated to indicate this. The joint source likelihood can be expressed as

$$Q(I, S) = \prod_{t=1}^{T} b_{S_t}(I_t) \, a_{S_{t+1}, S_t}.$$

3 Recognition: Finding the Best Hidden Sequence

In one formulation of HMM sequence recognition, a model is constructed for each observation class, and each of these models is used to score an unknown sequence. The unknown sequence is then identified according to which model gave it the maximum likelihood. For example, models $\{Q^i\}$ would be trained for a set of words $\{W^i\}$. An observation I would then be classified as an instance of a particular word W^j if $L_{Q^j}(I) \geq L_{Q^i}(I)\ \forall i$ according to some model-based likelihood criterion L_Q.

The scoring criterion considered here is the maximum likelihood Viterbi score $\max_R Q(I, R)$, so called because of the DP-based algorithm used in its computation [For67]. R is used to denote estimates of the hidden state sequence, S, to emphasize the distinction between the unobserved source hidden process that generated the observation and any estimate of it by the recognizer. For an observed sequence I, the *most likely state sequence* (MLSS) R_I is found. The joint likelihood $Q(I, R_I) = \max_R Q(I, R)$ is used to score the observation.

The Viterbi algorithm is a dynamic programming technique that solves $\max_R Q(I, R)$. For an observation sequence I, it directly produces the likelihood score $\max_R Q(I, R)$. Backtracking can then be used to find the MLSS R_I. If only the Viterbi score $\max_R Q(I, R)$ is desired, neural architectures are available that can compute this quantity [LG87].

This formulation is typical of maximum likelihood HMM–based recognizers. While it does not describe all neural network sequence recognition systems, it can be used to describe systems that use a DP algorithm to transform static pattern classifiers (i.e., feed-forward neural networks) into sequence recognizers. Such systems have been widely experimented with and have been termed *hidden control neural networks* [Lev93]. Neural networks have also been used in HMM hybrid systems that also employ the Viterbi algorithm [MB90, FLW90].

4 Controlled Sequence Recognition

If HMMs are considered as source models and inherently as models of sequence generation, they are easily understood as systems in which the hidden state process controls the production of the observation sequence. In recognition, however, control flows in the opposite direction: observed sequences control the formation of symbol sequences that are estimates of the source hidden state sequence. An architecture that models both sequence production and recognition should include mechanisms by which the observable and underlying events can control each other. The role of control processes in these two systems is presented in Figure 1. A complex control framework of this nature can be described using *controlled Markov models* (CMM) [Ros83]. The value of formulating both sequence production and recognition in terms of CMMs will be shown by using the same basic architecture in both problems. This differs from the usual HMM formalism in which a model is first trained in "source mode" and its parameters are then embedded in a recognition system of a different architecture.

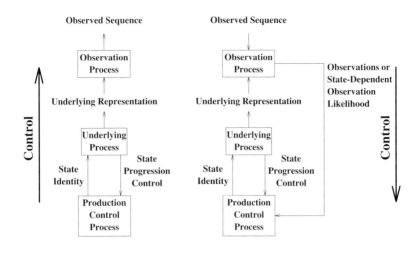

FIGURE 1. Aspects of control in sequence generation and recognition.

4.1 Controlled Markov Models

A CMM is a Markov process whose state transition probabilities can be modified by an applied control. The control is usually a function of the current model state and is applied to improve system performance as the process evolves. The CMM formalism can be used to describe both sequence generation and MLSS recognition by hidden Markov models.

Suppose a homogeneous Markov process X_t has the following transition probability:

$$P(X_{t+1} = x' \mid X_t) = a_{x,x'}, \quad X_t = x.$$

A CMM has a modified transition probability that depends upon an applied control process U:

$$P(X_{t+1} = x' \mid X_t; U_t) = a_{x,x'}(u), \quad X_t = x, \quad U_t = u.$$

U_t is called a stationary Markov control if it is a function of the process state X_t but depends only on the state identity and is not a function of time.

The choice of which control to apply when the system is in a given state is determined according to a control policy. If a policy is based upon stationary Markov controls, the resulting CMM will also yield a stationary Markov process [Mak91]. If such a policy, π, is chosen, the probability distribution it defines is denoted P^π. It will later be necessary to take expectations with respect to this distribution. Given that the process starts from a state x, expectation with respect to the distribution that arises from the control policy is denoted by E_x^π.

Source Models: A CMM Description

The HMM source model describes jointly the observed process I and the hidden process S involved in sequence production. The production of I and S in a left-to-right HMM will be described here as a CMM.

It is assumed that the progression of the hidden process is completely determined by a binary control signal U_t. Applying $U_t = 0$ forces S_{t+1} to equal S_t, i.e., there is no change in the hidden state from time t to time $t+1$. Conversely, applying $U_t = 1$ forces a change of state, so that if $S_t = n$, then $S_{t+1} = n + 1$.

The control U_t is a random process defined as

$$Pr(U_t = u \mid S_t) = \begin{cases} a_{S_t, S_t}, & u = 0, \\ a_{S_t, S_{t+1}}, & u = 1. \end{cases}$$

The original hidden process is effectively embedded in the control law. While the effect of an applied control is exact, the choice of control is random, and the choice is made in a way that duplicates the original hidden process. This describes how the hidden Markov process can be described as a CMM. The observations I_t are then generated as a function of S_t according to Equation 1.

While this may seem somewhat contrived, its value will be shown in the next section, in which this same CMM formalism will be used to describe sequence recognition.

MLSS Recognition: A CMM Description

As described earlier, the MLSS is obtained using the Viterbi algorithm. The observed sequence I is assumed to be generated by an HMM jointly with an unobserved sequence S. The log-likelihood of the observed sequence is computed as $\max_R \log Q(I, R)$. R is used to distinguish the recognizer state sequence from the source hidden state sequence S that was generated, but not observed, with I.

For any recognition strategy, including but not necessarily the Viterbi algorithm, the joint log-likelihood of the observed sequence and hidden state sequence estimated by the recognizer is

$$\log Q(I, R) = \sum_{t=1}^{T} \log b_{R_t}(I_t) a_{R_t, R_{t+1}}.$$

This sum can be accumulated by the recognizer while the sequence is observed, and it is possible to describe this as a controlled process. Suppose that at time t, the recognizer is in state $R_t = n$ and the symbol I_t is observed. The control $u_t = 0$ can be applied so that $R_{t+1} = n$, or $u_t = 1$ can be applied so that $R_{t+1} = n + 1$. The action of the applied control is summarized as

$$f(R; u) = \begin{cases} R, & u = 0, \\ R + 1, & u = 1. \end{cases} \tag{2}$$

The function f indicates the new recognizer state that results from applying the control.

At each time, the recognizer receives a reward that depends upon the observed symbol, the current recognizer state, and the chosen control. If at time t the recognizer state is $R_t = n$ and $I_t = i$ is observed, the reward received is

$$v(i, n; u_t) = \begin{cases} \log b_n(i)\, a_{n,n}, & u_t = 0, \\ \log b_n(i)\, a_{n,n+1}, & u_t = 1. \end{cases} \qquad (3)$$

The observations are scored under the state observation distribution that corresponds to the current recognizer state. Before the observation is scored, the observer chooses whether or not to advance the recognizer state at the next time instant. The contribution of the hidden state sequence likelihood is added accordingly. The accumulated reward is then the joint log-likelihood of the recognizer state sequence and the observation sequence

$$\sum_t v(I_t, R_t; U_t) = \log Q(I, R).$$

This is the cumulative score that the recognizer obtains by applying controls U_t to produce the hidden state sequence estimate R in response to the observations I.

Any control law could be used in recognition. While it would be unnecessarily complicated to formulate the Viterbi algorithm in this way, the recognition controls could be applied to obtain the Viterbi score and the corresponding Viterbi sequence if the entire observation sequence were known beforehand. However, this is not possible if the recognizer is not provided information from arbitrarily far into the future. In the next section, suboptimal but causal recognition strategies will be described that are based on providing limited future information to the recognizer.

As a technical note, the source emits an end-of-string symbol when the sequence ends. When this symbol is observed, the recognizer is driven into the final state N, and the recognition process terminates. If some future information is available, the sequence termination can be anticipated gracefully.

The Viterbi score has been described as a decision–reward process that occurs incrementally as estimates of the hidden state sequence are produced. In the next section, the choice of recognition control rules will be investigated.

4.2 Source-Driven Recognizers

When the recognizer is not provided complete information about the future, it is necessary to guess what the correct recognizer behavior should be. It is possible to describe this as a Markov decision problem [Ros83]. In this formulation the optimal DP search is approximated by a gambling strategy that uses estimates of the future based on the stochastic source model. To use Markov decision theory in finding a recognition control law, the entire process — which includes both the source and the recognizer — must be described as a Markov process.

The Joint Source–Recognizer Model

While the source process (I_t, S_t) is Markov, during recognition the process S_t is not available. It is not true in general that I_t is Markov, i.e., $Pr(I_{t+1}|I_1^t) \neq Pr(I_{t+1}|I_t)$ (where I_t^{t+h} denotes $\{I_t, \ldots, I_{t+h}\}$); however, it is possible to accumulate a statistic

$$\tilde{\alpha}_t(n) = Pr(S_t = n \mid I_1^t)$$

so that the joint process $(I_t, \tilde{\alpha}_t)$ is Markov. This state occupancy statistic is found by the forward part of the scaled forward–backward algorithm [Lev85] and is also well known in the literature on the control of partially observed Markov processes [Mon82].

More generally, it is also possible to compute state occupancy statistics that maintain some limited "future" information. Define a vector of conditional probabilities

$$\tilde{\alpha}_t^h(n) = Pr(S_t = n \mid I_1^{t+h}), \qquad n = 1, \ldots, N,$$

that maintains a current source state probability based on information that extends h observations into the future. It is not difficult to show (as in [KM93]) that $\tilde{\alpha}_t^h$ satisfies a recursion in I_t^{t+h} and $\tilde{\alpha}_{t-1}^h$. This recursion is denoted by $\tilde{\alpha}_t^h = T^h(I_t^{t+h}, \tilde{\alpha}_{t-1}^h)$. It is also straightforward to determine that because the hidden process is Markov, $\tilde{\alpha}_t^h$ is sufficient to determine $Pr(I_{t+1}^{t+1+h}|I_1^{t+h})$. This computation is denoted by $Pr(I_{t+1}^{t+1+h} \mid I_1^{t+h}) = \Psi^h(I_{t+1}^{t+1+h}, \tilde{\alpha}_t^h)$. It will be shown that by maintaining these statistics it is possible to describe a recognition decision process that at time t uses information from the future up to time $t+h$.

The first step in summarizing the joint source–recognizer process as Markov uses the following property of the source model:

Property 1 $(I_t^{t+h}, \tilde{\alpha}_t^h)$ *is a time-homogeneous Markov process.*

Proof

$\Pr\left(I_{t+1}^{t+1+h} = i, \tilde{\alpha}_{t+1}^h = a \mid I_1^{t+h}, \tilde{\alpha}_t^h, \ldots, \tilde{\alpha}_1^h\right)$
$= \Pr\left(\tilde{\alpha}_{t+1}^h = a \mid I_1^{t+1+h}, \tilde{\alpha}_t^h, \ldots, \tilde{\alpha}_1^h\right) \Pr\left(I_{t+1}^{t+1+h} = i \mid I_1^{t+h}, \tilde{\alpha}_t^h, \ldots, \tilde{\alpha}_1^h\right)$
$= \Pr\left(T(i, \tilde{\alpha}_t^h) = a \mid I_1^{t+1+h}, \tilde{\alpha}_1^t\right) \Psi^h(i, \tilde{\alpha}_t)$
$= \delta_a(T(i, \tilde{\alpha}_t^h)) \Psi^h(i, \tilde{\alpha}_t^h).$

□

The process $(I_t^{t+h}, \tilde{\alpha}_t^h) \to (I_{t+1}^{t+1+h}, \tilde{\alpha}_{t+1}^h)$ is therefore first-order Markov. The accumulated source statistics are fairly complex, however, consisting of the $(h+1)$-element observation vector I_t^{t+h} and the N-element probability vector $\tilde{\alpha}_t^h$.

The recognizer state R_t and the observed and accumulated source statistics $(I_t^{t+h}, \tilde{\alpha}_t^h)$ can be combined into a state $(R_t, I_t^{t+h}, \tilde{\alpha}_t^h)$ and treated jointly as a single process. This is termed the source–recognizer process. In a sense, the recognizer is modeled as a CMM driven by the observation process. Because the observations and the recognizer are Markov, the source–recognizer process is also Markov.

The source–recognizer process has the following CMM transition probability:

$$\Pr((R_{t+1}, I_{t+1}^{t+1+h}, \tilde{\alpha}_{t+1}^h) = (n, i, a) \mid R_t, I_t^{t+h}, \tilde{\alpha}_t^h; u)$$
$$= \Pr(R_{t+1} = n \mid R_t; u)\, \Pr((I_{t+1}^{t+1+h}, \tilde{\alpha}_{t+1}^h) = (i, a) \mid I_t^{t+h}, \tilde{\alpha}_t^h).$$

If u is a stationary Markov control, this defines a valid stationary Markov process [Mak91].

Note that while the control may be a function of the complete source–recognizer state $(R_t, I_t^{t+h}, \tilde{\alpha}_t^h)$, it appears only in the recognizer state transition probability. This reflects the separation between the source and the recognizer: the recognizer can be controlled, while the source statistics can only be accumulated.

For simplicity, $\Pr((R_{t+1}, I_{t+1}^{t+1+h}, \tilde{\alpha}_{t+1}) = (n, i, \tilde{\alpha}) \mid R_t, I_t^{t+h}, \tilde{\alpha}_t; u)$ is denoted by $p^h(n, i, \tilde{\alpha} \mid R_t, I_t^{t+h}, \tilde{\alpha}_t; u)$. Some portion of the state process is deterministic, so this probability simplifies to

$$p^h(n, i, \tilde{\alpha} \mid R_t, I_t^{t+h}, \tilde{\alpha}_t; u) = \Psi^h(i, \tilde{\alpha}_t^h)\, \delta_n(f(R_t; u))\, \delta_{\tilde{\alpha}}(T^h(i, \tilde{\alpha}_t^h)). \quad (4)$$

To completely specify the source–recognizer process, the initial source–recognizer state probability must also be defined. It must be consistent with the knowledge that the source starts in state $S_1 = 1$. This requires that $\tilde{\alpha}_1^h$ assign probability 1 to state 1. The initial state probability is

$$P_1((R_1, I_1^{1+h}, \tilde{\alpha}_1^h) = (n, i, \tilde{\alpha})) = \begin{cases} Q(I_1^{1+h} = i), & n = 1,\ \tilde{\alpha}(n) = \delta_1(n), \\ 0, & \text{otherwise.} \end{cases}$$

Recognition as a Markov Decision Problem

When a reward is associated with the observations and control policy in a CMM, maximizing the expected reward is termed a *Markov decision problem* (MDP). It will be shown here how MLSS recognition can be formulated as an MDP.

It is first necessary to specify the allowable control policies. The set of admissible recognition control laws will be determined by fixing $h \geq 0$. Fixing h specifies the amount of future information provided to the recognition decision process. For a fixed h, $\mathcal{U}_h$ will be the 0/1-valued control laws measurable with respect to the source–recognizer state process $(I_t^{t+h}, \alpha_t^h, R_t)$.

Policies that are restricted to using control laws from $\mathcal{U}_h$ are denoted by π^h.

Using the incremental reward given in Equation 3 for the sequential recognition problem, the expected discounted cost resulting from a policy can be given as

$$J^\pi(x) = E_x^\pi \sum_t \beta^t v(I_t^{t+h}, \tilde{\alpha}_t^h, R_t; U_t),$$

where β ($0 \leq \beta \leq 1$) is a discounting parameter. This is the expected reward that can follow from a source–recognizer state x under the policy π.

The goal is to find the optimum policy that maximizes the expected discounted reward. This optimum expected reward is termed the *value function* and is defined as

$$V^h(x) = \max_{\pi \in \{\pi^h\}} J^\pi(x).$$

This is the maximum reward that can be expected given a CMM state x. The value function satisfies [Ros83]

$$V^h(r, i, \tilde{\alpha}) = \max_{u=0,1} \{v(r, i, \tilde{\alpha}; u) + \beta \sum_{r', i', \tilde{\alpha}'} p^h(r', i', \tilde{\alpha}' \mid r, i, \tilde{\alpha}; u) V^h(r', i', \tilde{\alpha}')\}.$$

Using the simplified expression of the transition probability, Equation 4, this reduces to

$$V^h(r, i, \tilde{\alpha}) = \max_{u=0,1} \{v(r, i, \tilde{\alpha}; u) + \beta \sum_{i'} \Psi^h(i', \tilde{\alpha}) V^h(f(r; u), i', T(i', \tilde{\alpha}))\},$$
(5)

where f describes the action of the control law as defined in Equation 2. The corresponding optimum control for each state is [Ros83]

$$u^h(r, i, \tilde{\alpha}) = \arg\max_{u=0,1} \{v(r, i, \tilde{\alpha}; u) + \beta \sum_{i'} \Psi^h(i', \tilde{\alpha}) V^h(f(r; u), i', T(i', \tilde{\alpha}))\}.$$

This is a complete, exact description of the combined source–recognizer processes and the optimum control rules that maximize the expected reward following from any source–recognizer state.

As a technical note, β may equal 1 if the final state can be reached with probability 1 from any state in a finite number of transitions, regardless of the controls. This is called the terminating assumption [MO70, p. 42], which is satisfied here. All observation sequences are of finite length with probability 1, and the recognizer is forced to its final state when the end-of-string symbol is observed.

Any technical assumptions required for the MDP formulation are assumed to be met by placing restrictions on the source model. For example, the observation distributions b are assumed to be bounded away from 0 for all possible observations, so that $B \leq \log b_n(i) \leq 0$. However, B can be arbitrarily small, so imposing this constraint is not restrictive.

There are several problems with this formulation, however. Although the state space is countable, $\tilde{\alpha}$ can take an extremely large number of values — almost as many values as sequences that could be observed. The dimensionality of the value function and control laws therefore grows unmanageably large. If it is necessary to maintain the control law explicitly for each state, the computational advantages obtained by assuming that the source processes are Markov are lost.

Further, these optimal rewards and their associated decision rules are difficult to obtain from these equations. The equations are contractions, so they can be solved numerically. However, a different approach will be described here that is based on neural computation and control.

Relationship to the Viterbi Algorithm

While basing a recognizer on the optimum expected reward may be an unusual formulation, it is possible to compare it to the usual Viterbi score. When the amount of future information is unrestricted, choosing the control that optimizes this criterion leads to scoring all observation sequences according to the Viterbi algorithm. This will be shown here.

Consider the expected reward resulting from any of the valid initial, $t = 1$ source–recognizer states. For $\beta = 1$ the expected reward can be restated as

$$J^\pi(x) = E_x^\pi \log Q(I, R^\pi),$$

where R^π denotes the recognizer sequence produced by recognition control policy π in response to the observation sequence I. In this version of the expected reward, which is "pointwise" in I, the $\tilde{\alpha}$ are not required because they are functions of I.

When h is unrestricted, the maximization is performed over policies allowed to employ all possible controls $\mathcal{U} = \cup_h \mathcal{U}_h$, so that the optimum reward becomes

$$\max_\pi E \log Q(I, R^\pi).$$

Property 2

$$\max_\pi E \log Q(I, R^\pi) = E \max_R \log Q(I, R).$$

A sketch of a proof of this property is given for models that assign probability zero to infinite-length sequences, i.e., for which $Q(\{I : T = \infty\}) = 0$.

Proof
$\mathcal{U}_h \subset \mathcal{U}_{h+1}$ implies

$$\max_\pi E \log Q(I, R^\pi) = \lim_{h \to \infty} \max_{\pi \in \{\pi^h\}} E \log Q(I, R^\pi).$$

For a fixed h, the Viterbi algorithm is an allowable policy for all observations I with length $T \leq h$, so for such I, $\max_{\pi \in \{\pi^h\}} \log Q(I, R^\pi) = \max_R \log Q(I, R)$. Therefore

$$\max_{\pi \in \{\pi^h\}} E \log Q(I, R^\pi) = \sum_{I:T\leq h} Q(I) \max_R \log Q(I, R)$$
$$+ \max_{\pi \in \{\pi^h\}} \sum_{I:T>h} Q(I) \log Q(I, R^\pi)$$

and

$$\lim_{h\to\infty} \max_{\pi \in \{\pi^h\}} E \log Q(I, R^\pi) = \sum_{I:T<\infty} Q(I) \max_R \log Q(I, R)$$
$$+ \max_\pi \sum_{I:T=\infty} Q(I) \log Q(I, R^\pi).$$

Loosely, since $Q(\{I : T = \infty\}) = 0$, the sum over infinite-length sequences is negligible, so that

$$\lim_{h\to\infty} \max_{\pi \in \{\pi^h\}} E \log Q(I, R^\pi) = E \max_R \log Q(I, R).$$

□

In summary, for every possible observation sequence it is possible to pick a value of h that provides complete information about the future. Given unrestricted future information, the Viterbi algorithm is an admissible and, by design, optimum strategy for all possible sequences. This gives an intuitive motivation for the expected likelihood criterion. As the restrictions on the temporal locality of the decision-making process are removed, the Viterbi algorithm is recovered as the best recognition strategy.

Before investigating the application of the MDP sequence recognition formulation, a neural architecture that is particularly well suited for use as a source model or recognizer will be presented.

5 A Sequential Event Dynamic Neural Network

A neural network architecture is presented here that can be used to generate and recognize simple sequences of events. After the network is described, it will be shown that it can be embedded into the MDP recognition framework presented above.

The network is based on a single layer of units $n = 1, \ldots, N$ that inhibit each other with strength w_n ($w_n > 0$). The network operates in discrete time: each unit updates its potential $x(n)$ according to

$$x_t(n) = \sum_{\substack{j=1 \\ j \neq n}}^{N} [-w_j \, y_{t-1}(j) + c].$$

3. Neurocontrol in Sequence Recognition

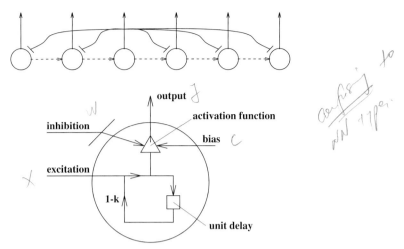

FIGURE 2. Dynamical network architecture: (top) lateral inhibitory network with directed excitation; (bottom) network unit schematic.

The unit output values $y(n)$ are $\{0,1\}$-valued with

$$y_t(n) = o(x_t(n)),$$

where o is the unit activation function. When $y(n) = 1$, unit n is on, or active. A bias term c is included so that uninhibited units will activate. The inhibition exceeds the bias: $w_n > c$.

As presented, this is a stable, lateral inhibitory network [MY72]. In particular, if the network reaches a state in which a single unit is active, that unit will remain active and prevent any other unit from activating.

The units can be made to activate sequentially by adding excitatory connections between units. While a unit n is active, it exhibits a slowly increasing weak excitatory effect upon its neighbor, unit $n+1$, so that this unit becomes less inhibited. The excitation of unit n by unit $n-1$ is given as

$$e_t(n) = (1-k)\, e_{t-1}(n) + g\, y_t(n-1).$$

This directed excitation channel is modeled as a connection of strength g followed by a leaky integrator with a decay factor of $1-k$. The result is that the excitation saturates at the value g/k. The lateral inhibitory network architecture with the directed excitation channels and the unit activity functions are presented in Figure 2. The unit states must be modified to include this excitation, so the network state vector is (x_t, e_t). The update equations for each unit are

$$x_t(n) = \sum_{\substack{j=1 \\ j \neq n}}^{N} [-w_j\, y_{t-1}(j) + e_{t-1}(n) + c],$$

$$e_t(n) = (1-k)\, e_{t-1}(n) + g\, y_t(n-1).$$

Suppose $k \approx 0$, i.e., the directed excitation grows linearly. If unit $n-1$ has been active for a period τ, the excitation of unit n is $e_t(n) = g\tau$ and all other excitations are zero. The unit states are then

$$x_t(n') = \begin{cases} c, & n' = n-1, \\ -w_{n-1} + g\tau + c, & n' = n, \\ -w_{n-1} + c, & \text{otherwise.} \end{cases} \quad (6)$$

If the activation function o is the unit-step function, unit n activates when x_n becomes nonnegative. When this happens, unit n shuts off unit $n-1$. After unit $n-1$ first activates, the time required for the directed excitation to overcome the inhibition of unit n is

$$\tau_{n-1} = \frac{w_{n-1} - c}{g}.$$

This determines the duration of unit $n-1$'s activity and leads to sequential behavior in that unit n activates only after unit $n-1$ has been active for a fixed duration.

A network can be constructed to represent events that occur sequentially for fixed durations. The parameters g, c, k, and w can be chosen to satisfy the above relationship so that each unit is active for a specified duration.

Under this updating rule the network activity sequence is fixed. Given an initial network state, each unit activates at a known time and remains active for a fixed period. The activity sequence of the network is denoted by S_t, where $S_t = n$ if $y_t(n) = 1$.

Such a network is not well suited to model sequences in which the event durations may vary. A simple way to model variable duration events is to randomize the unit activation function. Rather than mapping the unit activation to the unit output deterministically, suppose the activation function o is such that each unit activates randomly according to

$$\Pr(\,y_t(n) = 1 \mid x_t(n)\,) = \frac{1}{1 + e^{-x_t(n)}}.$$

The connectivities are chosen to satisfy $w_n \gg c \gg 0$, so that an inhibited unit will not activate, while a unit that is uninhibited will always activate. This is equivalent to activating the next unit in the sequence by flipping a biased coin whose bias towards activation increases with time.

Again consider the case when $k \approx 0$. If at time t unit $n-1$ has been active for a period τ, the unit states will be as in Equation 6. While unit $n-1$ is active and until unit n activates, under the assumption $w_{n-1} \gg c \gg 0$, the unit activation functions behave according to

$$\Pr(y(n') = 1 \mid x(n')) \approx \begin{cases} 1, & n' = n-1, \\ \frac{1}{1+e^{-[-w_{n-1}+g\tau+c]}}, & n' = n, \\ 0, & \text{otherwise.} \end{cases}$$

The probability that unit $n+1$ activates given that unit n has been active for a period τ is denoted by

$$a_n(\tau) = \frac{1}{1 + e^{-[-w_n + g\tau + c]}}.$$

Each unit remains active for a duration τ_n according to the distribution $\Pr(\tau_n = \tau) = d_n(\tau)$, where

$$d_n(\tau) = \prod_{t=1}^{\tau-1}(1 - a_n(t))\, a_n(\tau). \tag{7}$$

Without further modification, the network can be used to model sequences of the form

$$\{S_t : 1 \leq S_t \leq S_{t+1} \leq N\},$$

that is, ordered events of varying duration. The probability of a sequence S is found through the probabilities of its component events

$$\Pr(S) = \prod_{n=1}^{N} d_n(\tau_n), \tag{8}$$

where τ_n is the duration of the nth event in sequence S.

The hidden state process is not a simple first-order Markov process. Because the transition probabilities depend on the state duration, duration must be included in the process state. If duration information is retained, the state transition mechanism is described by a first-order Markov process (n, τ). If the process has value (n, τ), unit n has been active for a period τ. The process transition probability is

$$\Pr(\,(n,\tau)_{t+1} = (n',\tau') \mid (n,\tau)_t\,) = \begin{cases} a_{n_t}(\tau_t), & n' = n_t + 1,\ \tau' = 1, \\ 1 - a_{n_t}(\tau_t), & n' = n_t,\ \tau' = \tau + 1. \end{cases} \tag{9}$$

This is illustrated in Figure 3.

More general sequences can be modeled by adding another group of units to the network. The original sequential event units now form a hidden layer, and these new units are the visible network units. The visible units are also stochastic, and their behavior depends on the unit activity sequence in the hidden layer. These visible units are meant to represent observations of labels, such as vector quantized acoustic features or phoneme identities.

At each time, an observation I_t is generated by the visible units according to distribution b_{S_t}, which depends upon transitions in the hidden layer. The probability of an observation sequence I given the underlying sequence S is

$$\Pr(I|S) = \prod_{t=1}^{T} b_{S_t}(I_t).$$

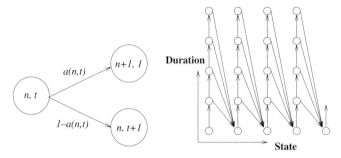

FIGURE 3. Duration-dependent transition probabilities: (left) Markov process defined by duration-dependent state transition probabilities. (right) Markov chain corresponding to duration-dependent transition probabilities.

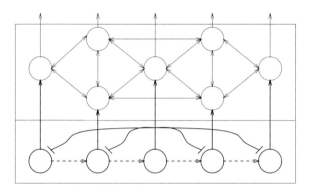

FIGURE 4. Network of visible units controlled by a sequential network.

An exact mechanism for the behavior of the visible units is not needed for this presentation; however, a possible architecture would be a Boltzmann machine whose units were influenced by the sequential units, as in Figure 4.

Alternatively, the observations could be generated according to state-dependent Gaussian distributions. While this is not covered by the current MDP formulation, which assumes discrete observations, the log-likelihood computation becomes a distance measurement between the observation and an exemplar feature. The interpretation of this process is that a state is represented by a single-feature vector, and the reward accrued in recognition is based on the distance from the observations to the exemplars.

The network can now be described as a probabilistic model with processes I and S that are the output sequence and unit activity sequence

$$\{(I_t, S_t) : 1 \leq S_t \leq S_{t+1} \leq N,\ t = 1, \ldots, T\}.$$

The joint distribution of the activity and observation sequences is

$$Q(S, I) \;=\; \Pr(I|S)\,\Pr(S)$$

$$= \prod_{t=1}^{T} b_{S_t}(I_t) \prod_{n=1}^{N-1} d_n(\tau_n).$$

The distribution has the form of a *hidden Markov model*, specifically, a *variable-duration hidden Markov model* (VDHMM) [Lev86], where the probability of leaving a state depends on how long the system has been in that state.

The duration distribution d_n determined by the network parameters has some attractive properties. When $k \approx 0$, it has a peak at $\tau_n \approx \frac{w_n}{g}$, which specifies the most likely duration. Additionally, for $\frac{w_n}{g}$ fixed, the variance of the unit activity duration decreases as g increases. This can be used to incorporate uncertainty about event duration in the network model.

In a non-variable-duration HMM, the state duration probability has the form $(1-a)^{\tau-1}a$, where a is the probability of remaining in a state. It has been argued that other distributions, such as Gaussian and gamma distributions, provide better temporal modeling. The distribution that arises here enjoys the two main features of the previously used distributions, namely the nonzero maximum likelihood duration and an adjustable variance. The difference between this model and other VDHMMs is that the duration distribution is not chosen beforehand — $d_n(\tau)$ doesn't have a closed-form expression — but arises from the state transition mechanism.

When k is not negligible, the potential of unit $n+1$ when excited by unit n eventually approaches

$$x_t(n+1) = -w_n + c + g/k,$$

so that $a_n(\tau)$, the probability of unit $n+1$ activating, approaches

$$K = \frac{1}{1 + e^{-[-w_n + c + g/k]}}.$$

Since $a_n(\tau)$ approaches K for large τ, the duration distribution d_n falls off as $(1-K)^{\tau}$ (Equation 7). This shows the importance of the excitation channel decay parameter k. It can be used to control the tail of the state duration distribution.

Two examples of model density durations are presented in Figure 5. The durations of 7500 instances of the phonemes /iy/ and /n/ were obtained from the TIMIT database. Model fits are plotted along with the sample densities. The parameter k is particularly valuable in fitting the exponential decay often described in phoneme duration histograms.

Training this network is discussed in Chapter 3 of [Byr93]. The EM algorithm is used to train it as a VDHMM [Lev86] under a maximum likelihood criterion. This training problem is developed in an information geometric framework similar to that used to describe Boltzmann machine learning in [Byr92]. Other neural training schemes based on sequential approximations to the EM algorithm are also possible [WFO90]. In general,

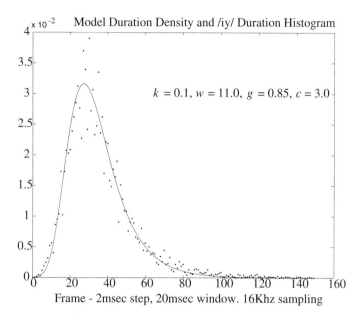

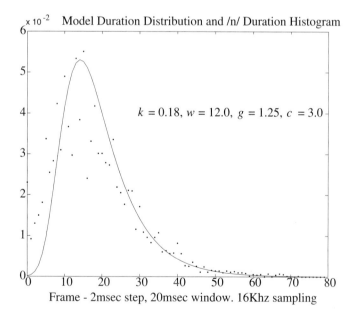

FIGURE 5. Modeling sample duration histograms computed from phoneme durations found in the TIMIT database: 6950 instances of /iy/ (top); 7068 instances of /n/ (bottom).

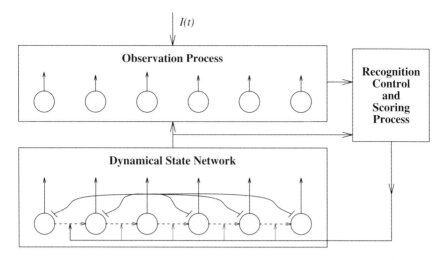

FIGURE 6. Dynamical network architecture embedded in a sequence recognizer.

modeling duration-dependent transition probabilities is difficult; however, there has been previous work using neural architectures to address this problem [GS89].

Similar networks have been presented elsewhere [RT91, BT91], and in general, dynamic neural networks intended for sequence recognition and production have been widely studied [DCN87, Kle86, SK86, BK91]. The network presented here has the benefit that individual parameters can be associated with desirable aspects of the network behavior. The gain parameter g determines the variance of the duration distribution; for example, the inhibition profile determines the relative duration of each unit's activity, and the decay factor k in the directed excitation channel is used in modeling the duration distribution tail, as described above.

In source mode, the gain function g is fixed and the state progression is a random function of the state (n, τ). In use as a recognizer, the gain function g is used to control the state progression. To force a change in state from (n, τ) to $(n+1, 1)$, g is set to a very large value, so that the directed excitation immediately activates the next unit in the sequence. Otherwise g is kept small so that the current unit remains active. This architecture is illustrated in Figure 6.

6 Neurocontrol in Sequence Recognition

Thus far, two topics have been discussed. MLSS sequence recognition with limited future information has been formulated as a Markov decision problem, and a stochastic neural architecture intended for modeling observation sequences has been introduced. In this section, a controlled sequence recognizer built on this network architecture will be described.

As described in the previous section, the hidden process of the dynamical network is a first-order Markov process with state $S_t = (n, \tau)_t$. While this is more complicated than the formulation of Section 4.2, which is based on a simple Markov process, the recognizer and control rules are formulated identically.

The following conditional probability can be computed recursively:

$$\tilde{\alpha}(n, \tau)_t^h = \Pr(S_t = (n, \tau)_t \mid I_1^{t+h}),$$

which is denoted by $\tilde{\alpha}_t^h = T^h(I_t^{t+h}, \tilde{\alpha}_{t-1}^h)$, as before. The statistics $(I_t^{t+h}, \tilde{\alpha}_t^h)$ again form a first-order Markov process. The joint source–recognizer description is as in Equation 4.

The specification of the optimum recognition control is as presented earlier (Equation 5). While the MDP formulation proves the existence of an optimum rule and provides methods to construct it, it is impractical to solve explicitly for the control rules. However, the MDP formulation describes the input to the control rule, i.e., how the observations should be transformed before presentation to the controller. According to this formulation, the optimum control U_t should be a function of $(\tilde{\alpha}_t^h, I_t^{t+h}, R_t)$. The control rule that is produced as a function of the source–recognizer state is unknown; however, the MDP formulation specifies that it does exist. Here, a neural network can be trained in an attempt to approximate it.

A set of training sequences is assumed to be available for the model. For example, if a network is to be trained to model the digit "nine," utterances of "nine" form the training set. After the network has been trained, the Viterbi algorithm is used to find the best hidden state sequence for each training sequence. The training sequences and their corresponding MLSSs form a training set that can be used to build a neural network to implement the recognizer control rule. The source–recognizer statistics are accumulated recursively, and the recognizer control rule neural network is trained to implement the control that generates the Viterbi sequence. This is illustrated in Figure 7.

Experimental Results

A small, speaker-independent, isolated-digit speech-recognition experiment was performed. The observations used were vector quantized features obtained from the cochlear model as described in [BRS89]. The features were computed at a frame rate of 20 msec and a step size of 2 msec and quantized using a 32 codeword vector quantizer. The speech was taken from the TI Connected Digits database, and networks were trained for each of the ten digits using ten utterances from ten different male speakers (10 utterances from each of 10 speakers). The recognition score on a test of ten utterances from each of ten other speakers, using the Viterbi algorithm,

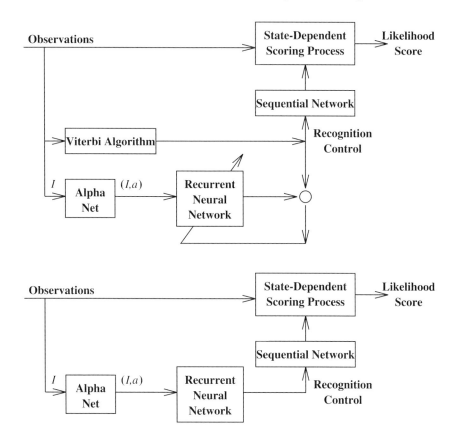

FIGURE 7. Viterbi algorithm supplying recognizer control law as training target (top); neurocontrol implementation of recognition control law (bottom).

was approximately 95% correct.

From the Viterbi segmentation, a recurrent neural network was trained to implement the recognition control law. The network consisted of two hidden layers with four units in the first hidden layer and two units in the second layer. Five frames ($h = 5$, 10.0 msec) of future information was provided to the recognizer, and a discounting parameter $\beta = 0.9$ was chosen. Using this neurally implemented recognition control law, recognition performance of approximately 93% correct was obtained.

This experiment is presented as an example of an application of the MDP formulation of MLSS sequence recognition and is far from conclusive. While promising, as currently implemented in software on conventional computers, the computational burden in training and testing prohibits evaluating the performance on large problems. However, it is hoped that this formulation might prove valuable both in investigations into the behavior of suboptimal approximations to the Viterbi algorithm and to prompt further investigation into applications of neurocontrol in sequence recognition.

7 Observations and Speculations

As MDP sequence recognition has been formulated, it has been assumed that the observations are produced by the source HMM Q that is also used to define the likelihood criterion L_Q. A more general formulation is possible. A more complex source could be used, for example, such as a mixture of HMM sources. The only restriction is that it must be possible to accumulate statistics so that the recognizer can be described as driven by a Markov process.

Model training and training of the neural network that implements the control law were presented as separate procedures. In many applications of neurocontrol, estimation of model parameters and the control law are performed simultaneously [BSS90]. Such a formulation is possible here, and could be based upon sequential versions of the EM algorithm [WFO90].

7.1 The Good Recognizer Assumption

An interesting simplification that follows from the MDP formulation of the MLSS sequence recognition problem arises from the following assumption. If the recognizer state R_t is assumed to be a very good estimate of the source hidden state S_t, the problem is greatly simplified. The assumption is

$$\Pr(S_t = n \mid I_1^{t+h}, R_t) = \begin{cases} 1, & R_t = n, \\ 0, & \text{otherwise.} \end{cases}$$

This assumption leads to a drastic reduction in the source–recognizer state dimension and to an interesting relationship between the source and recognition controls.

The first simplification due to the assumption is the elimination of the statistic $\tilde{\alpha}_t$. Because $\tilde{\alpha}_t^h(n) = \Pr(S_t = n \mid I_1^{t+h})$ and R_t is a function of I_1^{t+h}, it follows that $\tilde{\alpha}_t^h(n) = \delta_n(R_t)$. Therefore $\tilde{\alpha}_t$ is constant for fixed R_t. As a result, the optimum value function $V(r, i, \tilde{\alpha})$ is a function of r and i alone.

Similarly, the recursive computation of the likelihood term involving I_t^{t+h} is also simplified. Note that R_{t+1} is a function of I_t^{t+h} and R_t, since the control is applied at time t to determine the next recognizer state, so that

$$\begin{aligned}
\Pr(I_{t+1}^{t+1+h} = i_1^{h+1} \mid I_1^{t+h} = i_1^h) &= \Pr(I_{t+h+1} = i_{h+1} \mid I_1^{t+h} = i_1^h) \\
&= \sum_n \Pr(I_{t+1+h} = i_{h+1} \mid S_{t+1}, I_1^{t+h}) \Pr(S_{t+1} = n \mid I_1^{t+h} = i_1^h) \\
&= \sum_n \Pr(I_{t+1+h} = i_{h+1} \mid S_{t+1}, I_{t+2}^{t+h} = i_2^h) \Pr(S_{t+1} = n \mid R_{t+1}, I_1^{t+h}).
\end{aligned}$$

Using $\phi^h(i|S_{t+1})$ to denote the probability

$$\Pr(I_{t+1+h} = i_{h+1}|\, S_{t+1}, I_{t+2}^{t+h} = i_2^h),$$

the above reduces to

$$\Pr(\, I_{t+1}^{t+1+h} = i\,|\, I_1^{t+h} = i_1^h) = \phi^h(i\,|\, R_{t+1}).$$

The accumulated statistics $\tilde{\alpha}_t^h$ are no longer used in computing the term $\Pr(\, I_{t+1}^{t+1+h} = i\,|\, I_1^{t+h})$; instead, it is approximated by

$$\Pr(\, I_{t+1}^{t+1+h} = i\,|\, I_{t+2}^{t+h}, S_{t+1}),$$

where the recognizer state R_{t+1} is used as an estimate of the source state S_{t+1}.

Modifying the optimum value equations to include this simplification yields

$$V^h(r, i) = \max_u \{\, v(r, i_1; u) + \beta \sum_{i'} V^h(f(r; u),\, i')\, \phi^h(\, i'\,|\, f(r; u)\,)\,\}.$$

This assumption leads to interesting interpretations of the control rules. Because the recognition control u appears directly in the source observation statistics, the recognizer acts as if it directly controls the source sequence production. This suggests a close link between the production and recognition processes. When the recognition process is accurate, the recognition control law recovers the source control law best suited to produce the observation sequence. This suggests that this formulation of sequence recognition may provide an avenue for the use of speech production mechanisms, or articulatory models, in speech recognition.

Control Rules

The value functions that result from the simplifying assumption can be solved fairly easily. The *good recognizer* assumption removes the dependence upon the accumulated statistics $\tilde{\alpha}$, so that the dimensionality of the value functions is greatly reduced. It is possible to solve them in a left-to-right manner: $V(r, i)$ depends upon itself and $V(r+1, i)$. This is made particularly easy when the recognizer state r is expanded to include the state-duration τ. In this case, the Markov chain allows only the two transitions $(r, \tau) \to \{(r, \tau+1), (r+1, 1)\}$. In this case, $V(r, \tau, i)$ depends solely upon $V(r, \tau+1, i)$ and $V(r+1, 1, i)$. $V(N, \tau, i)$ is solved first, and then $V(r, \tau, i)$ is solved for decreasing r. In practice, this requires picking a maximum value of τ. The $V(N, \tau, i)$ can then be solved directly; an approximation is to pick $V(N, \tau_{\max}, i)$ at random and solve backwards for decreasing τ.

Consider the $h = 1$ case. Here, $i = (i_1, i_2)$ and

$$\phi^1(i|S_{t+1}) = \Pr(I_{t+2} = i_2|S_{t+1}).$$

The value functions are presented here with the reward expressed in likelihood form

$$V(N, \tau, i) = \log(1 - a_N(\tau)) b_N(i_1) + \beta \sum_{i_3} \phi^1((i_2, i_3)|N) V(N, \tau + 1, (i_2, i_3)),$$

$$V(r, \tau, i) =$$
$$\max \left\{ \log(1 - a_r(\tau)) b_r(i_1) + \beta \sum_{i_3} \phi^1((i_2, i_3) | r) V(r, \tau + 1, (i_2, i_3)), \right.$$
$$\left. \log a_r(\tau) b_r(i_1) + \beta \sum_{i_3} \phi^1((i_2, i_3) | r + 1) V(r + 1, 1, (i_2, i_3)) \right\}.$$

Denoting $\sum_{i_2} \phi^1((i_1, i_2)|r) V(r, \tau, (i_1, i_2))$ by $\bar{V}(r, \tau, i_1)$, the decision rule can be simplified. Suppose that at time t the recognizer is in state (r, τ) and the observation symbol I_{t+1} becomes available. The recognition control law is chosen according to

$$\bar{V}(r, \tau + 1, I_{t+1}) - \bar{V}(r + 1, 1, I_{t+1}) \begin{array}{c} u = 0 \\ \geq \\ < \\ u = 1 \end{array} \frac{1}{\beta} \log \frac{1 - a_r(\tau)}{a_r(\tau)}.$$

The recognition control law becomes a fairly simple table look-up that using the next available observation is based upon comparing expected rewards against a duration-dependent threshold. This could easily be implemented in a neurocontrol architecture.

Experiments have been carried out using recognition control rules based upon this simplification. However, the results were unsatisfactory. For whatever reasons, but most likely due to the overly optimistic nature of the assumption, the recognizers behaved poorly. Typical behavior was either to remain in the initial state or to move to the final state as quickly as possible. This assumption may yet prove valuable, however, as it suggests methods by which the dimensionality of the value functions and control rules may be reduced. A topic of future interest might be the investigation of other, not so drastic, approximating assumptions that might yield reductions in computational cost without too much loss in performance.

7.2 Fully Dynamic Sequential Behavior

As presented in Section 5, the dynamical network is suitable for recognizing individual sequences, such as words spoken in isolation. While determining the underlying event sequence in such applications is not of crucial importance, it has been given as a demonstration of the MDP recognition formulation in which local decisions can be made that avoid the use of noncausal dynamic programming algorithms.

The ultimate desired application of the dynamical network is in identifying subsequences in continuous observation streams as necessary for connected or continuous speech recognition. To be useful in such applications, the network architecture must be modified. By allowing the final network unit N to excite the first network unit, the network can be made to oscillate. After the Nth unit has been active, the first unit activates, and the network repeats its pattern sequence. In this way the network behaves as a controlled oscillator.

Oscillatory networks have been used as models of central pattern generation in simple neural systems [Mat87, Mat85, SK86, Kle86]. Such oscillatory behavior has been investigated in the network presented here and has been described elsewhere [BRS89]. As in the isolated sequence case, it is desirable to use the dynamical behavior of the network in these more complex sequence recognition problems.

In problems in which higher-level context must be modeled, such as when sequences can appear as substrings in other sequences, it is hoped that large networks with cyclic behavior might be built that would capture the complexity of the task. Ideally, a "grammar" that describes the problem would control the cyclic behavior of the network in much the same way that language models are currently used in HMM speech recognition systems to constrain the acoustic search.

In such an application, the dynamical network operates in either phase-locked or free-cycling mode. A recognition controller is used to vary the excitation gain g to induce either of these modes, as described earlier in Section 5. In free-cycling mode, the excitation gain is set to a high value so that the network progresses quickly through its state sequence. In phase-locked mode, the network progresses through its state sequence at a rate matched to the observed sequence. This is an indication that the observations agree with the network model. Because this behavior is indicated by the control law itself, the value of g serves as an indication of the match between the observations and the model.

An example of early experiments into this phase-locking behavior is described here. The dynamical network is intended to synchronize with the output of phoneme classifiers when the correct word is in the input stream. When incorrect words are in the input, the network should lose synchronization and free-cycle.

Feed-forward networks were trained to classify hand-labeled segments of spoken digits. The classifier outputs were thresholded to make a binary decision about the identity of the observation, so that the network is presented with a binary vector of classifier signals. The Hamming distance between the network activity vector and the classifier output is used as a measure of

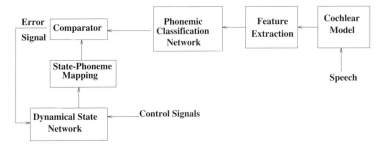

FIGURE 8. Dynamical network embedded in a simple phone classifier architecture.

instantaneous agreement between the observations and the recognizer. The network gain is obtained directly from this agreement measurement by simple low-pass filtering. If the network and classifier vectors are in agreement, the gain will decay, and the network state progression will slow. Conversely, if the agreement is poor, the error will drive up the gain, and the network will speed up. Ideally, the network will synchronize its progression to the rate of the input signal. This network architecture is presented in Figure 8, and preliminary experiments with this system are described in [Byr93]. While the example presented here is simple, it captures the formulation of the intended application of the dynamical network in sequence recognition.

Constructing a recognition control law to implement the desired control is a nontrivial task. A topic of future research is to extend the MDP formulation to describe recurrent behavior by multiple networks so that a rigorous framework for sequential decisions in connected and continuous speech recognition can be developed.

8 REFERENCES

[Bel57] R. Bellman. *Dynamic Programming*. Princeton Univ. Press, Princeton, New Jersey, 1957.

[BK91] W. Banzhaf and K. Kyuma. The time-into-intensity-mapping network. *Biological Cybernetics*, 66:115–121, 1991.

[BM90] H. Bourlard and N. Morgan. A continuous speech recognition system embedding MLP into HMM. In D. Touretzky, editor, *Advances in Neural Information Processing 2*, pages 186–193, Morgan Kaufmann, San Mateo, California, 1990.

[Bri90] J. Bridle. Alpha-nets: a recurrent neural network architecture with a Hidden Markov Model interpretation. *Speech Communication*, 9:83–92, 1990.

[BRS89] W. Byrne, J. Robinson, and S. Shamma. The auditory processing and recognition of speech. In *Proceedings of the Speech* and *Natural Language Workshop, Cape Cod, Massachusetts, October 1989*, pages 325–331, Morgan Kaufmann, San Mateo, California, 1989.

[BSS90]   A. Barto, R. Sutton, and C. Sutton. Learning and sequential decision making. In M. Gabriel and J. Moore, editors, *Learning and Computational Neuroscience: Foundations of Adaptive Networks*, chapter 13, pages 539–602. Bradford, Boston, 1990.

[BT91]   P. Bressloff and J. Taylor. Discrete time leaky integrator network with synaptic noise. *Neural Networks*, 4:789–801, 1991.

[Bur88]   D. Burr. Experiments on neural nets recognition of spoken and written text. *IEEE Transactions on Acoustics, Speech and Signal Processing*, 36:1162–1168, 1988.

[BW89]   H. Bourlard and C. Wellekens. Speech pattern discrimination and multilayer perceptrons. *Computer Speech and Language*, 3:1–19, 1989.

[Byr92]   W. Byrne. Alternating Minimization and Boltzmann Machine learning. *IEEE Transactions on Neural Networks*, 3(4):612–620, 1992.

[Byr93]   W. Byrne. *Encoding and representing phonemic sequences using nonlinear networks*. Ph.D. Thesis, University of Maryland, College Park, 1993.

[DCN87]   S. Dehaene, J.-P. Changeux, and J.-P. Nadal. Neural Networks that learn sequences by selection. *Proceedings of the National Academy of Science, U.S.A.*, 84:2727–2731, May 1987.

[Elm90]   J. Elman. Finding structure in time. *Cognitive Science*, 14:179–211, 1990.

[FLW90]   M. Franzini, K. F. Lee, and A. Waibel. Connectionist Viterbi training: a new hybrid method for continuous speech recognition. In *Proceedings of the ICASSP*, pp. 425–428, Albuquerque, New Mexico, April 1990.

[For67]   G. Forney. The Viterbi algorithm. *IEEE Transactions on Information Theory*, IT-13:260–269, April 1967.

[GS89]   S. Grossberg and N. Schmajuk. Neural dynamics of adaptive timing and temporal discrimination during associative learning. *Neural Networks*, 2(2):79–102, 1989.

[Kle86]   D. Kleinfeld. Sequential state generation by neural networks. *Proceedings of the National Academy of Science, U.S.A.*, 83:9469–9473, December 1986.

[KM93] V. Krishnamurthy and J. Moore. On-line estimation of Hidden Markov Models based on the Kullback-Leibler information measure. *IEEE Transactions on Signal Processing*, 41(8):2557–2573, August 1993.

[LMS84] S. Levinson, L. Rabiner, and M. Sondhi. An introduction to the application of the theory of probabilistic functions of a Markov process to automatic speech recognition. *The Bell System Technical Journal*, 64(4):1035–1074, April 1984.

[Lev85] S. Levinson. Structural methods in automatic speech recognition. *Proceedings of the IEEE*, 73(11):1625–1650, November 1985.

[Lev86] S. Levinson. Continuously variable duration Hidden Markov Models for automatic speech recognition. *Computer Speech and Language*, 1(1):29–46, March 1986.

[Lev93] E. Levin. Hidden Control neural architecture modeling of nonlinear time varying systems and its applications. *IEEE Transactions on Neural Networks*, 4(1):109–116, January 1993.

[LG87] R. Lippmann and B. Gold. Neural network classifiers useful for speech recognition. In *Proceedings of the First International Conference on Neural Networks, San Diego, June 1987*, pages 417–425. IEEE, 1987.

[Mak91] A. Makowski. ENEE 726. Stochastic Control Class Notes, Dept. Electrical Engineering, University of Maryland, College Park, Fall 1991.

[Mat85] K. Matsuoka. Sustained oscillations generated by mutually inhibiting neurons with adaptation. *Biological Cybernetics*, 52:367–376, 1985.

[Mat87] K. Matsuoka. Mechanisms of frequency and pattern control in the neural rhythm generators. *Biological Cybernetics*, 56:345–353, 1987.

[MB90] N. Morgan and H. Bourlard. Continuous speech recognition using multilayer perceptrons and Hidden Markov Models. In *Proceedings of the ICASSP*, pages 413–416, Albuquerque, New Mexico, April 1990.

[MO70] H. Mine and S. Osaki. *Markov Decision Processes*. American Elsevier, New York, 1970.

[Mon82]   G. Monaham. A survey of partially observable Markov decision processes: theory, models and applications. *Management Science*, 28(1):1–16, January 1982.

[MY72]   I. Morishita and A. Yajima. Analysis and simulation of networks of mutually inhibiting neurons. *Kybernetic*, 11, 1972.

[NS90]   L. Niles and H. Silverman. Combining Hidden Markov Models and neural network classifiers. In *Proceedings of the ICASSP*, pages 417–420, Albuquerque, New Mexico, 1990.

[Rab89]   L. Rabiner. A tutorial on Hidden Markov Models and selected applications in speech recognition. *Proceedings of the IEEE*, 77(2):257–286, February 1989.

[RBH86]   L. Rabiner and B.-H. Juang. An introduction to Hidden Markov Models. *IEEE ASSP Magazine*, pages 4–16, January 1986.

[Ros83]   S. Ross. *Introduction to Stochastic Dynamic Programming*. Academic Press, New York, 1983.

[RT91]   M. Reiss and J. Taylor. Storing temporal sequences. *Neural Networks*, 4:773–787, 1991.

[SIY$^+$89] H. Sakoe, R. Isotani, K. Yoshida, K. Iso, and T. Watanabe. Speaker independent word recognition using dynamic programming neural networks. In *Proceedings of the ICASSP*, pages 29–32, Glasgow, Scotland, May 1989.

[SK86]   H. Sompolinsky and I. Kanter. Temporal association in asymmetric neural networks. *Physical Review Letters*, 57(22):2861–2864, December 1986.

[SL92]   E. Singer and R. Lippmann. A speech recognizer using radial basis function neural networks in an HMM framework. In *Proceedings of the ICASSP*, pages 629–632, San Francisco, California, 1992.

[TH87]   D. Tank and J. Hopfield. Neural computation by concentrating information in time. *Proceedings of the National Academy of Science, U.S.A.: Biophysics*, 84:1896–1900, April 1987.

[UHT91]   K. P. Unnikrishnan, J. Hopfield, and D. Tank. Connected-digit speaker-dependent speech recognition using a neural network with time-delayed connections. *IEEE Transactions on Signal Processing*, 39(3):698–713, March 1991.

[WFO90] E. Weinstein, M. Feder, and A. Oppenheim. Sequential algorithms for parameter estimation based on the Kullback-Leibler information measure. *IEEE Transactions on Acoustics, Speech, and Signal Processing*, 38(9):1652–1654, September 1990.

[WHH+89] A. Waibel, T. Hanazawa, G. Hinton, K. Shikano, and K. Lang. Phoneme recognition using time-delay neural networks. *IEEE Transactions on Acoustics, Speech and Signal Processing*, 37(3):328–339, March 1989.

Chapter 4

A Learning Sensorimotor Map of Arm Movements: a Step Toward Biological Arm Control

Sungzoon Cho
James A. Reggia
Min Jang

ABSTRACT Proprioception refers to sensory inputs that principally regulate motor control, such as inputs that signal muscle stretch and tension. Proprioceptive cortex includes part of SI cortex (area 3a) as well as part of primary motor cortex. We propose a computational model of neocortex receiving proprioceptive input, a detailed map of which has not yet been clearly defined experimentally. Our model makes a number of testable predictions that can help guide future experimental studies of proprioceptive cortex. They are first, overlapping maps of both individual muscles and of spatial locations; second, multiple, redundant representations of individual muscles where antagonist muscle length representations are widely separated; third, neurons tuned to plausible combinations of muscle lengths and tensions; and finally, proprioceptive "hypercolumns," i.e., compact regions in which all possible muscle lengths and tensions and spatial regions are represented.

1 Introduction

It has long been known that there are multiple *feature maps* occurring in sensory and motor regions of the cerebral cortex. The term feature map refers to the fact that there is a systematic two-dimensional representation of sensory or motor features identifiable over the cortical surface. Generally, neurons close to one another in such maps respond to or represent features that are similar. In most cases, neurons or other supraneuronal processing units ("columns") have broadly tuned responses, and thus the receptive fields of neighboring units overlap.

Feature maps are conveniently classified as being either *topographic* or *computational*. A feature map is called topographic when the stimulus parameter being mapped represents a spatial location in a peripheral space, for instance, the location of a point stimulus on the retina or the location of a tactual stimulus on the skin. A feature map is called computational when the stimulus parameter represents an attribute value in a feature space, for instance, the orientation of a line segment stimulus or the spatial location of a sound stimulus [KdLE87, UF88].

Several computational models have been developed to simulate the self-organization and plasticity of these cortical maps, including topographic maps in somatosensory (SI) cortex [GM90, PFE87, Skl90, Sut92] and computational maps in visual cortex [BCM82, Lin88, MKS89, vdM73]. While not without their limitations, these and related models have shown that fairly simple assumptions, such as a Mexican hat pattern of lateral cortical interactions and Hebbian learning, can qualitatively account for several fundamental facts about cortical map organization.

To our knowledge, the goal of all past computational models of cortical maps has been to explain previously established experimental data concerning relatively well-defined maps. In contrast, in this chapter we develop a computational model of neocortex receiving proprioceptive input (hereafter called "proprioceptive cortex"), a detailed map of which has not yet been clearly defined experimentally. *Proprioception* refers to sensory inputs that principally regulate motor control, such as inputs that signal muscle stretch and tension. Proprioceptive cortex includes part of SI cortex (area3a) as well as part of primary motor cortex [Asa89]. Our model makes a number of testable predictions that can help guide future experimental studies of proprioceptive cortex. In addition, the results of our simulations may help clarify recent experimental results obtained from studies of primary motor cortex, a cortical region that is heavily influenced by proprioceptive inputs [Asa89]. To our knowledge, this is the first computational model of map formation in proprioceptive cortex that has been developed.

The overall concern in this chapter is with what sensory feature maps could emerge in cortex related to the control of arm movements. Insight into this issue is not only of interest in a biological context, but also to those concerned with control of robotic arms or other engineering control applications [WS92]. There have been several previous models of map formation with model arms [BG88, BGO$^+$92, Kup88, Mel88, RMS92]. These previous models are different from that described here in that they are usually concerned with visuomotor transformation process of a 3-D reaching arm movement taking place in motor cortex. Thus, they typically use spatial location such as xyz-coordinates of an arm's endpoint as input rather than only muscle length and tension as was done here.

Thus, these past models have not been concerned with proprioceptive map formation.

The role of primary motor (MI) cortex in motor control has been an area of intense research during the last several years [Asa89]. Recent studies have discovered a great deal about the encoding of movements in MI [CJU90, DLS92, GTL93, LG94, SSLD88], although many aspects of the organization of the MI feature map remain incompletely understood or controversial (see [SSLD88] for a cogent review). For example, experimental maps of MI muscle representations have revealed that upper extremity muscles are activated from multiple, spatially separated regions of cortex [DLS92]. It has been suggested that this organization may provide for local cortical interactions among territories representing various muscle synergies. While this may be true, the model proprioceptive cortex developed here offers another explanation: that such an organization may be secondary to multiple, spatially separated muscle representation in proprioceptive cortex. Proprioceptive input exerts a significant influence on motor cortex [Asa89]. Thus, this model of proprioceptive cortex may help clarify these and other organizational issues concerning primary motor cortex.

In our work, a model arm provides proprioceptive input to cortex. Our model arm is a substantial simplification of reality: there are only six muscles (or muscle groups), there are no digits, there is no rotation at joints, gravity is ignored, and only information about position is considered. Two pairs of shoulder muscles (flexor and extensor, abductor and adductor) and one pair of elbow muscles (flexor and extensor) control and move the model arm, which we study in a three-dimensional (3-D) space. Nevertheless, as will be seen, this simplified arm provides sufficient constraints for a surprisingly rich feature map in the cortex.

The resultant feature map consists of regularly spaced clusters of cortical columns representing individual muscle lengths and tensions. Cortical units become tuned to plausible combinations of tension and length, and multiple representations of each muscle group are present. The map is organized such that compact regions within which all muscle group lengths and tensions are represented could be identified. Most striking was the observation that although not explicitly present in the input, the cortical map developed a representation of the three-dimensional space in which the arm moved.

2 Methods

We first present a neural network model of proprioceptive cortex, its activation mechanism and learning rule. Secondly, the structure of the model arm and the constraints it imposes on input patterns are given. The model arm is *not* a neural model; it is a simple simulation of the physical constraints imposed by arm positioning. Finally, we describe how we generate the proprioceptive input patterns from the model arm.

2.1 Neural Network

The model network has two separate layers of units, the arm input layer and the proprioceptive cortex layer (or simply "cortical layer" from now on) (see Figure 1). Each unit in the arm layer competitively distributes its activation to every unit in the cortical layer. Each unit in the cortical

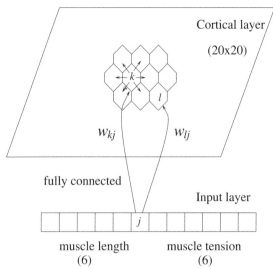

FIGURE 1. Neural network. The arm input layer contains six muscle length units and six muscle tension units, whose activation values represent the proprioceptive inputs of muscle length and tension values. The cortical layer consists of a grid of 20 × 20 units that represents proprioceptive cortex. Each unit is connected to its six immediate neighboring units (a hexagonal tessellation is used). To remove edge effects, units on the edges are connected with units on the opposite edges, so the cortical layer effectively forms a torus. The connection weights from input layer to cortical layer are initially randomly generated from a uniform distribution, then updated through training. The lateral connection weights between cortical units are constant.

layer also competitively distributes its activation to its neighbors through lateral connections. Competitive activation mechanisms have been shown to be quite effective in many different applications [RDSW92, RSC91]. With the recent development of learning algorithms for use with competitive activation mechanisms, these activation mechanisms can now be used in a wide variety of applications [CR92, CR93, RSC91]. One distinct feature of a competitive activation mechanism is its ability to induce lateral inhibition among units, and thus to support map formation, without using explicit inhibitory connections [CR92, RDSW92, Sut92, UF88]. Even with constant weight values for all corticocortical connections, a Mexican hat pattern of activation appears in the cortex [RDSW92]. It is this feature that we try to exploit in map formation at the cortical layer.

The activation level of unit k at time t, $a_k(t)$, is determined by[1]

$$\frac{da_k(t)}{dt} = c_s a_k(t) + (max - a_k(t))in_k(t), \qquad (1)$$

---

[1] Arm layer units are clamped to the length and tension values computed from random cortical signals to six muscles; thus the equation applies only to the cortical layer units.

where

$$in_k(t) = \sum_j o_{kj}(t) = \sum_j c_p \frac{(a_k^p(t) + q)w_{kj}}{\sum_l (a_l^p(t) + q)w_{lj}} a_j(t). \qquad (2)$$

This activation rule is the same as the rule used in [RDSW92]. The weight on the connection from unit j to unit k is denoted by w_{kj}, which is assumed to be zero when there is no connection between the two units. Although the weight variable is also a function of time due to learning, it is considered constant in the activation mechanism because activation levels change much faster than weights. The constant parameters c_s and c_p represent decay at unit k (with negative value) and excitatory output-gain at unit j, respectively. The value of c_s controls how fast activation decays, while that of c_p determines how much output a unit sends in terms of its activation level. The exponent parameter p determines how much competition exists among the units. The larger the value of p, the more competitive the model's behavior, and thus the greater the peristimulus inhibition. The parameter q (a small constant such as 0.0001) is added to $a_k(t)$ for all k to prevent division by zero (denominator term in Equation 2) and to influence the intensity of lateral inhibition. The parameter max represents the maximum activation level. The output $o_{kj}(t)$ from unit j to unit k is proportional not only to the sender's activation level, $a_j(t)$, but also to the receiver's activation level, $a_k(t)$. Therefore, a stronger unit receives more activation. Another unit, l, which also gets input from unit j, can be seen as competing against unit k for the output from unit j because the normalizing factor $\sum_{l \in N} (a_l(t) + q)w_{lj}$ in the denominator constrains the sum of the outputs from unit j to be equal to the unit's activation level, $a_j(t)$, when $c_p = 1$. The activation sent to unit k, therefore, depends not only on the activation values of the units from which it receives activation such as unit j, but also on the activation values of those of its competitors to which unit k has no explicit connections. Since competitive distribution of activation implicitly assumes that activation values are nonnegative, we used a hard lower bound of zero when we update activation values in (1) in order to prevent the activation values from ever going negative. The equation is approximated by a difference equation with $\Delta t = 0.1$. Other parameter values were determined empirically as follows. For cortical layer units, decay constant c_s and ceiling max values in (1) were set to -4.0 and 5.0, respectively. Their q and output gain parameter c_p values in (2) were set to 0.001 and 0.9, respectively. For arm layer units, q and c_p values in (2) were set to 0.1 and 0.8, respectively. Since arm layer units were clamped, their c_s and max values were not relevant. Further details of the activation mechanism can be found in [RDSW92].

2.2 Learning

Connection weights are modified according to competitive learning, a variant of Hebbian learning that tends to change the incoming weight vectors of the output units (cortical layer units here) into prototypes of the input patterns [RZ86]. The particular learning rule used here is adapted from [Sut92] and [vdM73]:

$$\Delta w_{kj} = \eta [a_j - w_{kj}] a_k^*, \qquad (3)$$

where

$$a_k^* = \begin{cases} a_k - \theta, & \text{if } a_k > \theta, \\ 0, & \text{otherwise,} \end{cases} \qquad (4)$$

and where parameters η and θ are empirically set to 0.1 and 0.32. Only the weights from the arm layer to the cortical layer are changed by (3); the corticocortical connections are constant. Before training, weights were randomly selected initially from a uniform distribution in the range [0.1, 1.0]. Updated weights were also normalized such that the 1-norm of the incoming weight vector of each cortical unit is equal to that of the input patterns (the average size of an input pattern was empirically found to be 7.45). Instead of checking at each iteration whether the network reached equilibrium, we ran the network for a fixed number of iterations, 32, which was found to approximate equilibrium empirically; at this point one step of learning was done according to (3).

2.3 Model Arm

Basically, the model arm consists of two segments, which we call the upper arm and lower arm, connected at the elbow. The model arm is fixed at the shoulder and has only six generic muscles or muscle groups. We assume that there are four muscles that control the upper arm and two muscles that control the lower arm. These "muscles" correspond to multiple muscles in a real arm. Abductor and adductor muscles move the upper arm up and down, respectively, through 180°, while flexor and extensor muscles move it forward and backward, respectively, through 180°. These four muscles are attached at points equidistant from the shoulder. The lower arm is moved up to 180° in a plane, controlled by closer (lower arm flexor) and opener (extensor) muscles as described in Figure 2.

This model arm is a great simplification of biological reality and is intended as only a first effort for modeling feature map formation in the proprioceptive cortex. Neither the dynamics of the arm movement nor the effects of gravity on the arm are considered. Also, the arm is assumed not to rotate around the elbow or shoulder joints. Only the positional information about the arm is part of the model.

4. A learning sensorimotor map of arm movements   67

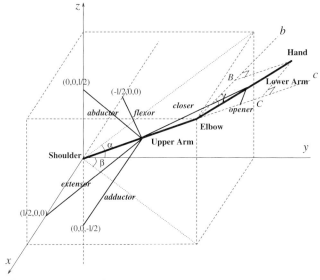

FIGURE 2. Schematic view of model arm. The model arm is considered as the right arm of a human facing the negative side of the x-axis. The pair of abductor and adductor muscles control the upper arm's vertical movement around the x-axis through contraction and stretch, with their joint angle denoted by α. The pair of flexor and extensor muscles control the arm's horizontal movement around the z-axis, with their angle denoted by β. All four muscles are attached to the midpoint of the upper arm and to imaginary points on either the x-axis or the z-axis. The upper arm can move up to $180°$ around the two axes, x and z, thus the possible positions of elbow E define a hemisphere. The pair of opener and closer muscles move the lower arm up to $180°$ around only one axis, a time-varying line perpendicular to the "hand plane" (the plane that is generated by the x-axis and elbow) and that passes through the elbow. Thus, the lower arm swings from a position collinear with the upper arm to a "folded" position where hand meets shoulder. Both muscles are attached to the midpoint of the lower arm and to imaginary points on the extended line of the upper arm, both length $l/2$ apart from the elbow. Their joint angle is denoted $\gamma(=\angle HEB)$, where H and E represent hand and elbow positions, respectively, and B is the projection of H onto the line segment that is perpendicular to the upper arm and on the hand plane. The possible positions of hand H define a semicircle with center E and radius l on the hand plane.

2.4 Proprioceptive Input

Since there is no motor cortex in our model, input activation to muscles must somehow be generated. We first generate six random numbers, which represent input activation to the six muscles that control the model arm. Given this input activation, we compute the expected proprioceptive information from muscles, i.e., muscle length and tension values. This information, consisting of twelve values, is used as input to the proprioceptive cortex in the model. The activation values of arm layer units are clamped to these values. Table 1 shows the formulae from which we compute proprioceptive input to the cortical layer.

Figure 3 shows a generic joint. First, we define the joint angle as a function of the difference between the input activation level of agonist and antagonist muscles.

Joint angle: *Let in_{ag} and in_{ant} denote the input activation levels of agonist and antagonist muscles, respectively. Then the joint angle θ is defined as $\theta = \frac{\pi}{2}(in_{\text{ag}} - in_{\text{ant}})$.* Note that value θ ranges from $-\pi/2$ to $\pi/2$, exclusive of the end points. In simulations, values of in are randomly generated from a uniform distribution in $[0, 1]$.

Muscle length units in the network model muscle spindle or stretch receptor inputs, which fire strongly when the muscle is mechanically stretched. We can derive the lengths of the muscles, l_1 $(=\overline{XZ})$ and l_2 $(=\overline{YZ})$, from the joint model shown in Figure 3.

Muscle length: *Given joint angle θ and appendage length l as in Figure 3, muscle lengths l_1 and l_2 are*

$$l_1 = l \cos \frac{1}{2}(\frac{\pi}{2} - \theta), \tag{5}$$

$$l_2 = l \sin \frac{1}{2}(\frac{\pi}{2} - \theta). \tag{6}$$

To see this, consider $\triangle OYZ$, an isosceles triangle with $\overline{OY} = \overline{OZ} = l/2$. Let W be on YZ such that $OW \perp YZ$, so $\triangle OWY$ is a right triangle with

$$\angle YOW = \frac{1}{2}(\frac{\pi}{2} - \theta) \tag{7}$$

TABLE 1. Proprioceptive input values for the network. The value in_M denotes the randomly generated neuronal input to muscle M. The values l_M and T_M respectively represent the length and tension input values of muscle M.

Joint	Angle	Muscle (M)	Length (l_M)	Tension (T_M)
α	$\frac{\pi}{2}(in_B - in_D)$	Abductor	$\sin\left(\frac{1}{2}\left(\frac{\pi}{2} - \alpha\right)\right)$	$in_B + 0.1 \cdot l_B$
		Adductor	$\cos\left(\frac{1}{2}\left(\frac{\pi}{2} - \alpha\right)\right)$	$in_D + 0.1 \cdot l_D$
β	$\frac{\pi}{2}(in_E - in_F)$	Extensor	$\sin\left(\frac{1}{2}\left(\frac{\pi}{2} - \beta\right)\right)$	$in_E + 0.1 \cdot l_E$
		Flexor	$\cos\left(\frac{1}{2}\left(\frac{\pi}{2} - \beta\right)\right)$	$in_F + 0.1 \cdot l_F$
γ	$\frac{\pi}{2}(in_O - in_C)$	Opener	$\sin\left(\frac{1}{2}\left(\frac{\pi}{2} - \gamma\right)\right)$	$in_O + 0.1 \cdot l_O$
		Closer	$\cos\left(\frac{1}{2}\left(\frac{\pi}{2} - \gamma\right)\right)$	$in_C + 0.1 \cdot l_C$

4. A learning sensorimotor map of arm movements 69

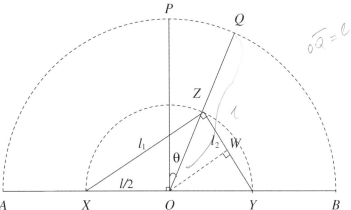

FIGURE 3. Generic joint of muscles XZ and YZ and arm segment OQ of length l. The pair of muscles XZ and YZ move the arm segment OQ from positions OA to OB through contraction and stretch. For example, contraction of muscle YZ and stretch of muscle XZ moves the arm segment to the right as shown in the figure. Thus, the possible positions of Q define a semicircle APB. Both muscles are attached to the midpoint Z of the arm segment, (i.e., $\overline{OZ} = l/2$). Muscle XZ is also attached to point X and muscle YZ to point Y, respectively, which are located distance $l/2$ apart from the joint O on opposite sides (i.e., $\overline{OX} = \overline{OY} = l/2$). Joint angle θ denotes the angle between OQ and OP.

and
$$\overline{YW} = \frac{l_2}{2}. \qquad (8)$$

From Equations 7 and 8 we get

$$\sin \frac{1}{2}(\frac{\pi}{2} - \theta) = \frac{\overline{YW}}{\overline{OY}} = \frac{l_2/2}{l/2} = l_2/l.$$

Thus, we have Equation 6.

Now consider $\triangle XZY$. Point Z is on a semicircle with center O and diameter l, so $\angle XZY = \frac{\pi}{2}$. Then

$$\overline{XZ}^2 + \overline{YZ}^2 = \overline{XY}^2$$
$$l_1^2 + l_2^2 = l^2$$
$$l_1 = \sqrt{l^2 - l_2^2}.$$

Substituting (6) for l_2, we have Equation 5 since $\frac{1}{2}(\frac{\pi}{2} - \theta) \in [0, \frac{\pi}{2}]$.

Because of their location terminal to muscle fibers, Golgi tendon organs strongly respond when the muscle actively contracts. Passive stretching of the muscle also activates the Golgi tendon organ but not as much [CG85]. These observations lead to the following definition.

Muscle Tension: *Let in_i denote the input activation to muscle i. Then the muscle tension T_i is defined as*

$$T_i = in_i + T\, l_i, \tag{9}$$

where T is a small constant.

The first term, in_i, at unit i represents the active portion of the total tension generated by the muscle. The second term, $T\, l_i$, represents the secondary sensitivity of the Golgi tendon organ to passive muscle stretching.

Input values to muscles in are uniform random variables to the neural network whose values range from 0 to 1. However, actual input values to the neural network such as joint angle, length, and tension are not uniform random variables. This is because any arbitrary transformation of uniform random variables does not usually result in another uniform random variable. This leads us to the observation that certain combinations of length and tension values are presented disproportionally more often during training. For instance, joint angle values near zero will be presented more often than other values.

3 Simulation Results

We present three types of results from this study of map formation in the proprioceptive sensory cortex. First, we show that both length and tension maps formed during training. Second, we characterize these maps by describing various redundancies and relationships that appear. Third, we describe the map of hand position in three-dimensional space that formed even though there was no explicit input of hand position.

3.1 Formation of Length and Tension Maps

To examine whether maps of muscle length and tension formed during training, we measured which muscle's length or tension each cortical unit responded to most strongly. Consider an input pattern where only one muscle length or tension unit (arm unit) is activated. There are 12 such input patterns, because we have six muscle lengths and six tension units. Since the arm units represent the length and tension of six muscles of the model arm (flexor and extensor, abductor and adductor in upper arm, and flexor and extensor in lower arm), each of these input patterns corresponds to the unphysiological situation where either length or tension of only one muscle is activated. For instance, an input pattern of $(P, 0, 0, \ldots, 0)$ represents the case where the upper arm extensor's length unit is activated, while all other units are not.[2] These input patterns were not used during training. Nev-

---

[2] A positive constant P is introduced to make the magnitude of a test input pattern similar to that of a normalized training pattern whose size was 7.45.

4. A learning sensorimotor map of arm movements   71

ertheless, they provide an unambiguous and simple method for measuring map formation.

A cortical unit is taken here to be "tuned" to an arm input unit if the sole excitation of the input unit produced activation larger than a threshold of 0.5 at that cortical unit. A cortical unit is "maximally tuned" to an arm input unit if it is tuned to that input unit and the activation corresponding to that input unit is largest. We determined to which of the six muscles each cortical unit was tuned maximally. This was done with respect to both the length and tension of each muscle independently.

Figure 4 shows the maximal tuning of cortical units, before (on the top) and after (at the bottom) training. Consider, for example, the unit displayed in the upper left corner of the cortical layer. After training (bottom figures), it was maximally tuned to "O" in the length tuning figure and "c" in the tension tuning figure. This translates into, This unit responded maximally to the opener with respect to the muscle length, but to the closer with respect to the muscle tension.[3] Cortical units marked with a "-" character were found to be not tuned to the length or tension of any muscle.

The number of untuned cortical units decreased 16% (length) and 30% (tension) with training. The number of cortical units tuned to multiple muscle lengths and multiple tension lengths after training were 46 and 27, respectively. The number of those units multiply tuned to either length or tension was 230.

Now compare the figures regarding length tuning before and after training (those on the left of Figure 4). Clusters of units responsive to the same muscle became more uniform in size after training. The size of clusters ranged from 2 to 10 before training, but ranged from 3 to 4 after training, and their shape became more regular. Clusters of units tuned to antagonist muscles were usually pushed maximally apart from each other during training. Many of these changes are more obvious visually if one considers a map of just two antagonist muscles. For instance, consider the clusters shown in Figure 5, where only those units in Figure 4 tuned to upper arm extensor ("E") and flexor ("F") muscles are displayed. After training, clusters of "E"s and "F"s are pushed maximally away from each other, evenly spaced, and more uniform in size. The network captures the mechanical constraint imposed by the model arm that two antagonist muscles cannot be stretched at the same time. This result is representative: clusters of antagonist muscles are pushed apart for other pairs, such as upper arm abductor and adductor muscles ("B" and "D") and opener and closer muscles ("O" and "C").

In the case of tension tuning figures, the results were similar. The size of

---

[3]Implications of this type of "multiple tuning to antagonists" will be explored in the next section.

(a) Length Tuning in
Untrained Cortical Layer

```
- - C - F - C B F D C C B - F C C C - -
- B B O - - - C F F D C - D F F C C C F F
- B O - - C E E - - - D O F C D D C F -
D F - E E C - - - - - - B B E E - - -
F F E E - B B - - - - B B E E - - O - -
D D - D O B B - F F - E E - - - O B - -
D D D O O E - C D E E E - D F O B E E -
B D D - E E - C D - - - D D F - E E C B
- - C B E - O O - - - - D C C - - C C B
E C C F F - O - B B - - - C - - - F - E
- C F F - - - - F O - E E - - - F F O E
- F O - - - - - - F - E E F - D D B O - -
- - O O O O C E D D - - O - - D - - - D D
- - - D O E E D D C C O E C - - - - D B
C - - D F F E - B B C E C C F F - - B B
C C B D F - E O B F F D - F F C - E E C
C B B D - - O O - F F E - - C C E E E D
F O - - C C E E - - E O - D D E O O D D
O O E - C D - - B - O - - D E - O B - F
O E E - F D - B B O O - B B - - B B - -
```

(b) Tension Tuning in
Untrained Cortical Layer

```
e c f f c c d d e e c c - f d b b b - e
o o f d c c - e e - c c - d d o o - f e
- o d - - b f f o b b - - e e o d f f e
- d - - b e c o d b - - c e - d d - - -
- - - - e e c - - f f c c f - c c - - -
- - - e e - - e e f o o f b c c d d - -
b d d o - - - b e c o o b b - - o o f b
b d f f - - b b d c - - - e e - o f c b
- - - c c - o d d - - - - e e c c e e -
- - - c - o o - - b - - - c c - d - -
- b b - o o - - b b f d d e e o d d c -
- b b e f - - - e e f d - b b o - f f -
d d e e f b - - - - o o - - - - - f o -
d o c f f b o - - - o - - - - - - e - -
o o c f d o o - - c e o f - - d c - - -
b f - - d o f f c c c f f - d d c - - b
f - - - - e f f c c - f b b d - e - - f
- - - - e e b - - - - b b b - e e - - -
- - - - - b - o f - - - o o o o - - - d
- c c - - - d d - e - - o f f o b b d d
```

(c) Length Tuning in
Trained Cortical Layer

```
O E - - - F - E E - - F F - O E D D C F - -
O - C C - O O - C B B O O - D C C - - O
- B C C D O O - C C - O F - - - E D D -
F B E D D F F - E E - F F B B E E D D F
- E E - F F - E D D C F B B O - - - C C F
O O - C C - O O D - C - - O O - - C C -
O - B B B - O - - - E E D D F B B - D D
F - B E - - - F F B - E D D F F B E E D D
F - E E - - F B B O - C C F - O E E - F
- O D D C C - - O O - - C C - O O - - C B
- O D - C C D D F F B B E D D - - B B -
- F F - E E D D F - B E E D D F - - E -
F B B - E - - C - - O E - C F F O E E -
C B - O O - C C - O O - C C - O O D - C
- - - F B B C E D D F - B B - O D D C C
D D F F B B E E - F F - E E - - F - - E
D - C - - O - - - F - O E - - F F - E E
- C C - O O - B C - O O - C C B B O O -
B C E D O F B B C - D D - C B B - O F B
B E E D F F - E E - F F - E E D - F F B
```

(d) Tension Tuning in
Trained Cortical Layer

```
c - - - - e - f f - - e e c f b b - e e c
- d d o o c c - d o o c c - - - o o c c
- d o o - c e d d o - - - e - - f - b -
e - f b b e e - f - b - d e c f f b b e
c f f b - e - f f b b o d c c - - b - e
c - - d o c c - - b o o - c - d d o o c
- - d d o c c e e - f f b b e d - o - -
e e f f - - - e e f f b b e e - f f b b
e - f f b - d d c c - - o e c c f - b -
c c - b b o o - c - d d o - c - - d o o
c c - - - - o - b - e d - - - b - d d o o
- d e e f f b b e e - f f b b e e f f -
d d e f f - b o e c c f - o e e c f b -
o o c c - d o o - c - - d o o c c b b -
o - c - d d f - b b e d d - - c - b - o
- b - e e f f - b e e f f - - - e e f f
b b o e - c - d - e c f f b d d e c f f
b o o - c c d d o o c c b d o o c c - -
- - f b c e d - o - b b - - - o - c - d d
- f b b e e f f - b b e e f f b b d e -
```

FIGURE 4. Tuning of cortical units to muscle length and tension. Labels E, F, B, D, O, and C represent length of upper arm extensor and flexor, upper arm abductor, adductor, lower arm opener and closer, respectively, while labels e, f, b, d, o, and c represent tension of corresponding muscle.

clusters became more uniform. However, the clusters of antagonistic muscles were not separated maximally. In fact, some antagonist clusters were located adjacent to each other. This is due to the fact that in contrast to what we see with muscle lengths, there are no constraints preventing two antagonist muscles from contracting at the same time. *Cocontraction* of antagonist muscles is employed when a stiffer joint is necessary, for instance to ensure the desired position when unexpected external forces are present [CG85].

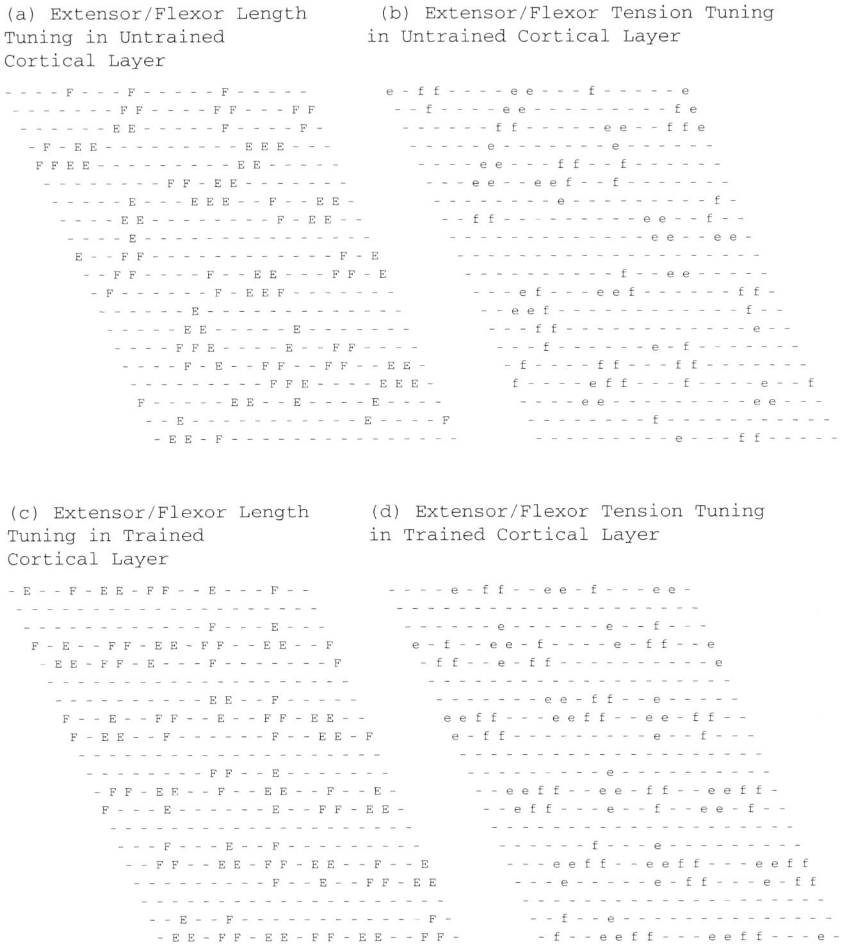

4. A learning sensorimotor map of arm movements 73

(a) Extensor/Flexor Length Tuning in Untrained Cortical Layer

(b) Extensor/Flexor Tension Tuning in Untrained Cortical Layer

(c) Extensor/Flexor Length Tuning in Trained Cortical Layer

(d) Extensor/Flexor Tension Tuning in Trained Cortical Layer

FIGURE 5. Tuning of cortical units to length and tension of upper arm extensor and flexor muscles only. The same set of labels defined in Figure 4 is used.

3.2 Relationships

Additional evidence of the trained network's capturing of the mechanical constraints imposed by the arm is found among those cortical units that are tuned to multiple proprioceptive inputs (i.e., activation over 0.5 for multiple test patterns, each of which corresponds to an input unit). Such multiple tuning could potentially not be compatible with physiological constraints. For instance, it seems unlikely that a cortical unit would be tuned to both a muscle's length and to its tension together since a muscle tends not to contract (high tension) and lengthen simultaneously. Another implausible case would be when a cortical unit is tuned to lengths of two antagonist muscles since they cannot be stretched at the same time.

Table 2 shows the number of implausible multiple tuning cases found in the network before and after training. For instance, pair (E, F) represents the number of cortical units that are tuned to both "E" (length of upper arm extensor) and "F" (length of upper arm flexor), and pair (B, b) represents the number of cortical units that are tuned to both length and tension of the upper arm abductor muscle. Each entry represents the number of cortical layer units that were tuned to a physiologically implausible pair of arm layer units. Entries in the top row show the number of units before training and those at the bottom row after training. Before training, a total of 69 cortical units were tuned to implausible pairs. After training, none of the cortical units had implausible tuning. This clearly shows that the trained network captured the physiological constraints imposed by the mechanics of the arm by eliminating implausible multiple tuning effects introduced by random initial weights.

Tuning of units to some multiple proprioceptive inputs, on the other hand, could be compatible with the constraints imposed by the mechanics of the model arm. For instance, in Section 3.1, we considered the unit shown on the upper left corner in Figures 4c and 4d, which is tuned to both the length of the opener and to the tension of the closer. This unit is, in that sense, tuned to the contraction of a single muscle, the closer. Contraction of this muscle increases its arm tension (c) and also increases the length of its antagonist muscle, the opener (O). Table 3 shows the number of the cortical units tuned to specific plausible tuning pairs, with the top row being before training and the bottom row being after training. The tuning pairs follow the same convention used in Table 2. The pair (E, f), for instance, represents the extensor's length and flexor's tension, thus contraction of the upper arm flexor. Those cortical units that were also tuned to implausible pairs were not counted here even though they might also be tuned to contraction of a plausible pair. The data "before training" show the effect of randomness of initial weights. Training increased the number of such cortical units by more than four times. This effect is clearly

TABLE 2. Numbers of implausibly tuned cortical layer units. Uppercase letters represent muscle length while lowercase letters represent muscle tension.

Tuning pairs	E, F	B, D	O, C	E, e	F, f	B, b	D, d	O, o	C, c	Total
Before training	7	5	6	6	10	7	9	9	10	69
After training	0	0	0	0	0	0	0	0	0	0

4. A learning sensorimotor map of arm movements

TABLE 3. Numbers of plausibly tuned cortical units.

Tuning pairs	E, f	F, e	B, d	D, b	O, c	C, o	Total
Before training	12	13	8	6	2	6	47
After training	42	37	18	35	35	33	200

illustrated in Figures 5c and d. (Compare the left (c) illustration with the corresponding right (d) illustration.)

After training, the map can be viewed as being organized into fairly compact contiguous regions where all possible features are represented in each region. For instance, the region of about 30 units in the lower left corner of the upper right quadrant (Figures 4c and d) illustrates this especially clearly: it has units tuned to every possible muscle length and tension. Such an organization is reminiscent of hypercolumns in visual cortex and quite different from that seen with past cortical maps of touch sensation [GM90, PFE87, Sut92].

3.3 Formation of Hand Position Map

Recall that the sole input information to the model cortex is length and tension information from each of the six muscle groups that control arm position. In other words, *there is no explicit input information about the "hand" position in the three-dimensional space in which it moves.* To assess what, if any, kind of map of three-dimensional hand position develops in cortex, we divided up the hand position space into 27 cubicles (three segments for each axis), computed an "average" hand position for each cubicle, presented the input patterns corresponding to the average hand positions, and determined to which of these 27 test input patterns each cortical unit is maximally tuned. We considered also for each cortical unit to which of the three segments of x-, y-, and z-axes it is tuned. In this scheme, the x, y, and z axes are divided into three equal-length segments (Figure 6). We chose this particular division of space based on the facts that a large number of the training patterns were covered by the resulting 27 cubicles (86%) and that every cubicle contains at least one training pattern.[4]

A cubicle is identified as a triplet (i, j, k), where values of i, j, and k

---

[4]The training patterns were not evenly spaced.

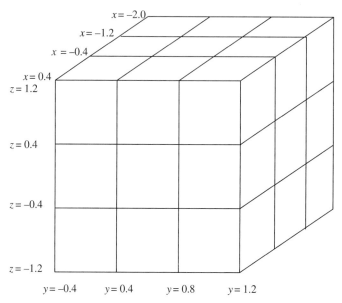

FIGURE 6. Division of hand position space into 27 cubicles. The x-axis was segmented into three sections, X_1, X_2, and X_3, of lengths $[-2, -1.2]$, $[-1.2, -0.4]$, and $[-0.4, 0.4]$, respectively. The y-axis was segmented into three sections, Y_1, Y_2, and Y_3, of lengths $[-0.4, 0.4]$, $[0.4, 1.2]$, and $[1.2, 2.0]$, respectively. The z-axis was segmented into three sections, Z_1, Z_2, and Z_3, of lengths $[-1.2, -0.4]$, $[-0.4, 0.4]$, and $[0.4, 1.2]$, respectively.

denote the location of the cubicle as

$$i = \begin{cases} 1, & \text{if } H_x \in [-2, -1.2], \\ 2, & \text{if } H_x \in [-1.2, -0.4], \\ 3, & \text{if } H_x \in [-0.4, 0.4]; \end{cases}$$

$$j = \begin{cases} 1, & \text{if } H_y \in [-0.4, 0.4], \\ 2, & \text{if } H_y \in [0.4, 1.2], \\ 3, & \text{if } H_y \in [1.2, 2.0]; \end{cases}$$

$$k = \begin{cases} 1, & \text{if } H_z \in [-1.2, -0.4], \\ 2, & \text{if } H_z \in [-0.4, 0.4], \\ 3, & \text{if } H_z \in [0.4, 1.2], \end{cases}$$

where hand position is (H_x, H_y, H_z). For each cubicle (i, j, k), the average hand position, was calculated from the training samples whose resultant hand positions were within the boundaries of the cubicle, and the corresponding muscle lengths and tensions were computed. Note, however, that only muscle lengths are determined uniquely, given hand positions: muscle tensions are not unique. For simplicity, we chose the tension values such that the total tension at each joint was either maximal or minimal. We ran the 27 resulting testing patterns with the already trained network and observed each cortical unit's activation. Since we got similar results from maximal tension and minimal tension patterns, we present the results from maximal tension patterns only from now on.

Figures 7 and 8 show the cortical units' spatial tuning to arm location before and after training, respectively. Tuning after training clearly shows map formation.

There are also clear relationships between spatial position and specific proprioceptive inputs in the map. To understand this, recall that muscle length and hand positions are jointly involved in a set of mechanical constraints imposed by the model arm. For example, the contraction of the adductor muscle, and thus the stretch of its antagonist abductor muscle, positions the elbow and hand below the shoulder. This translates into the hand position's z-coordinate being negative (namely, the segment Z_1 in Figure 6). In other words, a stretched abductor muscle is very likely to correlate with hand position being in Z_1.[5] Stretching of the adductor muscle, on the other hand, is very unlikely to place the hand in Z_1 but is very likely to be correlated with the hand position in Z_3 (i.e., a positive z-coordinate). Another similar constraint is that the contraction of the upper arm flexor muscle, and thus the stretching of its antagonist upper arm extensor muscle, tends to position the elbow in front of the body, resulting in the hand being placed very far in front of the body. This translates to the hand position's x-coordinate being very negative (i.e., in segment X_1, also defined in Section 3.3). Therefore, the stretch of the upper arm extensor is very likely to position the hand in X_1. In short, the mechanics of the model arm imposes constraints on the relations between muscle length and hand positions such that there are certain pairs of muscle and hand positions that are very likely to happen simultaneously and such that there are some other pairs that are not likely to happen simultaneously.

To see whether the network learned these types of constraints, we calculated the number of cortical units that were tuned both to the stretch of a muscle and to various segments of the hand positions of all three axes, both before and after training. Tables 4 and 5 show the number of cortical units that are maximally tuned both to the length (stretch) of a certain muscle and to a certain segment of hand positions before and after training, respectively. For instance, the entry "28" in the upper left corner of Table 4 represents the number of cortical units that were tuned to the stretch of the upper arm extensor muscle and to the hand position in segment X_1, before training. The entry "37" in the upper left corner of Table 5 represents the same thing after training. After training, the number of the cortical units tuned to plausible pairs of muscle stretch and hand position values increased significantly, while the number of cortical units tuned to implau-

---

[5] Recall that the model arm segments do not rotate.

x-direction Tuning in Untrained Cortical Layer

```
            2 1 1 3 3 3 3 1 3 3 3 - - 3 2 1 1 1 - 3
            1 1 3 1 1 1 1 1 1 2 2 - - 2 3 3 1 2 2 3
            3 2 1 1 1 1 2 2 1 2 3 3 3 3 3 1 1 1 - 2
            3 3 1 - 2 2 3 3 1 3 3 3 2 - 1 1 1 1 2 3
            2 1 - 2 2 3 3 1 1 3 2 2 2 1 1 2 2 2 3 2
            1 3 - 3 3 3 2 - 1 2 1 1 1 1 3 2 2 - 1 2
            3 3 2 3 1 2 2 1 1 1 1 2 2 3 3 2 2 1 1 3
            2 3 1 1 1 1 2 - - 1 1 2 2 2 1 1 1 2 2 3
            3 2 1 2 1 2 1 1 1 1 3 2 3 1 1 2 1 1 2 2
            1 2 2 2 3 1 1 2 3 3 3 2 1 1 2 2 2 2 3 2
            2 3 3 2 2 3 3 2 2 2 2 1 1 2 1 1 3 3 3 1
            3 3 3 2 1 1 2 2 2 2 1 1 2 1 1 1 3 3 2 1
            3 3 2 3 1 1 1 2 3 - 1 - - 2 1 1 3 2 1 1
            3 3 2 2 2 2 3 3 1 1 1 1 1 1 1 3 2 2 2 2
            - 3 3 2 3 3 3 1 3 3 1 1 1 - 3 3 2 2 2 2
            1 3 3 3 3 3 2 1 3 3 3 2 2 3 3 3 3 3 2 1
            2 3 3 - 3 1 1 2 2 3 3 2 1 3 3 2 3 3 2 1
            2 3 - - 2 1 2 1 1 3 1 1 1 1 2 2 2 2 2 2
            3 3 2 1 2 3 1 1 1 1 3 1 1 1 2 2 2 3 2 2
            3 1 1 1 1 3 3 3 1 1 3 3 - - 2 2 2 1 - - 3
```

y-direction Tuning in Untrained Cortical Layer

```
            2 1 2 3 2 1 1 3 1 1 2 - - 1 2 1 1 1 - 3
            2 3 2 2 1 2 3 2 1 2 1 - - 1 3 3 1 3 3 3
            2 3 2 2 1 2 3 2 1 1 2 2 1 2 1 3 3 2 - 3
            3 3 1 - 2 3 2 1 2 1 2 1 1 - 2 3 2 1 1 1
            3 1 - 1 1 2 1 3 3 3 1 1 1 2 2 1 1 1 1 2
            1 1 - 2 1 1 2 - 3 3 3 1 1 1 2 2 2 - 3 2
            1 1 1 2 2 1 1 1 1 3 1 1 1 1 3 3 3 2 2 2
            1 3 2 1 1 2 2 - - 1 1 1 1 3 3 3 2 3 3 2
            3 3 1 1 3 2 2 1 1 2 3 2 1 3 3 1 2 2 2 2
            2 2 1 1 2 3 1 1 1 2 3 3 3 2 1 1 2 2 1 2
            3 3 2 1 3 3 3 2 1 3 3 3 3 1 2 1 1 3 2 1 3
            3 3 1 1 2 2 3 2 1 3 1 1 1 1 1 1 3 3 1 1 3
            3 3 2 3 1 2 3 1 1 - 3 - - 3 3 1 3 2 1 3
            3 3 1 2 3 3 3 3 2 3 3 2 2 2 1 3 2 3 3 3
            - 3 2 2 2 3 3 3 3 1 2 1 1 - 3 1 2 3 3 3
            2 3 2 1 1 1 2 2 1 2 2 2 3 2 2 3 3 3 1
            3 1 2 - 1 3 2 3 3 1 1 1 1 3 1 2 1 2 2 1
            2 1 - - 3 3 1 2 2 1 1 2 3 3 1 2 2 1 1 2
            2 2 1 2 2 3 1 3 3 1 2 1 3 1 3 2 1 1 1 1
            2 1 1 1 2 3 1 3 3 1 2 - - 2 3 3 1 - - 1
```

z-direction Tuning in Untrained Cortical Layer

```
            2 2 1 1 2 2 2 2 1 1 2 - - 2 1 1 3 3 - 2
            3 1 1 1 2 2 2 1 1 3 3 - - 1 1 3 3 1 1 3
            2 1 1 1 3 3 1 2 1 3 2 1 1 1 3 3 1 1 - 3
            1 1 1 - 3 1 2 2 1 3 2 1 1 - 3 3 1 1 2 2
            2 1 - 3 3 2 2 1 2 1 2 1 1 3 3 2 1 1 2 2
            2 2 - 3 3 2 1 - 3 2 1 1 3 3 2 2 1 - 3 3
            2 2 1 3 1 1 1 3 3 2 2 3 3 3 3 1 1 1 3 3
            2 2 2 1 2 2 3 - - 1 2 3 3 3 2 1 1 1 2 3
            1 2 3 3 3 2 1 1 1 2 2 1 1 2 1 2 1 2 2 1
            2 3 3 2 2 1 1 1 2 1 1 1 3 3 1 2 2 2 1 1
            2 3 3 2 3 1 1 1 2 2 3 2 2 3 3 2 2 1 2 1
            3 2 1 3 3 1 1 3 3 3 1 2 2 3 3 2 2 3 2 3
            1 1 1 2 1 2 2 3 3 - 1 - - 1 1 2 3 1 3 3
            1 1 1 1 2 2 2 3 3 1 1 1 1 1 1 2 1 1 2 3 1
            - 1 1 2 2 1 2 3 2 1 1 1 1 - 1 1 2 3 1 1
            1 2 3 2 1 1 1 2 2 1 1 1 1 1 1 2 3 1 1 1
            1 3 3 - 1 1 1 3 3 3 3 2 2 1 2 3 3 1 2 1
            1 3 - - 3 1 3 3 3 3 2 2 2 1 3 3 1 1 2 2
            1 2 3 3 3 3 3 3 3 2 1 1 1 3 3 1 1 2 1 1
            2 1 2 2 3 3 2 3 3 1 2 - - 2 2 1 1 - - 1
```

FIGURE 7. Tuning of untrained cortical units to hand position in each direction, x, y, and z. Each unit is labeled such that the corresponding element of cubicle (i, j, k) is displayed when the unit is maximally tuned to the hand position from the cubicle.

sible pairs decreased. For example, as discussed above, the number of units tuned to abductor-Z_1 pair and to adductor-Z_3 pair (i.e., likely pairs) has increased from 15 to 40 and 19 to 39, respectively, while the number of units tuned to adductor-Z_1 pair and abductor-Z_3 pair (i.e., unlikely pairs) has decreased from 11 to 0 and 8 to 1, respectively. Figure 9 illustrates that

4. A learning sensorimotor map of arm movements   79

x-direction Tuning in Trained Cortical Layer

```
2 - 2 3 3 2 2 - - 3 3 2 2 2 1 1 3 3 2 2
3 - 2 2 2 2 2 - 2 2 2 2 2 2 2 1 2 - 2 2
3 2 1 1 1 2 2 2 1 1 2 2 2 2 2 1 1 1 2 2
3 1 1 1 2 2 2 2 1 1 2 3 3 2 1 1 1 1 3 3
2 2 - 2 3 3 2 2 1 1 2 3 2 2 2 2 1 1 3 3
2 2 - 2 3 2 2 2 2 2 1 2 - 2 3 2 1 2 2 2
2 2 1 1 1 - 3 2 2 1 1 1 2 2 3 3 1 1 1 2
3 2 2 1 1 - 3 2 2 1 1 1 2 3 3 2 1 1 1 3
2 2 1 1 - 3 3 2 2 2 - 1 2 3 3 2 2 1 1 3
2 2 2 2 2 2 2 - 3 3 - 2 1 - 2 2 2 1 2 3
2 3 2 2 1 1 1 2 2 2 2 1 1 1 2 2 - 1 1 1
3 3 2 2 1 1 1 2 3 3 2 1 1 1 3 3 2 1 1 1
3 3 2 2 2 - - 3 3 - 2 - 1 2 3 2 2 1 1 1
2 2 2 2 3 - 2 1 1 2 3 - 1 2 2 2 2 2 1 2
1 2 3 2 3 1 1 1 2 2 2 2 1 1 - 2 2 2 2 1
1 2 3 3 2 1 1 - 2 3 3 2 1 1 - 3 3 2 1 1
1 3 3 3 - 2 2 - 3 3 2 2 1 1 1 3 2 2 2 1
2 2 2 2 2 2 - 2 2 2 2 2 1 1 2 2 2 2 2 -
1 1 1 2 3 2 2 1 1 1 2 2 - 1 1 - 2 2 2 3
1 1 1 3 3 3 2 1 1 - 3 2 - 1 1 1 2 3 3 2
```

y-direction Tuning in Trained Cortical Layer

```
1 - 3 3 3 2 1 - - 3 2 1 1 1 2 2 3 3 2 1
1 - 3 3 2 2 1 - 3 3 1 2 1 1 1 2 3 3 - 1 1
2 2 2 2 1 1 2 1 2 2 2 2 2 2 2 2 1 1 2 2
2 1 1 2 2 2 1 1 1 1 2 3 2 2 1 1 1 2 2 3
1 1 - 3 3 2 1 1 1 2 3 3 2 1 1 1 2 3 3 3
1 1 - 3 3 2 1 1 2 3 3 3 - 1 1 2 3 3 3 2
2 2 3 2 1 - 1 2 2 1 2 1 2 1 2 1 2 2 1 1
2 2 3 1 1 - 2 2 1 1 1 2 3 3 2 1 1 1 1 1
2 1 1 1 - 3 3 2 1 1 - 3 3 3 2 1 1 2 3 3
1 2 1 1 2 3 3 - 1 1 - 3 3 - 1 1 1 3 3 3
2 1 2 1 1 2 1 1 2 2 2 2 1 1 1 2 - 3 1 1
1 2 2 1 1 1 1 2 3 2 1 1 1 2 2 2 1 2 1 1
3 2 1 1 1 - - 3 3 - 1 - 2 3 3 2 1 1 2 3
3 2 1 1 1 - 3 2 1 1 1 - 3 3 2 1 1 1 3 3
1 2 1 2 1 2 2 1 1 1 2 1 2 1 - 1 2 2 3 2
1 3 2 2 2 1 1 - 2 2 1 1 1 1 - 2 2 2 1 1
2 3 3 3 - 1 1 - 3 3 2 1 1 2 3 3 2 1 1 1
3 3 3 2 1 1 - 3 3 3 1 1 2 3 3 2 1 1 1 -
2 2 1 2 1 2 1 2 2 1 1 2 - 3 1 - 1 1 2 1
1 1 2 1 2 2 2 1 2 - 2 2 - 1 1 1 1 2 2 1
```

z-direction Tuning in Trained Cortical Layer

```
2 - 3 2 2 1 2 - - 2 1 1 2 2 3 3 3 1 1 1
2 - 2 3 3 3 2 - 2 2 1 3 2 3 3 2 1 - 3 2
2 1 1 2 3 3 2 1 1 2 3 3 2 2 1 1 2 3 3 2
1 1 2 3 3 2 1 1 2 2 3 3 2 1 1 1 2 3 3 2
1 2 - 3 3 1 1 2 3 3 3 1 1 1 2 2 2 2 2 1
2 2 - 2 1 1 2 2 3 3 2 1 - 3 2 2 2 2 3 3
2 2 2 1 1 - 2 2 2 1 1 2 3 3 2 1 1 1 3 3
2 2 3 2 2 - 3 2 1 1 2 3 3 2 1 1 1 2 3 3
1 1 3 3 - 3 1 1 1 2 - 3 2 1 1 2 2 2 3 2
1 3 3 3 3 1 1 - 2 2 - 1 1 - 2 2 2 2 2 1
3 3 2 3 1 2 3 3 3 2 1 1 2 3 3 2 - 1 1 1
3 2 2 3 1 2 3 3 2 1 1 2 2 3 3 2 1 1 2 2
3 1 1 3 2 - - 3 1 - 2 - 3 3 1 1 1 3 3 3
1 1 3 2 2 - 1 3 3 2 2 - 2 1 1 1 2 2 3 2
1 3 3 2 1 1 1 2 3 3 2 1 1 1 - 3 3 2 3 1
3 3 3 1 1 1 2 - 3 3 1 1 1 2 - 3 2 2 1 2
2 3 2 1 - 2 2 - 1 2 1 2 3 3 3 2 1 1 2 2
2 2 3 3 3 2 - 1 2 3 3 2 3 3 1 1 1 2 2 -
1 1 3 3 3 2 1 1 2 3 3 2 - 1 1 - 3 3 2 1
1 2 3 3 3 1 1 2 3 - 3 2 - 1 2 2 3 3 1 1
```

FIGURE 8. Tuning of trained cortical units to hand position in directions x, y, and z. Each unit is labeled such that the corresponding element of cubicle (i, j, k) is displayed when the unit is maximally tuned to the hand position from the cubicle. In the x-axis tuning, stripes of 1s, 2s and 3s in the orientation of northwest to southeast appear. Also in the y-axis and z-axis tuning shown are similar stripes of 1s, 2s and 3s in the orientation of northeast to southwest. There were no such tuning stripes found in the untrained cortical layer (Figure 7). A careful examination of the spatial location of stripes formed reveals that their orientation does not match the hexagonal tessellation of the underlying network, and thus it is not an artifact of the particular tessellation used in the model.

TABLE 4. Number of cortical units maximally tuned to length and hand position (before training).

	X_1	X_2	X_3	Y_1	Y_2	Y_3	Z_1	Z_2	Z_3	Total
Extensor (E)	28	17	5	19	17	14	17	17	16	150
Flexor (F)	5	13	24	15	13	14	21	13	8	126
Abductor (B)	6	13	15	12	10	12	15	11	8	102
Adductor (D)	14	12	15	19	9	13	11	11	19	123
Opener (O)	11	10	15	13	17	6	22	9	5	108
Closer (C)	23	11	8	13	15	14	15	11	16	126
Total	87	76	82	91	81	73	101	72	72	

TABLE 5. Number of cortical units maximally tuned to length and hand position (after training).

	X_1	X_2	X_3	Y_1	Y_2	Y_3	Z_1	Z_2	Z_3	Total
Extensor (E)	37	7	0	39	5	0	6	30	8	132
Flexor (F)	0	15	37	2	33	17	18	24	10	156
Abductor (B)	17	21	5	21	16	6	40	2	1	129
Adductor (D)	18	23	4	22	21	2	0	6	39	135
Opener (O)	0	38	9	41	6	0	2	33	12	141
Closer (C)	19	23	5	1	11	35	16	19	12	141
Total	91	127	60	126	92	60	82	114	82	

cortical units representing a stretched, longer abductor muscle are overwhelmingly embedded in the stripes representing hand position Z_1. The other constraints we discussed above also seemed to be learned, as shown in the significant change between before and after training of the entries in the upper left box and the lower middle box of Tables 4 and 5.[6] In addition, these tables show more instances of interesting tuning such as in the upper middle box, where the entries in upper arm extensor-Y_1 and upper arm flexor-Y_2 greatly increased while those in upper arm extensor-Y_2, upper arm extensor-Y_3, and upper arm flexor-Y_1 significantly decreased. This is due to the fact that the stretch of the upper arm extensor and stretch of the upper arm flexor tend to place the hand toward the negative side of the y-axis (i.e., Y_1) and toward the positive side of the y-axis (i.e., Y_2 and Y_3), respectively. Comparison of the two tables shows that the network learned the constraint that the contraction/stretch of certain muscles positions the hand in certain locations in space. Since the hand position was not explicitly provided as input, the network seems to learn to encode the "interrelationship" among the muscle lengths. The spatial map of hand position that the model developed can be considered as a higher-order map

---

[6] Entries in the tables are divided up into nine boxes, excluding the "total" column and row. Each box is associated with one set of antagonist muscles and one axis of hand positions.

4. A learning sensorimotor map of arm movements

```
Units tuned to the abductor          Units tuned to Z_1
- - - - - - - - - - - - - - - - - -  - - - - - 1 - - - - 1 1 - - - - - 1 1 1
- - - - - - - - - 1 1 - - - - - - -  - - - - - - - - - - 1 - - - - - 1 - - -
- 1 - - - - - - - - - - - - - - - -  - 1 1 - - - - 1 1 - - - - - 1 1 - - - -
- 1 - - - - - - - - - - 1 1 - - - -  1 1 - - - - 1 1 - - - - 1 1 1 - - - - -
- - - - - - - - - - - 1 1 - - - - -  1 - - - - 1 1 - - - - 1 1 1 - - - - - 1
- - - - - - - - - - - - - - - - - -  - - - - 1 1 - - - - - 1 - - - - - - -
- - 1 1 1 - - - - - - - - - 1 1 - -  - - - 1 1 - - - - 1 1 - - - - 1 1 1 - -
- - 1 - - - - - 1 - - - - - 1 - - -  - - - - - - - - 1 1 - - - - 1 1 1 - - -
- - - - - - - 1 1 - - - - - - - - -  1 1 - - - - 1 1 1 - - - - 1 1 - - - - -
- - - - - - - - - - - - - - - - - 1  1 - - - - 1 1 - - - - 1 1 - - - - - - 1
- - - - - - - - - 1 1 - - - - - 1 1  - - - - 1 - - - - - 1 1 - - - - - 1 1 1
- - - - - - - - - - - 1 - - - - - -  - - - - 1 - - - - 1 1 - - - - - 1 1 - -
- 1 1 - - - - - - - - - - - - - - -  - 1 1 - - - - 1 - - - - - - 1 1 1 - - -
- 1 - - - - - - - - - - - - - - - -  1 1 - - - - 1 - - - - - - 1 1 1 - - - -
- - - - 1 1 - - - - - 1 1 - - - - -  1 - - - 1 1 1 - - - - 1 1 1 - - - - - 1
- - - - 1 1 - - - - - - - - - - - -  - - - 1 1 1 - - - 1 1 1 - - - - - - 1 -
- - - - - - - - - - - - - - - - - -  - - - 1 - - - - 1 - 1 - - - - - 1 1 - -
- - - - - - - - 1 - - - - - 1 1 - -  - - - - - - - - 1 1 - - - - - 1 1 1 - -
1 - - - - - 1 1 - - - - - - 1 1 - - 1  1 1 - - - - 1 1 - - - - - 1 1 - - - - 1
1 - - - - - - - - - - - - - - - - 1  1 - - - - 1 1 - - - - - - 1 - - - - 1 1

Units tuned to both                  Units tuned to either
the abductor and Z_1                 the abductor or Z_1
- - - - - - - - - - - - - - - - - -  - - - - - 1 - - - - 1 1 - - - - - 1 1 1
- - - - - - - - - 1 - - - - - - - -  - - - - - - - - - - 1 1 - - - - - 1 - -
- 1 - - - - - - - - - - - - - - - -  - 1 1 - - - - 1 1 - - - - - 1 1 - - - -
- 1 - - - - - - - - - - 1 1 - - - -  1 1 - - - - 1 1 - - - - 1 1 1 - - - - -
- - - - - - - - - - - 1 1 - - - - -  1 - - - - 1 1 - - - - 1 1 1 - - - - - 1
- - - - - - - - - - - - - - - - - -  - - - - 1 1 - - - - - 1 - - - - - - -
- - 1 1 - - - - - - - - - - 1 1 - -  - - 1 1 1 - - - - 1 1 - - - - 1 1 1 - -
- - - - - - - - 1 - - - - - 1 - - -  - - 1 - - - - - 1 1 - - - - 1 1 1 - - -
- - - - - - - 1 1 - - - - - - - - -  1 1 - - - - 1 1 1 - - - - 1 1 - - - - -
- - - - - - - - - - - - - - - - - 1  1 - - - - 1 1 - - - - 1 1 - - - - - - 1
- - - - - - - - - 1 1 - - - - - 1 1  - - - - 1 - - - - - 1 1 - - - - - 1 1 1
- - - - - - - - - - - 1 - - - - - -  - - - - 1 - - - - 1 1 - - - - - 1 1 - -
- 1 1 - - - - - - - - - - - - - - -  - 1 1 - - - - 1 - - - - - - 1 1 1 - - -
- 1 - - - - - - - - - - - - - - - -  1 1 - - - - 1 - - - - - - 1 1 1 - - - -
- - - - 1 1 - - - - - 1 1 - - - - -  1 - - - 1 1 1 - - - - 1 1 1 - - - - - 1
- - - - 1 1 - - - - - - - - - - - -  - - - 1 1 1 - - - 1 1 1 - - - - - - 1 -
- - - - - - - - - - - - - - - - - -  - - - 1 - - - - 1 - 1 - - - - - 1 1 - -
- - - - - - - - 1 - - - - - 1 1 - -  - - - - - - - - 1 1 - - - - - 1 1 1 - -
1 - - - - - 1 1 - - - - - - 1 - - - 1  1 1 - - - - 1 1 - - - - - 1 1 1 - - - 1
1 - - - - - - - - - - - - - - - - 1  1 - - - - 1 1 - - - - - - 1 - - - - 1 1
```

FIGURE 9. Relation between the tuning to abductor length and tuning to hand position Z_1. Units tuned to abductor length comprise a subset of units tuned to hand position Z_1.

than muscle length or tension maps.

Finally, the cortical units inside a compact contiguous region mentioned in the last paragraph of Section 3.2 also contained the cortical units tuned to all three segments of three axes. This particular region of about 30 units located in the lower left corner of the upper right quadrant, for instance, contains those cortical units tuned to hand positions from 24 out of all possible 27 cubicles.[7]

3.4 Variation of Model Details

The results reported above are from the network trained with arbitrary model arm positions. Quantitatively identical results were also obtained when the network was trained with *equilibrium* model arm positions [JC93].

---

[7]Very few training samples were picked from the three cubicles that were not represented in that region, but were represented in another area of the cortical layer.

The model arm is in equilibrium if at each joint the total tension (active and passive) of the agonistic and antagonistic muscles is the same. Given two different neuronal input values, the two muscles generate the same total tension as the muscle with less neuronal input (therefore with less active tension) becomes stretched, thus generating passive tension. The network trained with equilibrium model arm positions produced almost identical maps as in the case of arbitrary model arm positions. Both length and tension maps were qualitatively identical. So were the spatial hand position maps. Also, the mechanical constraints of the model arm were learned.

In addition, we have done simulations to identify the possible role of some model parameters in shaping the computational maps. In particular, the lateral connection radius (LCR), cortical layer size, and competition parameter value were altered and the resulting maps examined [CJR]. First, the average size of the length clusters grew proportionally to the square of the LCR value, while the number of clusters remained the same. Second, as the cortical layer size increased, the number of clusters increased, while the size of clusters stayed almost constant. Finally, a small change in the competition parameter value made an enormous change in the qualitative behavior of length maps, ranging from total inactivity of units to full saturation.

4 Discussion

To the authors' knowledge, this is the first attempt to develop a computational model of primary proprioceptive cortex. Input to our model cortex consists of length and tension signals from each of six muscle groups that control arm position. Although this model arm is greatly simplified from reality, it still leads to formation of a remarkably rich feature map with an unexpected representation of external three-dimensional spatial positions.

Our results can be summarized as follows. First, cortical units became tuned to length or tension of a particular muscle during map formation. The units tuned to the same muscle, be they units of length or tension, tended to group together as clusters, and the size of these clusters became more uniform with training. In particular, the clusters of cortical units tuned to antagonistic muscle lengths were pushed far apart from each other, thus implying learning by the network of the constraints imposed by the mechanics of arm movement (antagonistic muscles do not become stretched simultaneously; usually only one tends to be highly activated; etc.).

Second, many cortical units were tuned to multiple muscles. Among the cortical units that were initially tuned to more than one arm layer unit, some did not follow the constraints of the arm movement mechanics (implausible tuning), while some did (plausible tuning). It was found that training eliminated the implausibly tuned cortical units, while it increased the number of the cortical units that were tuned to plausible pairs of arm layer units. The map self-organized so that redundant length and tension clusters exist. These regularly spaced clusters are reminiscent of clusters of orientation-sensitive cells in primary visual cortex.

A spatial map of hand positions was also found in the cortical layer. Units tuned to one area of hand position were located in the cortical layer near those units tuned to adjacent areas of hand location. The units tuned to certain segments of axes formed stripes that ran in different orientations from the hexagonal tessellation. To the authors' knowledge, there has been no report of finding a spatial map of hand position in the somatosensory cortex, so this represents a testable prediction of our model. Further, the physical constraints involving muscle length and hand position were also learned by the network. The number of cortical units tuned to plausible pairs of muscle stretch and hand position values increased, while that of cortical units tuned to less plausible pairs decreased significantly. Another characteristic is that when multiple parameters were mapped onto the same 2-D surface, they tended to organize in such a way that there is maximum overlap between the parameters (muscle vs. spatial in our case). Thus muscle tuning forms a fine-grained map within a coarse-grained map of spatial segments. Many of these results from the computational model can be viewed as testable predictions about the organization of primary proprioceptive cortex. Our model predicts that experimental study of proprioceptive regions of cortex should find the following: 1) overlapping maps of both individual muscles and of spatial locations; 2) multiple, redundant representations of individual muscles where antagonist muscle length representations are widely separated; 3) neurons tuned to plausible combinations of muscle lengths and tensions; and 4) proprioceptive "hypercolumns," i.e., compact regions in which all possible muscle lengths and tensions and spatial regions are represented.

Acknowledgments: This work was supported by POSTECH grant P93013 to S. Cho and NIH awards NS-29414 and NS-16332 to J. Reggia.

5 References

[Asa89] H. Asanuma. *The Motor Cortex.* Raven, New York, 1989.

[BCM82] E. Bienenstock, L. Cooper, and P. Munro. Theory for the development of neuron selectivity: Orientation specificity and binocular interaction in visual cortex. *Journal of Neuroscience*, pages 32–48, 1982.

[BG88] D. Bullock and S. Grossberg. Neural dynamics of planned arm movements: Emergent invariants and speed-accuracy properties during trajectory formation. *Psychological Review*, 95:49–90, 1988.

[BGO+92] Y. Burnod, P. Grandguillaume, I. Otto, S. Ferraina, P. Johnson, and R. Caminiti. Visuomotor transformations underlying arm movements toward visual targets: A neural network model of cerebral cortical operation. *Journal of Neuroscience*, 12:1435–1453, 1992.

[CG85] T. Carew and C. Ghez. Muscles and muscle receptors. In E. Kandel and J. Schwartz, editors, *Principles of Neural Science*, pages 443–456. Elsevier, New York, 1985.

[CJR] S. Cho, M. Jang, and J. Reggia. Effects of varying parameters on properties of self-organizing feature maps. *Neural Processing Letters*, 7:129-147, 1996.

[CJU90] R. Caminiti, P. Johnson, and A. Urbano. Making arm movements within different parts of space: Dynamic aspects of the primate motor cortex. *Journal of Neuroscience*, 10:2039–2058, 1990.

[CR92] S. Cho and J. Reggia. Learning visual coordinate transformations with competition. In *Proceedings of the International Joint Conference on Neural Networks*, volume 4, pages 49–54. IEEE, 1992.

[CR93] S. Cho and J. Reggia. Learning competition and cooperation. *Neural Computation*, 5(2):242–259, 1993.

[DLS92] J. Donoghue, S. Leibovic, and J. Sanes. Organization of the forelimb area in primate motor cortex: Representation of individual digit, wrist, and elbow muscles. *Experimental Brain Research*, 89:1–19, 1992.

[GM90] K. Grajski and M. Merzenich. Hebb-type dynamics is sufficient to account for the inverse magnification rule in cortical somatotopy. *Neural Computation*, 2:71–84, 1990.

[GTL93] A. Georgeopoulos, M. Taira, and A. Lukashin. Cognitive neurophysiology of the motor cortex. *Science*, 260:47–51, 1993.

[JC93] M. Jang and S. Cho. Modeling map formation in proprioceptive cortex using equilibrium states of model arm. In *Proceedings of the 20th Korean Information Science Society Conference*, pages 365–368, 1993.

[KdLE87] E. Knudsen, S. du Lac, and S. Esterly. Computational maps in the brain. *Annual Review of Neuroscience*, 10:41–65, 1987.

4. A learning sensorimotor map of arm movements

[Kup88]   M. Kuperstein. Neural model of adaptive hand-eye coordination for single postures. *Science*, 239:1308–1311, 1988.

[LG94]   A. Lukashin and A. Georgeopoulos. A neural network for coding of trajectories by time series of neuronal population vectors. *Neural Computation*, 6:19–28, 1994.

[Lin88]   R. Linsker. Self-organization in a perceptual network. *Computer*, pages 105–117, 1988.

[Mel88]   B. Mel. MURPHY: A robot that learns by doing. In *Neural Information Processing Systems*, pages 544–553. American Institute of Physics, New York, 1988.

[MKS89]   K. Miller, J. Keller, and M. Stryker. Ocular dominance column development: Analysis and simulation. *Science*, 245:605–615, 1989.

[PFE87]   J. Pearson, L. Finkel, and G. Edelman. Plasticity in the organization of adult cerebral cortical maps: A computer simulation based on neuronal group selection. *Journal of Neuroscience*, 7:4209–4223, 1987.

[RDSW92] J. Reggia, C. L. D'Autrechy, G. Sutton, and M. Weinrich. A competitive distribution theory of neocortical dynamics. *Neural Computation*, 4(3):287–317, 1992.

[RMS92]   H. Ritter, T. Martinez, and K. Schulten. *Neural Computation and Self-organizing Maps*. Addison-Wesley, Reading, Massachusetts, 1992.

[RSC91]   J. Reggia, G. Sutton, and S. Cho. Competitive activation mechanisms in connectionist models. In M. Fraser, editor, *Advances in Control Networks and Large Scale Parallel Distributed Processing Models*. Ablex, Norwood, New Jersey, 1991.

[RZ86]   D. Rumelhart and D. Zipser. Feature discovery by competitive learning. In D. Rumelhart, J. McClelland, and the PDP Research Group, editors, *Parallel Distributed Processing, volume 1: Foundations*, pages 151–193. MIT Press, Cambridge, Massachusetts, 1986.

[Skl90]   E. Sklar. A simulation of cortical map plasticity. In *Proceedings of International Joint Conference on Neural Networks, volume 3*, pages 727–732. IEEE, 1990.

[SSLD88] J. Sanes, S. Suner, J. Lando, and J. Donoghue. Rapid reorganization of adult rat motor cortex somatic representation patterns after motor nerve injury. *Proceedings of the National Academy of Science, U.S.A.*, 85:2003–2007, 1988.

[Sut92] G. Sutton. *Competitive Learning and map formation in artificial neural networks using competitive activation mechanisms.* Ph.D. Thesis, University of Maryland, College Park, 1992.

[UF88] S. Udin and J. Fawcett. Formation of topographic maps. *Annual Review of Neuroscience*, 11:289–327, 1988.

[vdM73] C. von der Malsburg. Self-organization of orientation sensitive cells in the striate cortex. *Kybernetic*, pages 85–100, 1973.

[WS92] D. White and D. Sofge. *Handbook of Intelligent Control.* Van Nostrand Reinhold, Princeton, New Jersey, 1992.

Chapter 5

Neuronal Modeling of the Baroreceptor Reflex with Applications in Process Modeling and Control

Francis J. Doyle III
Michael A. Henson
Babatunde A. Ogunnaike
James S. Schwaber
Ilya Rybak

> ABSTRACT Biological control systems exhibit high performance and robust control of highly complex underlying systems; on the other hand, engineering approaches to robust control are still under development. This situation motivates neuromorphic engineering: the reverse engineering of biological control structures for applications in control systems engineering. In this work, several strategies are outlined that exploit fundamental descriptions of the neuronal architectures that underly the baroreceptor vagal reflex (responsible for short-term blood pressure control). These applications include process controller scheduling, nonsquare controller design, and dynamic process modeling. A simplified neuronal model of the baroreflex is presented, which provides a framework for the development of the process tools.

1 Motivation

The biological term *homeostasis* refers to the coordinated actions that maintain the equilibrium states in a living organism. A control engineer can readily associate this term with the systems engineering concept of "regulation." In each case, a variety of tasks are performed that include the collection, storage, retrieval, processing, and transmission of data, as well as the generation and implementation of appropriate control action. In the engineering context, these tasks are accomplished by "hard-wired" networks of devices whose tasks are typically coordinated by distributed computer controllers. In the biological context, there are analogous devices and architectures, the most important of which is the brain. Comprising a vast network of "microprocessors" (neurons), this "central controller" simultaneously coordinates many complex functions.

Consider the regulation of arterial blood pressure. The mean blood pressure is controlled around a setpoint dictated by cardiovascular system demands. The pressure is a function of the cardiac output and the resistance of the blood vessels. However, the blood volume is an order of magnitude less than that of the blood vessels. Thus, in order to optimize circulating blood weight and pumping requirements, the distribution of blood to specific vascular beds varies as a function of: (i) demand (e.g., eating, exercise); (ii) external influence (e.g., cold weather); (iii) emotional state (e.g., joy, anger); and (iv) anticipated action (e.g., postural adjustment). Because the major objective in maintaining blood pressure (and thus blood flow) is the exchange of gases in the tissues, the respiratory and cardiovascular systems are intimately linked. Consequently, blood gas composition and respiratory action modulate cardiovascular function.

The regulation of blood pressure in response to changing requirements and external disturbances is accomplished by a complex network of processing elements in the central nervous system. This control system performs a wide variety of tasks, which include:

1. integration of multiple inputs from pressure sensors, chemosensors, and other brain systems;

2. noise filtering of the sensory inputs;

3. provision of control that is robust to sensor drift and loss;

4. compensation for nonlinear, interacting features of cardiovascular function.

Clearly, these functions have direct parallels in engineering applications. Our long-term objectives are therefore to understand the mechanisms behind the control of blood pressure and cardiovascular function and to "reverse engineer" the relevant attributes of the baroreceptor reflex for process engineering applications.

This chapter contains a summary of some preliminary results; it is organized as follows. In Section 2, we provide an overview of the baroreceptor reflex, including a description of its key processing elements. In Section 3, simplified neuron models are used as the basis for constructing a network model of the overall reflex. A potential application of this structure to scheduled process control is then described. In Section 4, blood pressure control architectures are examined from a systems perspective, and applications to the control of "nonsquare" process systems are discussed. In Section 5, a simplified, "biologically inspired" dynamic processing element is presented for process modeling using network architectures. These models are used to develop a model-based control strategy for a simple

5. Modeling of the Baroreceptor Reflex with Applications   89

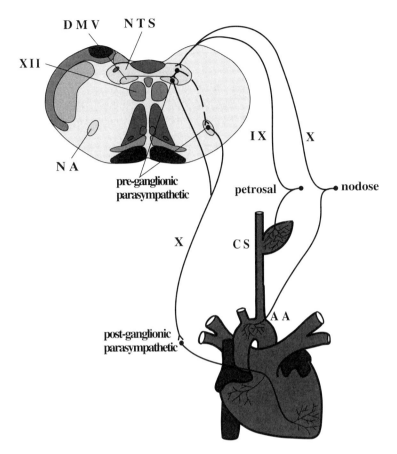

FIGURE 1. Schematic diagram of the baroreceptor reflex.

reactor problem. Finally, some conclusions and directions for future work are discussed in Section 6.

2 The Baroreceptor Vagal Reflex

2.1 Background

The baroreceptor reflex (baroreflex) performs adaptive, nonlinear control of arterial blood pressure. Its components are pressure transducers in major blood vessels, a central processing network in the brain, and actuators in the heart and vessels. A schematic diagram of the baroreceptor reflex circuit is shown in Figure 1. Arterial pressure is transduced by stretch receptors (baroreceptors) located in the major blood vessels. These "first-order" neurons project their input onto "second-order" neurons in a specific "cardiorespiratory" subdivision of the nucleus tractus solitarii (crNTS), where they are integrated with other sensory signals that reflect demands on cardiorespiratory performance [Sch87, Spy90]. Control signals are sent to the heart to regulate its rate, rhythm, and force of contraction. Other limbs of the baroreflex send signals to the individual vascular beds to determine flow and resistance. For example, if the blood pressure rises above its desired setpoint, the heart rate is slowed, thereby reducing cardiac output and increasing total peripheral resistance, with a consequent reduction in blood pressure.

The underlying signal processing mechanism appears to be more complex then mere linear filtering of input signals. For instance, following the elimination of the baroreceptor inputs, rhythmic output activity and stability in the heart rate and blood pressure are observed, although "reflex" adjustments to pressure perturbations are lost. In addition, there is a central processing delay (typically in the 100 ms range) that is an order of magnitude larger than would be anticipated for a straight-through transmission of input signals. Finally, the activity in the reflex oscillates at the cardiac frequency, and it is plausible that this behavior is due to reflex computation. In short, the processing of inputs by second-order NTS neurons is a remarkably complex operation.

We are interested in the baroreflex not only because it exhibits interesting behavior, but also because it offers important advantages for analysis: (i) the input and output are nerves, and are therefore easily accessible for morphological and physiological study; (ii) the circuit (in its simplest form) may be restricted to a single level of the brainstem, and thus may be studied (at least partially) *in vitro* using transverse slices of the brainstem; (iii) in principle, it is possible to delineate the complete reflex connectional circuit at the cellular level; (iv) the total number of neurons is small enough to allow system simulations that incorporate neuronal dynamics; and (v) the location of the NTS is highly advantageous for whole cell patch studies *in vivo*.

2.2 Experimental Results

In an effort to develop accurate network models of the baroreflex, we have performed a variety of experiments to understand the computational mechanisms carried out by its individual elements. The work discussed here will focus on the processing of inputs from the baroreceptors by second-order neurons in the NTS. By focusing on the interactions taking place within the NTS at the initial stage of the processing, we aim to determine the circuit architectures and the basis for the nonlinear, dynamical, adaptive signal processing it performs.

The first-order baroreceptors are highly sensitive, rapidly adapting neurons that encode each pressure pulse with a train of spikes on the rising phase of pressure, with activity that is sensitive to dP/dt [AC88, SvBD$^+$90]. A typical response of the baroreceptor to rhythmic changes of blood pres-

5. Modeling of the Baroreceptor Reflex with Applications

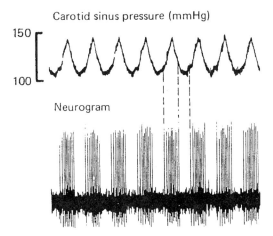

FIGURE 2. Recording of natural waveform pulses into the isolated carotid sinus (top trace) and associated activity of a single baroreceptor sensory neuron in the carotid sinus nerve (bottom trace) (Data provided courtesy of M. Chapleau and F. Abboud, 1991; cf. [AC88]).

sure is shown in Figure 2. There are approximately 100 baroreceptor afferent fibers per nerve. Variations in the pressure thresholds of these fibers are considerably more than a scattering around the mean pressure, but rather cover a range from well below (approximately 35 mmHg) to well above (approximately 170 mmHg) resting pressure. We have studied the connections between the first- and second-order neurons in neuroanatomical experiments using a virus that crosses synapses [SES94]. The results of this work suggest the possibility of a topographic organization of the crNTS, such that there is a spatial arrangement of the first-order inputs by their pressure thresholds [BDM+89, MRSS89].

The second-order neurons are of interest not only because it is among them that the first synaptic processing of pressure information in the NTS takes place, but also because this processing creates an activity pattern that is not well understood but appears important. In order to analyze the processing characteristics, we have conducted single-neuron recording experiments in the NTS of anesthetized rats. In initial experiments we have recorded from second-order neurons and characterized their responses to naturalistic changes in arterial pressure. Although the first-order neurons have ongoing bursting activity patterns at the cardiac rhythm (Figure 2), this pattern is not observed in the relatively low-rate, irregular spiking activity of second-order neurons (Figure 3). In addition, our results show that second-order neurons exhibit nonlinear responses to changes in blood pressure and seem to encode both mean arterial blood pressure and the rate of pressure change. Figure 3 shows a typical second-order neuron that initiates its response as pressure rises but decreases its firing frequency at

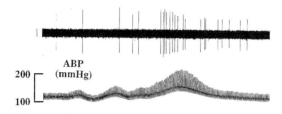

FIGURE 3. Typical response of an NTS neuron to an arterial pressure step change. Unpublished data of Rogers and Schwaber, cf. [RPS93].

higher pressures. This is difficult to interpret because the sign and strength of the synaptic connection from first- to second-order neurons is strong and positive.

In order to develop conductance-based Hodgkin-Huxley neuron models [HH52] for the second-order neurons, we have performed *in vitro* experiments [FPSU93, FUS93, PFS93, SGP93]. These experiments aimed: (1) to characterize the voltage dynamics of the NTS neuronal population; and (2) to determine whether (and in what approximate amount) candidate conductances that might contribute to the voltage dynamics are present in various neuron types. The *in vitro* work showed that NTS neuronal responses to current steps fall into three broad classes that depend on the relative abundance of conductance channels: (i) single spike response; (ii) rapidly adapting, delayed response; and (iii) adapting but repetitive response. It is not known at this time whether baroreceptor inputs land haphazardly on neurons of each of these response types or whether these different neural types represent the front ends of different information channels for NTS processing.

2.3 Nonlinear Dynamical Processing

The role of *nonlinear* neuronal mechanisms is highlighted by our *in vitro* observations of dynamical behavior of baroreceptive NTS neurons arising from their active membrane properties, in particular the large potassium conductances and the calcium dependent potassium channels. This behavior presents the interesting possibility that neuronal dynamics play an important role in the signal processing performed by the network of first-order inputs to second-order neurons. Thus, one of our strong interests is to explore whether or not nonlinearities in cellular input–output functions play an important signal-processing role in baroreceptive NTS neurons and to extend this work to explore the interaction of cell properties with synaptic inputs for network processing and parallel processing in this system.

We use computational models to explore the contribution of neuron dynamics and specific baroreceptor circuitry to the function of the baroreceptor vagal reflex [GSP+91]. The model circuitry is composed of specific classes of neurons, each class having unique cellular–computational properties. Focusing on the interactions taking place within the NTS at the input synaptic stage of the processor, we aim to determine the circuit architectures and single-neuron functionality that contribute to the complex signal processing in the reflex. Our work suggests that biological neural networks compute by virtue of their nonlinear dynamical properties. Individual neurons are intrinsically highly nonlinear due to active processes inherent in their membrane biophysics. Collectively, there is even more opportunity for nonlinearity due to the connectivity patterns between neurons.

Characterizing the behavior of this sort of system is a difficult challenge, as a neuronal system constantly receives many parallel inputs, executes some dynamic computation, and continuously generates a set of parallel outputs. The relationship between inputs and outputs is often complex, and the first task in emulating biological networks is to find this relationship, and then to understand the dynamical computational mechanisms underlying it. If this functionality can be captured mathematically in a model, one has a powerful tool for investigating mechanisms and principles of computation that cannot be explored in physiological experiments. The work presented in this chapter represents a preliminary step in this process.

3 A Neuronal Model of the Baroreflex

In this section, a simple closed-loop model of the baroreflex is presented. This network model serves a dual purpose: (i) it provides information about the network-level computations that underlie the control functions of the baroreflex; and (ii) it provides the basis for "reverse engineering" the scheduled transitions in neuronal activity that occur in response to blood pressure changes for applications in scheduling the action of a process controller.

3.1 Background

In the previous section, we described some of the relevant experimental results on the dynamics of the second-order NTS neurons (Figure 3), which were used as a basis for the development of a neural network model of the baroreceptor reflex. An analysis of these results (see Figure 3) reveals the following dynamic properties of the second-order neurons:

1. The second-order NTS neurons respond to a change in mean blood pressure with a burst of activity whose frequency is much lower than the frequency of the cardiac cycle;

2. The responses suggest that NTS neurons are inhibited immediately before and immediately after the bursts;

3. It is reasonable to assume that this bursting activity is the source of regulatory signals that are relayed to, and cause the compensatory changes at, the heart;

4. It is plausible that each NTS neuron responds to pressure changes and provides this regulation in a definite static and dynamic range of pressure.

These observations, combined with other physiological data and general principles of sensory system organization, suggest the following hypotheses, which have been used to construct a simple baroreflex model:

1. The first hypothesis, *barotopical organization*, as explained previously in [SPRG93, SPR+93], proposes that: (a) the thresholds of the baroreceptors are topographically distributed in pressure space; and (b) each second-order neuron receives inputs from baroreceptors with thresholds belonging to a narrow pressure range. There are anatomical [BDM+89, DGJS82] and physiological [RPS93] data that support these suppositions.

2. The second hypothesis proposes that projections of the first-order neurons onto the second-order neurons are organized like "ON–center–OFF–surround" receptive fields in the visual sensory system [HW62]. Each group of second-order neurons receives "lateral" inhibition from neighboring neuron groups, which respond to lower and higher levels of blood pressure (compared to the center group). This supposition results from the second experimental observation listed above and corresponds to a general organizational principle of sensory systems.

3.2 Model Development

Structure of the Model

A diagram of the proposed network model for the closed-loop baroreflex is shown in Figure 4. The first-order neurons, which are arranged in increasing order of pressure threshold, receive an excitatory input signal that is proportional to the mean blood pressure. The second-order neurons receive both synaptic excitation and inhibition from the first-order neurons as depicted in Figure 4. The lateral inhibition of the second-order neurons is achieved by direct synaptic inhibition from the neighboring off-center, first-order neurons (i.e., the periphery of the receptive field [HW62]). A more biologically accurate mechanism would employ inhibitory interneurons and reciprocal inhibition between the second-order neurons. An investigation of these more complex inhibition mechanisms is left for future work; here we consider only the simple mechanism shown in Figure 4. The outputs of the second-order neurons are summed and, via an intermediate dynamic subsystem, are used as an input to a model of the the heart. This model

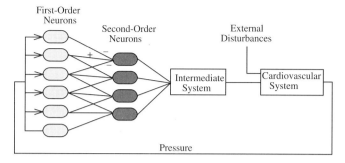

FIGURE 4. Schematic of the simplified baroreflex model.

receives inputs from both the neural feedback subsystem and an external disturbance signal. The output of this model is fed back to the neural control system as the blood pressure signal.

Model of a Single Neuron

Detailed conductance-based neuron models of first- and second-order baroreflex neurons show ([SPRG93, SPR+93]) close correspondence to experimental observations. However, the complexity of these models poses a difficult problem for *efficient* network-level simulations. In this case, a simplified model of a spiking neuron is preferred. A summary of the single-neuron model used in the baroreflex network (based on previously described neuron models [Get89, Hil36, Mac87]) is given below.

Following the Hodgkin-Huxley formalism, the dynamics of a neuron's membrane potential can be described by the following differential equation:

$$c\dot{V} = \sum_i g_{iabs}(E_i - V) + I,$$

where c is the membrane capacitance, V is the membrane potential, g_{iabs} is the conductance of the ith ionic channel, E_i is the reversal potential of the ith ionic channel, and I is the input current.

Following a long period that is devoid of excitatory and inhibitory signals ($I = 0$), the neuron will cease to generate action potentials, and the variables will attain the following "resting" or steady-state values: $V = V_r$ and $g_{iabs} = g_{ir}$. The conductances can be represented as "deviation variables" by defining

$$g_i = g_{abs} - g_{ir},$$

so that g_i is the relative change of the ith conductance. The deviation form of the membrane potential equation is

$$c\dot{V} = g_0(V_r - V) + \sum_i g_i(E_i - V) + I,$$

where the resting membrane potential, V_r, and generalized conductance, g_0, are defined by the following expressions:

$$V_r = \frac{\sum_i g_{ir} E_i}{\sum_i g_{ir}} \; ; \qquad g_0 = \sum_i g_{ir}.$$

Three types of conductances (g_i) are used in the current model. They include conductances for excitatory and inhibitory synaptic currents (g_{esyn} and g_{isyn}), which are opened by action potentials (AP) coming from other neurons, and a g_{AHP} conductance for the potassium current, which is opened by AP generation in the neuron itself. There are, in fact, several potassium channel types [CWM77], and the AHP notation identifies the specific class considered here.

With this assumption, the membrane potential can be represented in the following form:

$$\begin{aligned} c\dot{V} &= g_0(V_r - V) + g_{esyn}(E_{syn} - V) + g_{isyn}(E_{isyn} - V) \\ &+ g_{AHP}(E_K - V) + I. \end{aligned} \qquad (1)$$

Because the first-order baroreflex neurons do not receive synaptic inputs, they can be described by the following simplified expression:

$$c\dot{V} = g_0(V_r - V) + g_{AHP}(E_K - V) + I, \qquad (2)$$

where the input signal I is proportional to the blood pressure. The membrane potential of second-order neurons is described as in Equation (1) without the input I.

In models of this type [Get89, Hil36, Mac87] it is generally assumed that g_0 is constant and that g_{esyn}, g_{isyn} and g_{AHP} depend on time, but not on the membrane potential. It is also assumed that the neuron generates an action potential at the moment of time when its membrane potential reaches, or exceeds, a threshold value. The dynamic behavior of the threshold value (H) is described as follows:

$$\tau_{H_o}\dot{H}_o = -H_o + H_r + Ad\,(V - V_r), \qquad (3)$$

$$H = H_o + (H_m - H_o)\exp\left(-\frac{t - t_0}{\tau_H}\right). \qquad (4)$$

Equation (4) describes the fast changes of the threshold immediately following an AP that is generated in the neuron at time t_0. The threshold (H) jumps from the current level to the higher level H_m at t_0 and then decays exponentially to H_0 with time constant τ_H. Equation (3) describes the slow adaptive dynamics of the current threshold level (H_0). The degree of adaptation is determined by the coefficient Ad. The resting level of the threshold is denoted by H_r, and τ_{H_0} denotes the time constant of adaptation.

5. Modeling of the Baroreceptor Reflex with Applications

The dynamics of the g_{AHP} conductance are described as follows:

$$g_{AHP} = g_{mAHP} \sum_{t_i \leq t} \exp\left(-\frac{t - t_i}{\tau_{AHP}}\right). \tag{5}$$

The conductance increases from the current level by the constant value g_{mAHP} at each time t_i when an AP is generated in the neuron, and then decays back to zero with time constant τ_{AHP}. These changes in g_{AHP} cause the short-time hyperpolarization that occurs after each AP. Equations (3)–(5) define slow and fast interspike dynamics of the neuron excitability. A more realistic description of neuron dynamics can be obtained by considering the dynamics of Ca^{++}, as well as the voltage and Ca^{++} dependencies of the conductances. Nevertheless, our results have shown that the simplified model describes the behavior of the baroreflex neurons with sufficient accuracy for the purpose of network modeling.

The connections between neurons are captured in the model by the changes of synaptic conductances in target neurons caused by each AP coming from source neurons. The transmittance of the action potential is captured in the output activity of a neuron (Y):

$$Y = V + (Am - V) f_1(t - t_0),$$

where Am is the amplitude of the action potential, and $f_1=1$ if $t = t_0$ and 0 otherwise. Synaptic potentials in a *target* neuron, which cause its excitation or inhibition, result from changes of g_{esyn} and g_{isyn} conductances in that neuron. These changes are modeled using the output variable of the *source* neuron (y), which causes the inhibition or excitation:

$$y = y_m \sum_{t_i \leq t} \exp\left(-\frac{t - t_i}{\tau_y}\right),$$

where t_i is the time at which an action potential is generated and y_m and τ_y are the parameters that define the normalized amplitude and decay time constant, respectively. The synaptic conductances in the *target* neuron are generated by the weighted sum of the respective output signals from the *source* neurons:

$$g_{esyn} = k_e \sum_j a_{ej} y_j,$$

$$g_{isyn} = k_i \sum_j a_{ij} y_j,$$

where a_{ej} and a_{ij} are weights associated with the excitatory and inhibitory synapses, respectively, from the neuron j; and k_e and k_i are tuning parameters.

A Simplified Model of the Baroreflex Control System

Let us now consider how the single-neuron model is used in the baroreflex control model depicted in Figure 4. The first-order neurons are arranged in increasing order of threshold rest levels (H_r) using a constant threshold difference of ΔH_r. The input signal to the first-order neurons depends on the pressure P via the amount of stretch in the blood vessels, modeled simply here as $I = f_P(P)$. As a first approximation, a linear relationship is assumed: $f_P(P) = k_P P$, where k_P is a "tuning" coefficient. The synaptic inputs from the first-order neurons to the second-order neurons are sketched in Figure 4. The weighted sum of the outputs from the second-order neurons forms the input for an intermediate subsystem, which is modeled as a simple linear filter:

$$\tau_{int} \dot{I}_{int} = -I_{int} + k_{int} \sum_j y_j .$$

This dynamical system captures the effects of the interneurons and motor neurons that lie between the second-order baroreflex neurons and the heart. (Note: in this model we have focused on the *vagal* motor neurons that affect the cardiac output and have ignored the effects of the sympathetic system on the peripheral resistance in the vascular bed.)

A first-order approximation of the blood pressure dynamics is described below. The pressure decays exponentially from a current level to the level P_0 with the time constant τ_P. At selected time points, denoted by t_1, the pressure responds with a "jump" to the level P_m in response to the pumping action of the heart:

$$P = P_0 + (P_m - P_0) \exp\left(-\frac{t - t_1}{\tau_P}\right) .$$

This pressure jump occurs at the moment when P_{min} exceeds P_0, where P_{min} is modeled by a first-order differential equation with time constant T_P and rest level P_{min0} (in the absence of inputs):

$$T_P \dot{P}_{min} = -P_{min} + P_{min0} + P_i - k_{fb} I_{int} .$$

One of the driving forces in this equation is the disturbance P_i, which represents the effects of an external agent (e.g. drug infusion). The second input is the feedback signal from the neural mechanism (I_{int}) multiplied by a constant feedback gain (k_{fb}).

Computer Simulation Results

The responses of four first-order neurons (the four upper rows) with distributed blood pressure thresholds (increasing from the bottom to the top) to increasing mean blood pressure (the bottom row) are shown in Figure 5. The neurons exhibit a spiking response to each pressure pulse, and the

TABLE 1. Baroreflex model parameter values.

Parameter	Value
c	1.0 μF
g_0	0.5 mS (1.0 for second-order neurons)
V_r	-60 mV
E_k	-70 mV
E_{esyn}	20 mV
E_{isyn}	-70 mV
H_r	-48 to -56 mV
ΔH_r	1 mV
H_m	-10 mV
Ad	0.6
g_{mAHP}	0.12 mS
Am	45 mV
y_m	0.5 mV
τ_{H_0}	30 ms
τ_H	10 ms
τ_{AHP}	10 ms
τ_y	60 ms
k_e	1.0
k_i	20.0
a_{ej}	3.6 $\frac{mS}{mV}$
a_{ij}	1.6 $\frac{mS}{mV}$
k_P	0.038 $\frac{mA}{mmHg}$
k_{int}	1.0 $\frac{1}{mV}$
τ_{int}	700 ms
P_0	50 $mmHg$
P_{min0}	120 $mmHg$
P_m	30 $mmHg$
τ_P	400 ms
T_P	3000 ms
k_{fb}	600 $mmHg$

neurons with lower thresholds exhibit increased activity. The values of the model parameters are shown in Table 1. These values are consistent with the physiological results described in the previous section.

Figures 6 and 7 show the responses of the four first-order neurons (the 2nd–5th rows) and one second-order neuron (the upper row) to a fluctuating pressure signal (the bottom row). Due to the barotopical distribution of thresholds, the first-order neurons respond sequentially to increasing mean blood pressure. Hence, the neuron with the lowest threshold (2nd row) displays the greatest amount of activity. The middle pair of first-order neurons (3rd and 4th rows) *excite* the second-order neuron, while the other

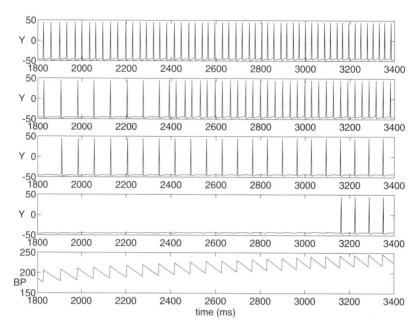

FIGURE 5. Responses of four first-order neurons (rows 1–4) with different blood pressure thresholds to increasing mean blood pressure (row 5).

two first-order neurons (2nd and 5th rows) are *inhibitory*.

In Figure 6, the feedback loop is disabled ($k_{fb} = 0$), and mean pressure increases in response to a persistent external signal P_i. It is clear that the first-order neurons respond sequentially with increasing activity in direct proportion to the pressure signal, while the second-order neuron is only active in a narrow pressure range.

In Figure 7, the feedback loop is closed, and the second-order neuron participates in pressure control. As the pressure enters the sensitive range of the second-order neuron, a signal burst is sent to the intermediate block. This block drives the heart with a negative feedback signal, leading to a temporary decrease in the pressure level. The persistent external signal drives the pressure up again, and the trend is repeated. Note that the second-order neuron exhibits low frequency bursts in a manner similar to that of its real counterpart (Figure 3).

Observe therefore that the network behavior of the proposed baroreflex model is a reasonable approximation of the experimentally recorded neuronal behavior. Refinements to the current model will be the subject of future work; in particular, the structural organization of the first- and second-order network will be modified to match the experimental data. As the sophistication of the model increases, we anticipate a commensurate increase in our understanding of the role of the second-order neurons in blood pressure control.

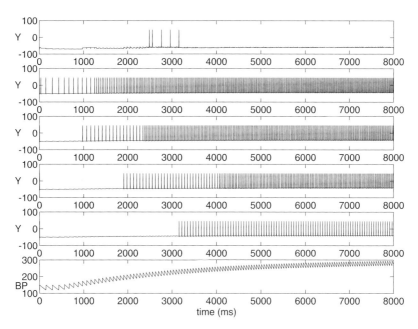

FIGURE 6. Open-loop responses of four first-order neurons (rows 2–5) and one second-order neuron (row 1) to a blood pressure signal (row 6).

3.3 Application to Scheduled Process Control

From a control perspective, an interesting feature of the proposed model is that individual second-order neurons are active *in a narrow static and dynamic range of pressure changes*. In effect, second-order neurons regulate the pressure through a sequence of adaptive control actions in response to the dynamics of pressure change. Thus, the second-order neurons may be considered as a set of interacting controllers that are active in a specific range of the controlled variable.

This behavior can be exploited in the formulation of scheduling algorithms for controller design [DKRS94]. Just as competition between second-order neurons leads to a selective dynamic response, a selectively scheduled nonlinear controller can be designed for a process system. Two paradigms for achieving this functionality are proposed:

1. In the *implicit* formulation, a control architecture consisting of a number of individual dynamic elements is designed to provide effective compensation over a wide operating regime. The second-order network structure is employed to provide the scheduling between these dynamic components. The individual entities do not represent distinct control laws; they represent basis elements of a larger dynamic structure. In this case, the network must be "trained" to learn the proper control strategies over the operating regime.

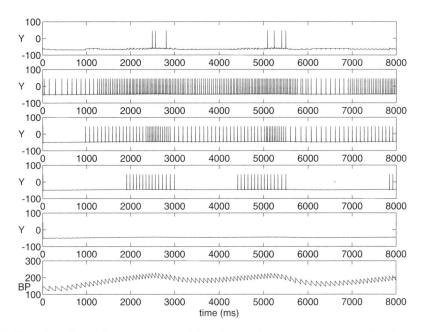

FIGURE 7. Closed-loop responses of four first-order neurons (rows 2–5) and one second-order neuron (row 1) to a blood pressure signal (row 6).

2. An *explicit* control formulation can be achieved by using the second-order network to model the open-loop response of a nonlinear system. Individual components of the second layer are trained to emulate the open-loop system behavior over a limited operating regime. In this case, the biological scheduling mechanism is used for transitions between different open-loop dynamic behaviors. A control law can be synthesized using traditional model-based control techniques [MZ89] (e.g., model predictive control (MPC), internal model control (IMC)).

Additional details of the control algorithm and simulations with chemical process examples are presented in [DKRS94].

4 Parallel Control Structures in the Baroreflex

In this section, two parallel control architectures in the baroreceptor reflex are described. Also discussed are two novel process control strategies that have been abstracted from these biological control architectures. Simplified block-diagrammatic representations of the reflex control structures are shown in Figure 8. In each case, the system is regulated by two controllers that operate in parallel. The two control systems, which differ according to the number of manipulated inputs and measured outputs, can be inter-

5. Modeling of the Baroreceptor Reflex with Applications

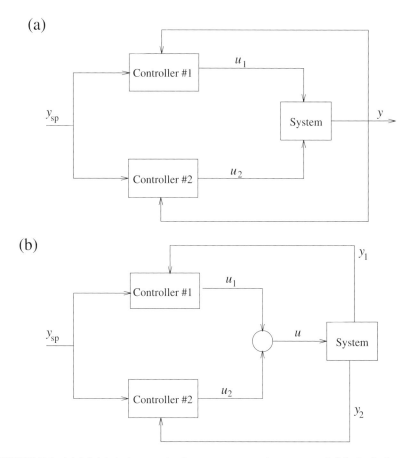

FIGURE 8. (a) Multiple-input, single-output control system and (b) single-input, multiple-output control system.

preted as *duals*.

1. *Multiple-Input, Single-Output (MISO) Control System.* The control system consists of two manipulated inputs (u_1, u_2) and a single measured output (y). The objective is to make y track the setpoint y_{sp}. The ith parallel controller $(i = 1, 2)$ receives y and y_{sp} and computes the manipulated input u_i.

2. *Single-Input, Multiple-Output (SIMO) Control System.* The control system consists of a single manipulated input (u) and two measured outputs (y_1, y_2). The objective is to make y_1 track y_{sp}. The ith parallel controller receives y_i and y_{sp} and computes the value u_i. The manipulated input u is the sum of the u_1 and u_2 values.

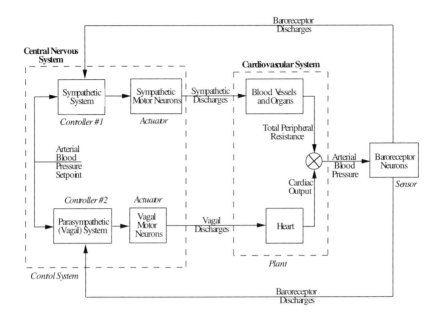

FIGURE 9. Simplified representation of a MISO control structure in the baroreflex.

4.1 MISO Control Structure

Baroreceptor Reflex

A simplified block-diagrammatic representation of a MISO control architecture employed in the baroreceptor reflex is shown in Figure 9. The baroreceptor discharges are processed by two parallel controllers in the central nervous system: the sympathetic and parasympathetic systems. The controllers compare the baroreceptor discharges to a desired blood pressure signal, which is determined by a variety of factors that affect cardiorespiratory performance [Spy90]. The sympathetic and parasympathetic systems affect the cardiovascular system via sympathetic and vagal postganglionic motor neurons, respectively. For simplicity, the effects of the sympathetic system on the heart have been neglected. Hence, the only couplings considered are those between the parasympathetic system and cardiac output, and between the sympathetic system and total peripheral resistance.

The effect of the parasympathetic system on arterial pressure is quite rapid, while that of the sympathetic system is comparatively slow. In modeling the closed-loop response of each control system to a step disturbance in the carotid sinus pressure of the dog, Kumada et al. [KTK90] reported the following results. Using a first-order-plus-deadtime model structure, the time constant and time delay for the sympathetic system response were estimated respectively as: $10 \leq \tau_1 \leq 80$ s, $2 \leq \theta_1 \leq 4.5$ s; for the parasympathetic response, the corresponding estimates ($7 \leq \tau_2 \leq 25$ s, $0.6 \leq \theta_2 \leq 1.2$ s) are comparatively small. Although the parasympathetic system is able to affect the arterial pressure quite rapidly, sustained variations in the cardiac output are undesirably "expensive," whereas long-term variations in the peripheral resistance are more acceptable [SKS71].

Cardiac output is therefore an expensive manipulated variable as compared to the peripheral resistance. The brain coordinates the use of the sympathetic and parasympathetic systems in order to provide effective blood pressure control while minimizing the long-term cost of the control actions. For instance, consider a blood pressure decrease caused by an external disturbance (e.g., standing up). The parasympathetic system induces a rapid increase in blood pressure by enhancing cardiac output, while a significantly slower increase in blood pressure is caused by the sympathetic system raising peripheral resistance. As the effects of increased peripheral resistance on the blood pressure become more pronounced, the parasympathetic controller *habituates* by returning cardiac output to its initial steady-state value.

Process Control Applications

The baroreceptor reflex provides an excellent biological paradigm for the development of control strategies for multiple-input, single-output (MISO) processes. As indicated in italics in Figure 9, the components of the system have well-defined control analogues: the central nervous system is the "controller," the sympathetic and vagal postganglionic motor neurons are the "actuators," the cardiovascular system is the "plant," and the baroreceptors are the "sensors." More importantly, many processes have manipulated inputs that differ in terms of their dynamic effects on the outputs and relative costs.

For example, consider the polymerization process depicted in Figure 10. The process consists of a continuous stirred tank polymerization reactor and an overhead condenser. The feed to the reactor consists of monomer, initiator, and solvent. The condenser is used to condense solvent and monomer vapors, and a cooling water jacket is available to cool the reactor contents. The process also includes a vent line for condensibles and a nitrogen admission line that can be used to regulate the reactor pressure P. One of the control objectives is to control the reactor temperature (T); the cooling water flow rate (F_j) and P (which can be changed almost instantaneously via nitrogen admission) are the potential manipulated variables. The reactor pressure P has a much more rapid and direct effect on T than does F_j. However, because significant and/or extended pressure fluctuations affect the reaction kinetics adversely, it is desirable to maintain P near its setpoint. It is therefore desirable to develop a control strategy in which P (the *secondary* input) is used track setpoint changes and reject disturbances rapidly. As F_j (the *primary* input) begins to affect T, P can

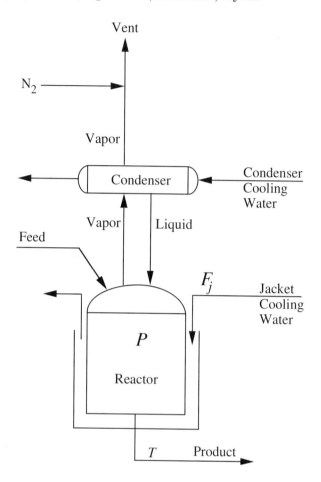

FIGURE 10. Polymerization process.

"habituate" by returning to its previous steady-state value.

Henson et al. [HOS95] have developed a habituating controller design methodology for two-input, single-output systems such as the polymerization process by reverse engineering the parallel control structure of the baroreceptor reflex. The approach is beneficial for processes with the following characteristics: (i) control system performance is limited by the nature of the dynamic effect exerted on the output by the primary manipulated input; (ii) a secondary input is available whose effect on the output is characterized by superior dynamics; and (iii) the long-term cost associated with the secondary input is greater than that associated with the primary input. There are several techniques that are similar to the habituating control strategy, including valve position control [Luy90, Shi78], coordinated control [CB91, PMB86], parallel control [BM88], and variants of H_∞ control [Med93, WHD+92]. These control strategies also employ more manipulated

5. Modeling of the Baroreceptor Reflex with Applications

inputs than controlled outputs. However, there are several important differences between the habituating control strategy and these related control schemes.

1. Our primary objective is to understand, and then to mimic, the functions of a biological system for process control applications. The habituating control strategy therefore is a translation of a biological control solution to a particular process control problem. By contrast, these other techniques are direct control solutions to control problems.

2. The habituating control strategy is formulated to exploit *specific* characteristics and operating objectives of processes with two *different* types of manipulated variables: (i) a slow, cheap type; and (ii) a fast, expensive type. By contrast, H_∞ control techniques were developed for a considerably more general class of systems, and therefore fundamental differences in the dynamic effects and costs of the manipulated inputs are not easily exploited. This point is illustrated quite clearly in [WHD+92]. In order to obtain an acceptable H_∞ controller for a system with one slow, cheap input and one fast, expensive input, significant design effort is required to select appropriate frequency domain weighting functions used in the H_∞ cost function.

3. The habituating control architectures are generalizations of the series [Luy90] and the parallel [BM88, CB91, PMB86] control structures employed in other techniques.

4. The habituating control strategy is supported by a *systematic* controller synthesis methodology. By contrast, the design procedures proposed for the other control techniques (valve position, coordinated, and parallel) are largely ad hoc, especially for nonminimum phase systems.

5. The effects of controller saturation and actuator failure on the habituating control strategy are considered explicitly, while these important issues are neglected in most other studies.

Habituating Controller Design

The following is a controller design methodology for habituating controllers based on the direct synthesis approach. An alternative technique based on model predictive control is discussed by Henson et al. [HOS95]. The discussion is restricted to transfer function models of the form

$$y(s) = g_1(s)u_1(s) + g_2(s)u_2(s) + g_3(s)d(s),$$

where y is the controlled output, u_1 and u_2 are the primary and secondary inputs, respectively, and d is an unmeasured disturbance. Because u_2 is chosen as a result of its favorable dynamic effects on y, the transfer function g_2 is assumed to be stable and of minimum phase. By contrast, the transfer function g_1 may be unstable and/or of nonminimum phase.

Because there are two manipulated inputs and one controlled output, the combination of control actions that produce the desired output y_{sp} at steady-state is nonunique. An additional objective is therefore required to obtain a well-defined control problem. In habituating control problems such as the polymerization process, the secondary input u_2 should also track a desired value $u_{2_{sp}}$. The desired control objectives are therefore as follows:

1. Obtain the transfer function $g_{y_d}(s)$ between y_{sp} and y.

2. Obtain the transfer function $g_{u_d}(s)$ between $u_{2_{sp}}$ and u_2.

3. Obtain a decoupled response between $u_{2_{sp}}$ and y.

4. Ensure nominal closed-loop stability.

5. Achieve asymptotic tracking of y_{sp} and $u_{2_{sp}}$ in the presence of plant-model mismatch.

The closed-loop transfer function matrix should therefore have the form

$$\begin{bmatrix} y \\ u_1 \\ u_2 \end{bmatrix} = \begin{bmatrix} g_{y_d} & 0 & * \\ * & * & * \\ * & g_{u_d} & * \end{bmatrix} \begin{bmatrix} y_{sp} \\ u_{2_{sp}} \\ d \end{bmatrix},$$

where g_{y_d} and g_{u_d} have the property that $g_{y_d}(0) = g_{u_d}(0) = 1$ and each asterisk (*) denotes a stable transfer function.

A parallel architecture for habituating control is shown in Figure 11. The term "parallel" is used because the input to both controllers is the error between y and y_{sp}, and each controller responds to setpoint changes and disturbances independently of the other controller. Note that this control structure is analogous to parallel architecture employed in the baroreceptor reflex (Figure 9). The parallel controllers have the form

$$u_1(s) = g_{c_{11}}(s)[y_{sp}(s) - y(s)] + g_{c_{12}}(s)u_{2_{sp}}(s),$$
$$u_2(s) = g_{c_{21}}(s)[y_{sp}(s) - y(s)] + g_{c_{22}}(s)u_{2_{sp}}(s).$$

If the transfer function g_1 associated with the primary input is of minimum phase, the control objectives can be satisfied by designing the primary and secondary controllers as [HOS95]:

$$g_{c_{11}} = \frac{g_{y_d} - (1 - g_{y_d})g_2 g_{c_{21}}}{(1 - g_{y_d})g_1}, \quad g_{c_{12}} = -\frac{g_2}{g_1}g_{c_{22}},$$
$$g_{c_{22}} = g_{u_d},$$

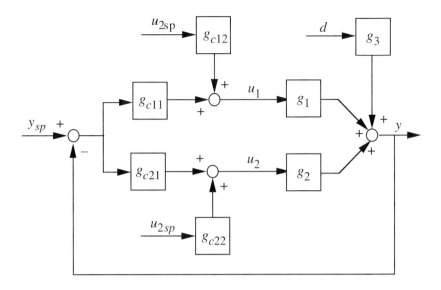

FIGURE 11. Parallel control architecture for habituating control.

where the Laplace variable s has been omitted for convenience. The free transfer function $g_{c_{21}}$ can be used to tune the responses of the two manipulated inputs. The transfer function g_{y_d} is tuned according to the dynamics of the secondary transfer function g_2, while g_{u_d} is chosen according to the dynamics of g_1. If the manipulated inputs are constrained, the habituating control approach offers the possibility of significantly improved performance as compared to conventional SISO control schemes, which only employ the primary input [HOS95].

If g_1 is of nonminimum phase, the primary and secondary controllers are chosen as [HOS95],

$$g_{c_{11}} = \frac{g_{u_d} g_{y_d}}{(1 - g_{y_d}) g_1^*}, \qquad g_{c_{12}} = -\frac{g_2}{g_1^*} g_{u_d},$$

$$g_{c_{21}} = \frac{(g_1^* - g_1 g_{u_d})}{(1 - g_{y_d}) g_1^* g_2}, \qquad g_{c_{22}} = \frac{g_1}{g_1^*} g_{u_d},$$

where g_1^* is the minimum phase approximation of g_1 [MZ89]. In the nonminimum phase case, a free controller transfer function is not available, and the u_2 tracking objective is only approximately satisfied:

$$\frac{u_2}{u_{2_{sp}}} = \frac{g_1}{g_1^*} g_{u_d} .$$

However, the undesirable effects of the nonminimum phase transfer function g_1 have been "transferred" from the output to the secondary input u_2. This property clearly demonstrates the advantage of habituating control as compared to conventional SISO control techniques. The transfer functions g_{y_d} and g_{u_d} can be tuned as in the minimum phase case.

Simulation Example

Consider the following process model:

$$y(s) = \frac{-2s+1}{(2s+1)^2} u_1(s) + \frac{1}{2s+1} u_2(s) + \frac{1}{s+1} d(s).$$

The transfer function g_1 contains a right-half-plane zero that limits the performance achievable with u_1 alone. An IMC controller [MZ89] and a habituating controller based on direct synthesis have been compared for this example [HOS95]. The IMC controller employs only the primary input u_1, while the habituating controller coordinates the use of the two available inputs. Therefore, this comparison demonstrates the performance enhancements that can be achieved by manipulating both the primary and secondary inputs. As discussed above, the habituating control strategy also offers important advantages over alternative control schemes that employ more manipulated inputs than controlled outputs. In the IMC design, a first-order filter with time constant $\lambda = 1$ and an additional setpoint filter with the same time constant are employed. The habituating controller is designed as

$$g_{y_d}(s) = \frac{1}{\epsilon_y s + 1}, \quad g_{u_d}(s) = \frac{1}{\epsilon_u s + 1},$$

with $\epsilon_y = \epsilon_u = 1$. An additional setpoint filter with the same time constant is also used.

Setpoint responses for IMC (dashed line) and habituating control (solid line) are shown in Figure 12. By using the secondary input u_2, habituating control yields excellent performance without an inverse response in the output. The secondary control returns to its setpoint ($u_{2_{sp}} = 0$) once the setpoint change is accomplished. By contrast, IMC produces very sluggish setpoint tracking with a significant inverse response. In Figure 13, the closed-loop responses of the two controllers for a unit step change in the unmeasured disturbance d are shown. Habituating control provides excellent performance, while the response of the IMC controller is very sluggish. The performance of the habituating controller for a setpoint change in the secondary input is shown in Figure 14. Note that the deleterious effects of the nonminimum phase element have been transferred to the $u_{2_{sp}}/u_2$ response, which is less important than the y_{sp}/y response. Moreover, the output is not affected by the $u_{2_{sp}}$ change. Additional simulation studies are presented by Henson et al. [HOS95].

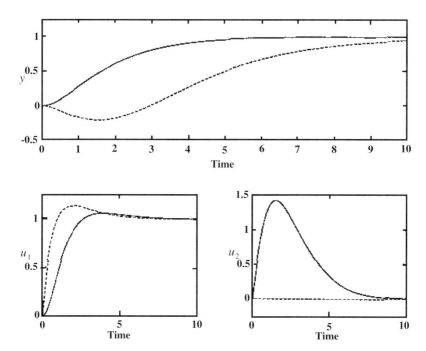

FIGURE 12. Direct synthesis and IMC control for an output setpoint change.

4.2 SIMO Control Structure

Baroreceptor Reflex

Carotid sinus baroreceptors have been classified as *Type I* or *Type II* receptors according to their firing patterns in response to slow ramp increases in pressure [SvBD+90]. Type I receptors exhibit the following characteristics: hyperbolic response patterns with sudden onset of firing at a threshold pressure, high sensitivities, and small operating ranges. By contrast, Type II receptors exhibit sigmoidal response patterns with spontaneous firing below a threshold pressure, low sensitivities, and large operating ranges. Type I and Type II baroreceptors also exhibit significant differences in acute resetting behavior [SGHD92], which is defined as a short-term (5–30 minutes) shift of the activity response curve in the direction of the prevailing pressure. Type I receptors acutely reset in response to mean pressure changes, while Type II receptors do not exhibit acute resetting. These firing characteristics indicate that Type I and Type II baroreceptors primarily measure rate of change of pressure and mean pressure, respectively [SGHD92]. Type I receptors generally have large myelinated fibers with high conduction velocities (2 – 40 m/s), while Type II baroreceptors have unmyelinated and small myelinated fibers with comparatively low conduction velocities (0.5 – 2 m/s). These physiological data suggest a differential role for Type I and Type II baroreceptors in dynamic and steady-state control of arterial

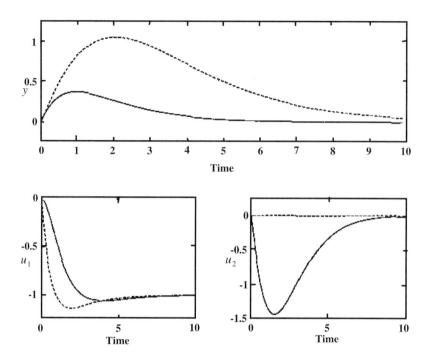

FIGURE 13. Direct synthesis and IMC control for an unmeasured disturbance.

blood pressure. Due to their high conduction velocities and measurement properties, Type I receptors may contribute primarily to dynamic control of blood pressure.

By contrast, Type II receptors may be effectively used for steady-state pressure control because they provide accurate, but slow, measurements of mean blood pressure. Seagard and coworkers [SHDW93] have verified this hypothesis by selectively blocking Type I and Type II receptors and examining the effects on dynamic and steady-state pressure control.

Coleman [Col80] has conducted an analogous investigation on the differential roles of the parasympathetic and sympathetic nervous systems in heart rate control. By selectively blocking the parasympathetic and sympathetic heart rate responses, Coleman has demonstrated that the parasympathetic and sympathetic systems are primarily responsible for dynamic and steady-state control of heart rate, respectively. Neglecting reflex manipulation of stroke volume and peripheral resistance, the results of Seagard [SHDW93] and Coleman [Col80] suggest a differential central nervous system pathway in which Type I and Type II baroreceptors preferentially affect the parasympathetic and sympathetic systems, respectively. Under this hypothesis, depicted in Figure 15, the heart rate is determined by two parallel controllers that selectively process input from Type I and Type II baroreceptors.

5. Modeling of the Baroreceptor Reflex with Applications

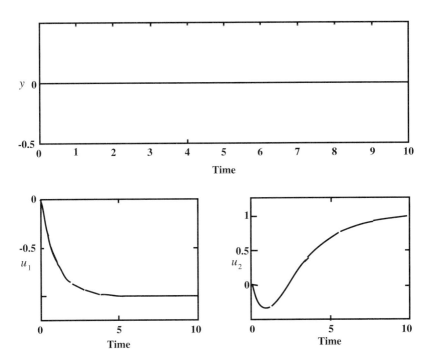

FIGURE 14. Direct synthesis control for an input setpoint change.

Process Control Applications

Many chemical processes contain output measurements that are analogous to the Type I and Type II baroreceptors. For example, consider the distillation column shown in Figure 16. Suppose that the objective is to control the composition of the product leaving the top of the column, and measurements of the top composition and an upper tray temperature are available. The top composition is the output variable to be controlled, but the inherent dynamics of typical on-line composition analyzers are such that such measurements are only available after a significant delay. By contrast, the tray temperature, measured by a thermocouple, is available without delay; it is, however, not always an accurate indication of the top composition. Observe that in this example the composition analyzer is analogous to the Type II receptor, while the thermocouple is analogous to the Type I receptor.

Hence, it is desirable to use the tray temperature for dynamic control and the top composition for steady-state control.

Pottmann et al. [PHOS96] have proposed a controller design methodology for single-input, two-output processes (such as this distillation column example) by reverse engineering the contributions of Type I and II receptors to blood pressure control. The approach is beneficial for processes that

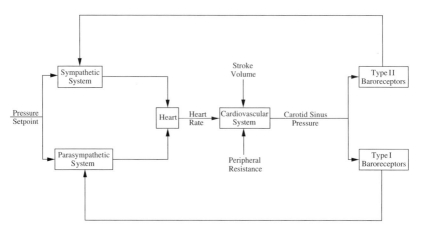

FIGURE 15. Simplified representation of a SIMO control structure in the baroreflex.

have two output measurements:

1. *Primary measurement* — a measurement of the process output to be controlled that has unfavorable (e.g., delayed) dynamic responses to changes in manipulated input and disturbance variables.

2. *Secondary measurement* — a measurement of a *different* process output that has more favorable dynamic responses to changes in manipulated input and disturbance variables.

Several related control schemes, including cascade control [Luy73, MZ89, SEM89, Yu88] have been proposed. In the most general sense the so-called "inferential control" schemes, as well as feedback control schemes incorporating state-estimation, may also be considered as related. In these instances, available "secondary" measurements are used to "infer" the status of the "primary" measurement. The novel feature of the strategy proposed by Pottmann et al. [PHOS96] is its control architecture, in which the controllers act in *parallel*; this offers the potential of superior performance and significantly improved robustness to controller and sensor failure as compared to cascade control approaches in which the controllers are in series.

Parallel Control Architecture

The process model is assumed to have the following parallel form:

$$y_1(s) = g_{11}(s)u(s) + g_{12}(s)d(s),$$
$$y_2(s) = g_{21}(s)u(s) + g_{22}(s)d(s),$$

where y_1 and y_2 are the primary and secondary measurements, respectively, u is the manipulated input, and d is an unmeasured disturbance. It is

5. Modeling of the Baroreceptor Reflex with Applications 115

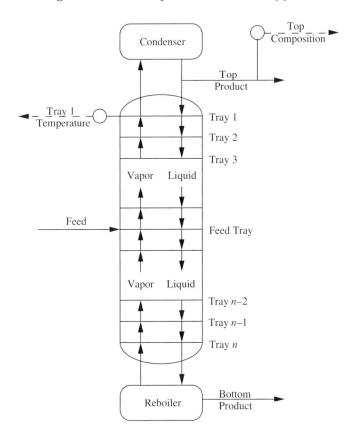

FIGURE 16. A distillation column.

easy to show that the parallel structure is more general than the cascade process structure used in most cascade control schemes [PHOS96]. Because the secondary output is assumed to exhibit favorable dynamic responses to input changes, the transfer functions g_{21} and g_{22} are assumed to be stable and of minimum phase. By contrast, the transfer functions g_{11} and g_{12} associated with the primary measurement may be of nonminimum phase.

The control objective is to make the primary output y_1 track its setpoint $y_{1_{sp}}$. In analogy to the baroreceptor reflex depicted in Figure 15, the parallel control architecture in Figure 17 is proposed. The controller has the form

$$u(s) = g_{c_1}(s)[y_{1_{sp}}(s) - y_1(s)] + g_{c_2}(s)[y_{2_{sp}}(s) - y_2(s)],$$

where $y_{2_{sp}}$ is the setpoint for y_2. Because y_2 is not a controlled output, the secondary setpoint is chosen as $y_{2_{sp}}(s) = g_{sp}(s)y_{1_{sp}}(s)$. The controller design problem is to select the transfer functions g_{c1}, g_{c2}, and g_{sp}.

For process control applications, the proposed architecture has two disadvantages: (i) it does not provide a convenient parameterization for controller design; and (ii) it is difficult to reconfigure the control system in

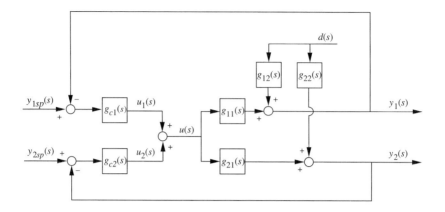

FIGURE 17. Parallel control architecture for SIMO control.

the event of a measurement failure. In order to overcome these shortcomings, the parallel controllers are *reparameterized*, and the resulting parallel control architecture is employed for controller design and implementation. Pottmann et al. [PHOS96] demonstrate that the parallel control strategy can yield superior performance and robustness as compared to a conventional cascade control scheme.

5 Neural Computational Mechanisms for Process Modeling

In this section, the neural computational mechanisms in the baroreflex are shown to have direct applications in the nonlinear modeling of chemical process systems. A brief description of a simplified conductance model will be presented, with special emphasis on the autoregulatory role played by the calcium channel. A novel processing element abstracted from the nonlinear dynamic nature of the neuron is then described, prior to discussing a chemical process modeling example. Finally, we outline a model-based control technique that employs the proposed dynamic processing element as a key component.

5.1 Neuron-Level Computation

As discussed earlier, the neurons in the cardiovascular NTS exhibit a wide range of complex nonlinear dynamic behavior. NTS neuron responses can be a function of time, voltage, and Ca^{++} concentration; and neurons in different regions of the baroreflex architecture display widely varying dynamic characteristics. These dynamic features are represented in Hodgkin-Huxley models by specific ion channels. For instance, *accommodation* (the lengthening of interspike intervals) is captured by the calcium channel. From a process modeling perspective, this suggests that neuronal elements used for computational modeling may be "tailored" to exhibit particular dynamic characteristics (e.g., asymmetric responses, oscillatory behavior, large deadtime), and incorporated in a suitable network architecture to yield desired input–output behavior.

As part of our research program, we seek to exploit these dynamic neuronal characteristics to develop tools for nonlinear process modeling. The approach discussed makes use of the biologically inspired neuron models (i.e., based on biologically plausible constitutive relations) for process applications. However, these detailed models will be reduced to a simpler form to facilitate network computation.

Role of Calcium in Autoregulation

The simplified model presented in Section 3 omitted the effect of calcium in modifying neuronal behavior. However, calcium plays an integral role in conductance-based neuron models, as it contributes to interspike interval modulation and accommodating responses [SGP93]. The intracellular calcium concentration has been proposed as an agent that regulates the maximal conductances [AL93]. This mechanism is described by modeling the maximal conductances of the membrane channels ($\bar{g}_i$) as a function of the calcium concentration:

$$\tau_i([Ca])\frac{d\bar{g}_i}{dt} = F_i([Ca]) - \bar{g}_i, \tag{6}$$

where [Ca] is the intracellular calcium concentration and F_i is the limiting value of the conductance. The function F_i is taken to be a rising or falling sigmoidal function in the original work [AL93]. In the context of dynamic chemical process models, Equation 6 may be recognized as a first-order system with variable time constant and steady-state gain; the process input is the calcium concentration; the process output is the maximal conductance. The incorporation of the simple mechanism in Equation 6 into a conductance model can lead to a broad range of dynamic behavior, including bursting activity, tonic firing, silent behavior, or "locked-up" (e.g., permanently depolarized) responses. Consequently, this mechanism was chosen as the basis for the development of a canonical element for dynamic process modeling.

A Canonical Dynamic Element

Calcium autoregulation suggests a simple computational element for process modeling: a first-order dynamic operator with a nonlinear time constant and an independent, nonlinear gain (cf. the Hopfield neuron model

[Hop90] where the gain and time constant share the same nonlinear dependence on the state). It should be noted that a *fixed* time constant and a *sigmoidal* gain function were used in [AL93]. In this work, we choose a more general formulation and employ Taylor series approximations of the nonlinear gain and time constant. Furthermore, the functional dependence of the time constant and gain are restricted to the operator output (y) to facilitate the numeric computations. By introducing first-order Taylor series approximations for the gain and time constant, one obtains

$$\mathcal{N}_i : \quad (\tau_0 + \tau_1 y)\frac{dy}{dt} = (K_0 + K_1 y)u - y. \tag{7}$$

Previous approaches for empirical nonlinear process modeling have employed similar mathematical forms to Equation 7 in an effort to capture the nonlinear dynamics of such chemical processes as distillation [CO93]. The present work differs from these earlier results by considering *network* arrangements of these processing elements.

Although the interconnection of these processing elements can take a variety of forms, we examine a fully recurrent Hopfield network [Hop90] in this work. The range of dynamic behavior of a Hopfield network composed of the biologically inspired neurons may be demonstrated by a simple interconnection of linear first-order systems. If the elements are connected in a feedback configuration with one system in the forward path and one system in the feedback path, a second-order transfer function is obtained. The coefficients of the first-order elements can be chosen to give *general* second-order responses between the input and output variables of the overall system.

This *cannot* be accomplished with many of the time series neural network techniques proposed for process modeling. For example, consider the approach in [MWD+91], where a first-order dynamic element is introduced at the output of a feedforward network. Such an architecture falls into the general class of Hammerstein dynamic systems (i.e., a static nonlinearity followed by a linear dynamic system). It is straightforward to show [SDS96] that such structures lead to nonlinear dynamic systems with relative degree one, underdamped responses, and (possibly) input multiplicity. By contrast, the architecture we propose yields dynamic systems that can have the following properties:

- arbitrary relative degree;

- arbitrary placement of the eigenvalues of the Jacobian matrix in the left-half (stable) complex plane;

- output and input multiplicity.

Clearly, the range of dynamic behavior that can be produced with the structure we propose is rather broad. In both [MWD+91] and the present case,

5. Modeling of the Baroreceptor Reflex with Applications

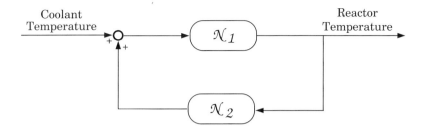

FIGURE 18. Dynamic model architecture.

an arbitrary system order can be achieved by employing an appropriate number of hidden layers.

Simulation Example

To demonstrate the effectiveness of the proposed structure, we now examine the problem of modeling a nonlinear continuous stirred-tank reactor (CSTR). The system considered is a stirred-tank jacketed reactor in which a simple first-order irreversible reaction occurs. This is a realistic example of practical significance, and it will serve as a preliminary test bed for the proposed modeling strategy. The dimensionless mass and energy balances for this system are given by [URP74]:

$$\dot{x}_1 = -x_1 + \mathcal{D}a(1-x_1)\exp(\frac{x_2}{1+x_2/\gamma}),$$
$$\dot{x}_2 = -x_2 + \mathbf{B}\mathcal{D}a(1-x_1)\exp(\frac{x_2}{1+x_2/\gamma}) + \beta(u-x_2).$$

The physical parameters chosen for this study are identical to those considered in [HS93]. The identification problem is to model the effect of coolant temperature (u) on the reactor temperature (x_2).

In Figure 18, the construction of a network model consisting of two fully interconnected dynamic processing elements is presented.

Additional dynamic elements can be added at the lower summation junction. Using a first-order Taylor series approximation for the nonlinear elements (i.e., gain, time constant), a model structure with eight parameters is obtained. The parameters of the network model were identified using a random search procedure [SGF90] because of the presence of multiple local minima in the solution space. The responses of the network model, an approximate linear model, and the actual CSTR to symmetric step changes in the input (±4 degrees) are shown in Figure 19. As can be seen in the figure, the system behavior is extremely nonlinear. While the linear model fails to track the reactor temperature accurately, the proposed network model exhibits excellent tracking over the range of these simulations. Additional details on the simulation results are contained in [SDS96].

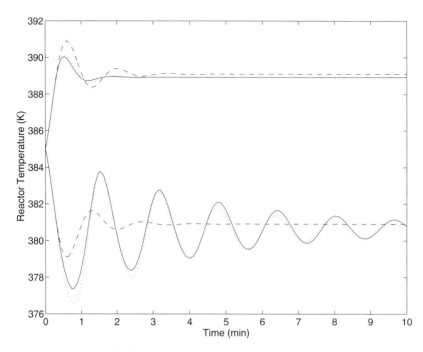

FIGURE 19. Process model dynamic response.

5.2 Model-Based Control Application

The biologically motivated dynamic network (BDN) model derived in the previous section can be directly incorporated in control schemes that depend explicitly upon a process model (e.g., internal model control (IMC) or model predictive control (MPC) [MZ89]). In this section, a direct synthesis approach to controller design will be presented that utilizes the BDN model as a key component. Such schemes typically rely on a model inverse for control move computations. However, recent results presented for Volterra-series-based models [DOP95] reveal a straightforward method for constructing a nonlinear model inverse that only requires linear model inversion. The details of this approach are omitted here; the interested reader is referred to the original reference. The resultant control structure is displayed in Figure 20, where it can be seen that the controller is composed of two components:

1. the proposed dynamic model (BDN), which contributes to a feedback signal representing the difference between the true process output and the modeled output; and

2. a model inverse loop, which contains the BDN model, a linear approximation to the BDN model, and a linear IMC controller.

5. Modeling of the Baroreceptor Reflex with Applications 121

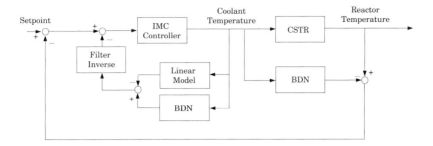

FIGURE 20. Closed-loop control structure.

Simulation Results

The reactor example from the previous section is considered, where the control objective is the regulation of the reactor temperature using the coolant temperature. Simulations were carried out for two control schemes: (i) a standard *linear* IMC controller that utilizes a linear model and its inverse; and (ii) the nonlinear controller depicted in Figure 20. In both cases, the desired closed-loop time constant was chosen to be 0.5 minutes. The closed-loop responses to a sequence of step changes in the temperature setpoint are shown in Figure 21. The setpoint is raised from 385°K to 400°K at $t = 0$ and back down to 380°K at $t = 25$. The dashed line represents the response of the linear controller, the dotted line represents the response of the nonlinear controller, and the solid line represents the *ideal* reference trajectory that would be achieved with perfect control. The nonlinear controller achieves vastly superior trajectory following. In fact, the linear controller response is unstable for the lower setpoint change. This demonstrates the improved performance that can be attained with a more accurate nonlinear model (such as the BDN) in a model-based control scheme.

6 Conclusions and Future Work

The neural circuitry in the baroreceptor reflex — the control system responsible for short-term regulation of arterial blood pressure — is a rich source of inspiration for process modeling and control techniques. Neuronal modeling has revealed some of the underlying principles that are responsible for the robust, nonlinear, adaptive, multivariable control functions that are utilized by the reflex. Preliminary results "reverse engineered" from this biological control system have been presented for scheduled control, parallel control, and nonlinear modeling strategies. Future work will focus on further development and industrial applications of the approaches described in this chapter.

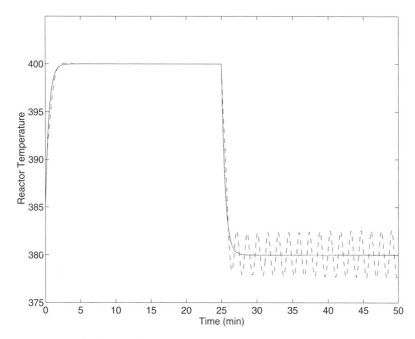

FIGURE 21. Closed-loop response to setpoint changes.

Acknowledgments: FJD would like to acknowledge funding from an NSF NYI award (CTS-9257059) and from an NSF grant (BCS-9315738). JSS acknowledges support from the following organizations: ONR (N00014-90-C-0224), NIH (NIH-MH-43787), NSF (IBN93-11388, BIR-9315303), and AFOSR (F49620-93-1-0285).

7 REFERENCES

[AC88] F. M. Abboud and M. W. Chapleau. Effects of pulse frequency on single-unit baroreceptor activity during single-wave and natural pulses in dogs. *Journal of Physiology (London)*, 401:295–308, 1988.

[AL93] L. F. Abbott and G. LeMasson. Analysis of neuron models with dynamically regulated conductances. *Neural Computation*, 5:823–842, 1993.

[BDM+89] J. Bradd, J. Dubin, B. Dueand, R. R. Miselis, S. Monitor, W. T. Rogers, K. M. Spyer, and J. S. Schwaber. Mapping of carotid sinus inputs and vagal cardiac outputs in the rat. *Neuroscience Abstracts*, 15:593, 1989.

[BM88] J. G. Balchen and K. I. Mumme. *Process Control: Structures and Applications.* Van Nostrand Reinhold, New York, 1988.

5. Modeling of the Baroreceptor Reflex with Applications

[CB91] T. L. Chia and C. B. Brosilow. Modular multivariable control of a fractionator. *Hydrocarbon Processing*, pages 61–66, June 1991.

[CO93] G. R. Srinivas, Y. Arkun, I-L. Chien and B. A. Ogunnaike. Nonlinear identification and control of a high-purity distillation column: a case study. *Journal of Process Control* 5(3):149–162, 1995.

[Col80] T. G. Coleman. Arterial baroreflex control of heart rate in the conscious rat. *American Journal of Physiology (Heart Circulation Physiology)*, 238:H515–H520, 1980.

[CWM77] J. A. Conner, D. Walter, and R. McKown. Neural repetitive firing: modifications of the Hodgkin-Huxley axon suggested by experimental results from crustacean axons. *Journal of Biophysics*, 18:81–102, 1977.

[DGJS82] S. Donoghue, M. Garcia, D. Jordan, and K. M. Spyer. Identification and brainstem projections of aortic baroreceptor afferent neurons in nodose ganglia of cats and rabbits. *Journal of Physiology (London)*, 322:337–352, 1982.

[DKRS94] F. J. Doyle III, H. Kwatra, I. Rybak, and J. S. Schwaber. A biologically-motivated dynamic nonlinear scheduling algorithm for control. In *Proceedings of the American Control Conference*, pages 92–96, 1994.

[DOP95] F. J. Doyle III, B. A. Ogunnaike, and R. K. Pearson. Nonlinear model-based control using second-order volterra models. *Automatica*, 31:697–714, 1995.

[FPSU93] W. R. Foster, J. F. R. Paton, J. S. Schwaber, and L. H. Ungar. Matching neural models to experiment. In F. Eeckman and J. M. Bower, editors, *Computation in Neural Systems*, pages 81–88. Kluwer Academic Press, Boston, 1993.

[FUS93] W. R. Foster, L. H. Ungar, and J. S. Schwaber. Significance of conductances in Hodgkin-Huxley models. *Journal of Neurophysiology*, 70:2502–2518, 1993.

[Get89] P. A. Getting. Reconstruction of small neural networks. In C. Koch and I. Segev, editors, *Methods in Neuronal Modeling*, pages 171–194. MIT Press, Cambridge, Massachusetts, 1989.

[GSP+91] E. B. Graves, J. S. Schwaber, J. F. R. Paton, K. M. Spyer, and W. T. Rogers. Modeling reveals mechanisms of central computation in the baroreceptor vagal reflex. *Society for Neurosciences Abstracts*, 17:993, 1991.

[HH52] A. L. Hodgkin and A. F. Huxley. A quantitative description of membrane current and its application to conduction and excitation in nerve. *Journal of Physiology (London)*, 117:500–544, 1952.

[Hil36] A. V. Hill. Excitation and accommodation in nerve. *Proceedings of the Royal Society (London)*, B119:305–355, 1936.

[Hop90] J. J. Hopfield. Dynamics and neural network computation. *International Journal of Quantum Chemistry: Quantum Chemistry Symposium 24*, pages 633–644, 1990.

[HOS95] M. A. Henson, B. A. Ogunnaike, and J. S. Schwaber. Habituating control strategies for process control. *AIChE Journal*, 41:604–618, 1995.

[HS93] M. A. Henson and D. E. Seborg. Theoretical analysis of unconstrained nonlinear model predictive control. *International Journal of Control*, 58(5):1053–1080, 1993.

[HW62] D. H. Hubel and T. N. Wiesel. Receptive fields, binocular integration and functional architecture in the cat's visual cortex. *Journal of Physiology (London)*, 160:106–154, 1962.

[KTK90] M. Kumada, N. Terui, and T. Kuwaki. Arterial baroreceptor reflex: Its central and peripheral neural mechanisms. *Progress in Neurobiology*, 35:331–361, 1990.

[Luy73] W. L. Luyben. Parallel cascade control. *Industrial & Engineering Chemistry Fundamentals*, 12:463–467, 1973.

[Luy90] W. L. Luyben. *Process Modeling, Simulation, and Control for Chemical Engineers*. McGraw-Hill, New York, 1990.

[Mac87] R. J. MacGregor. *Neural and Brain Modeling*. Academic Press, New York and London, 1987.

[Med93] J. V. Medanic. Design of reliable controllers using redundant control elements. In *Proceedings of the American Control Conference*, pages 3130–3134, San Diego, 1993.

[MRSS89] R. R. Miselis, W. T. Rogers, J. S. Schwaber, and K. M. Spyer. Localization of cardiomotor neurones in the anaesthetized rat; cholera-toxin HRP conjugate and pseudorabies labeling. *Journal of Physiology (London)*, 416:63P, 1989.

[MWD+91] G. A. Montague, M. J. Willis, C. DiMassimo, J. Morris, and M. T. Tham. Dynamic modeling of industrial processes with artificial neural networks. In *Proceedings of the International*

Symposium on Neural Networks and Engineering Applications, 1991.

[MZ89] M. Morari and E. Zafiriou. *Robust Process Control.* Prentice-Hall, Englewood Cliffs, New Jersey, 1989.

[PFS93] J. F. R. Paton, W. R. Foster, and J. S. Schwaber. Characteristic firing behavior of cells in the cardiorespiratory region of the nucleus tractus solitarii of the rat. *Brain Research*, 604:112–125, 1993.

[PHOS96] M. Pottmann, M. A. Henson, B. A. Ogunnaike, and J. S. Schwaber. A parallel control strategy abstracted from the baroreceptor reflex. *Chemical Engineering Science*, 51:931–945, 1996.

[PMB86] L. Popiel, T. Matsko, and C. Brosilow. Coordinated control. In M. Morari and T. J. McAvoy, editors, *Proceedings of the 3rd International Conference of Chemical Process Control*, pages 295–319, Elsevier, New York, 1986.

[RPS93] R. F. Rogers, J. F. R. Paton, and J. S. Schwaber. NTS neuronal responses to arterial pressure and pressure changes in the rat. *American Journal of Physiology*, 265:R1355–R1368, 1993.

[Sch87] J. S. Schwaber. Neuroanatomical substrates of cardiovascular and emotional-autonomic regulation. In A. Magro, W. Osswald, D. Reis, and P. Vanhoutte, editors, *Central and Peripheral Mechanisms in Cardiovascular Regulation*, pages 353–384. Plenum, New York, 1987.

[SDS96] A. M. Shaw, F. J. Doyle III, and J. S. Schwaber. A dynamic neural network approach to nonlinear process modeling. *Computers and Chemical Engineering, in press*, 1996.

[SEM89] D. E. Seborg, T. F. Edgar, and D. A. Mellichamp. *Process Dynamics and Control.* Wiley, New York, 1989.

[SES94] A. Standish, L. W. Enquist, and J. S. Schwaber. Innervation of the heart and its central medullary origin defined by viral tracing. *Science*, 263:232–234, 1994.

[SGF90] R. Salcedo, M. J. Goncalves, and S. Feyo de Azevedo. An improved random-search algorithm for non-linear optimization. *Computers and Chemical Engineering*, 14(10):1111–1126, 1990.

[SGHD92] J. L. Seagard, L. A. Gallenburg, F. A. Hopp, and C. Dean. Acute resetting in two functionally different types of carotid baroreceptors. *Circulation Research*, 70:559–565, 1992.

[SGP93] J. S. Schwaber, E. B. Graves, and J. F. R. Paton. Computational modeling of neuronal dynamics for systems analysis: Application to neurons of the cardiorespiratory NTS in the rat. *Brain Research*, 604:126–141, 1993.

[SHDW93] J. L. Seagard, F. A. Hopp, H. A. Drummond, and D. M. Van Wynsberghe. Selective contribution of two types of carotid sinus baroreceptors to the control of blood pressure. *Circulation Research*, 72:1011–1022, 1993.

[Shi78] F. G. Shinskey. Control systems can save energy. *Chemical Engineering Progress*, pages 43–46, May 1978.

[SKS71] R. M. Schmidt, M. Kumada, and K. Sagawa. Cardiac output and total peripheral resistance in carotid sinus reflex. *American Journal of Physiology*, 221:480–487, 1971.

[SPR+93] J. S. Schwaber, J. F. R. Paton, R. F. Rogers, K. M. Spyer, and E. B. Graves. Neuronal model dynamics predicts responses in the rat baroreflex. In F. Eeckman and J. M. Bower, editors, *Computation in Neural Systems*, pages 89–96. Kluwer Academic Press, Boston, 1993.

[SPRG93] J. S. Schwaber, J. F. R. Paton, R. F. Rogers, and E. B. Graves. Modeling neuronal dynamics predicts responses in the rat baroreflex. In F. Eeckman and J. M. Bower, editors, *Computation in Neural Systems*, pages 307–312. Kluwer Academic Press, Boston, 1993.

[Spy90] K. M. Spyer. The central nervous organization of reflex circulatory control. In A. D. Loewy and K. M. Spyer, editors, *Central Regulation of Autonomic Functions*, pages 168–188. Oxford Univ. Press, New York, 1990.

[SvBD+90] J. L. Seagard, J. F. M. van Brederode, C. Dean, F. A. Hopp, L. A. Gallenburg, and J. P. Kampine. Firing characteristics of single-fiber carotid sinus baroreceptors. *Circulation Research*, 66:1499–1509, 1990.

[URP74] A. Uppal, W. H. Ray, and A. B. Poore. On the dynamic behavior of continuous stirred tanks. *Chemical Engineering Science*, 29:967–985, 1974.

[WHD+92] S. J. Williams, D. Hrovat, C. Davey, D. Maclay, J. W. V. Crevel, and L. F. Chen. Idle speed control design using an H-Infinity approach. In *Proceedings of the American Control Conference*, pages 1950–1956, Chicago, 1992.

[Yu88] C.-C. Yu. Design of parallel cascade control for disturbance rejection. *AIChE Journal*, 34:1833–1838, 1988.

Chapter 6

Identification of Nonlinear Dynamical Systems Using Neural Networks

A. U. Levin
K. S. Narendra

ABSTRACT This chapter is concerned with the identification of a finite-dimensional discrete-time deterministic nonlinear dynamical system using neural networks. The main objective of the chapter is to propose specific neural network architectures that can be used for effective identification of a nonlinear system using only input–output data. Both recurrent and feedforward models are considered and analyzed theoretically and practically. The main result of the chapter is the establishment of input–output models using feedforward networks. Throughout the chapter, simulation results are included to complement the theoretical discussions.

1 Introduction

System theory provides a mathematical framework for the analysis and design of dynamical systems of various types, regardless of their special physical natures and functions. In this framework a system may be represented as an operator σ that belongs to a class Σ of operators that map an input space $\mathcal{U}$ into an output space $\mathcal{Y}$. The inputs $u \in \mathcal{U}$ comprise the set of all external signals that influence the behavior of the system and the outputs $y \in \mathcal{Y}$ comprise the set of dependent variables that are of interest and that can be observed by an external observer. To analyze any system σ we need to select a model $\bar{\sigma}$ that approximates σ in some sense. The model $\bar{\sigma}$ is an element of a parameterized family of operators $\bar{\Sigma} \subset \Sigma$. To be able to find a model that approximates any $\sigma \in \Sigma$ as closely as desired, $\bar{\Sigma}$ must be dense in Σ. For example, in the celebrated Weierstrass theorem, Σ is the class of continuous functions on a compact set, while $\bar{\Sigma}$ is the class of polynomial functions. In this chapter Σ represents a class of finite-dimensional discrete-time nonlinear systems, while $\bar{\Sigma}$ is the class of discrete dynamical systems generated by neural networks.

An extensive literature exists on linear system identification (a comprehensive list of references is given in [LL91]). For such systems, transfer functions, linear differential equations, and state equations have been used as models. In some cases, the class of systems Σ may itself be the class of nth order transfer functions or n-dimensional state equations, and in such cases the model $\bar{\Sigma}$ is also chosen to have the same form. We shall assume in this chapter that the class of interest, Σ, is the class of discrete-time finite-dimensional systems of the form

$$\Sigma: \begin{array}{rcl} x(k+1) &=& f[x(k), u(k)], \\ y(k) &=& h[x(k)], \end{array} \qquad (1)$$

where $x(k) \in \mathcal{X} \subset \mathbb{R}^n$ is the state of the system, $u(k) \in \mathcal{U} \subset \mathbb{R}^r$ is the input to the system, $y(k) \in \mathcal{Y} \subset \mathbb{R}^m$ is the output of the system, and f and h are smooth functions.[1] Based on some prior information concerning the system (1), our objective is to identify it using neural network–based models. In particular, the following class of identification models will be considered:

(i) state space (recurrent) models,

(ii) input–output (feedforward) models.

The structure of the neural networks used to identify the system is justified using results from analysis and differential topology. The relative merits of the models are compared and simulation results are presented wherever necessary to complement the theoretical developments.

Notation

The space of input and output sequences of length l will be denoted by $\mathcal{U}_l$ and $\mathcal{Y}_l$, respectively.

Input and output sequences of length l starting at time k will be denoted respectively by

$$U_l(k) \triangleq [u(k), u(k+1), \ldots u(k+l-1)]$$

and

$$Y_l(k) \triangleq [y(k), y(k+1), \ldots y(k+l-1)].$$

By definition of the state, it follows that $x(k+l)$ can be represented as

$$x(k+l) \triangleq F_l[x(k), U_l(k)],$$

---

[1] For clarity of exposition, we will state all results for SISO systems. Extension of these to MIMO systems is quite straightforward. Also, without loss of generality, an equilibrium point x_0, u_0, y_0 will always be assumed to be $(0, 0, 0)$.

where $F_l : \mathcal{X} \times \mathcal{U}_l \to \mathcal{X}$. Similarly, the output at time $k+l$ can be expressed as
$$y(k+l) = h[F_l(x(k), U_l(k)), u(k)] \triangleq h_l[x(k), U_{l-1}(k)],$$
where $h_l : \mathcal{X} \times \mathcal{U}_l \to \mathcal{Y}$; and $Y_l(k)$ can be expressed as
$$Y_l(k) \triangleq H_l[x(k), U_{l-1}(k)],$$
where $H_l : \mathcal{X} \times \mathcal{U}_{l-1} \to \mathcal{Y}_l$. When no confusion can arise, the index k will be omitted, e.g., $U_l \triangleq U_l(k)$.

Following the notation introduced in [NP90], an L-layer neural network with n_l neurons at the lth layer will be denoted by
$$\mathcal{N}^L_{n_0, n_1, n_2, \ldots n_L}.$$

For example, a network with two inputs, three neurons in the first hidden layer, five in the second, and one output unit will be described by $\mathcal{N}^3_{2,3,5,1}$. The set of weights of a network NN will be denoted by $\Theta(NN)$, and a generic weight (or parameter) will commonly be denoted by θ.

Organization of the Chapter

The chapter is organized as follows: Section 2 presents mathematical preliminaries and is devoted to concepts and definitions as well as mathematical theorems that will be used throughout the chapter. Section 3 deals with identification using state space models. Using the dynamic backpropagation algorithm, it is shown how a recurrent structure can be used to identify a system. In Section 4 the problem of identification using input–output models is considered. First, the simpler problem of constructing a local input–output model around an equilibrium state is considered, and then conditions for the existence of a global model are derived. In all cases the theoretical basis is stated for the architectures chosen, and simulation results are presented to complement the theoretical discussions.

2 Mathematical Preliminaries

This section is intended to serve as a concise introduction to some of the notions that this chapter relies upon. First, in Section 2.1 we give a brief summary of neural networks as they will be used in the chapter. The establishment of input–output models will rely on the concept of observability, which is presented in Section 2.2. Finally, in Section 2.3 some definitions and results from differential topology, which will be used to establish the global existence of input–output realizations of nonlinear systems, are introduced.

2.1 Neural Networks

In the current work, neural networks are treated merely as conveniently parameterized nonlinear maps, capable of approximating arbitrary continuous functions over compact domains. Specifically, we make use of sigmoidal feedforward networks as components of dynamical systems. The algorithms presented rely on supervised learning. Since the main objective of this work is to propose a general methodology by which identification based on neural networks can be made more rigorous, no particular effort is made to optimize the computation time, and training relies on the standard backpropagation and dynamic backpropagation algorithms. These could be easily replaced by any other supervised learning method. Also, all results are presented in such a way that they can be implemented by any feedforward architecture capable of universal approximation.

In the following, the term *neuron* will refer to an operator that maps $\mathbb{R}^n \to \mathbb{R}$ and is explicitly described by the equation:

$$y = \Gamma(\sum_{j=1}^{n} w_j u_j + w_0), \qquad (2)$$

where $U^T = [u_1, u_2, \ldots u_n]$ is the input vector, $W^T = [w_1, w_2, \ldots w_n]$ is referred to as the weight vector of the neuron, and w_0 is termed its bias. $\Gamma(\cdot)$ is a monotone continuous function $\Gamma : \mathbb{R} \to (-1, 1)$ (commonly referred to as a "sigmoidal function" e.g., $\tanh(\cdot)$). The neurons are organized in a feedforward layered architecture ($l = 0, 1 \ldots L$), and a neuron at layer l receives its inputs only from neurons in the layer $l - 1$.

A neural network, as defined above, represents a specific family of parameterized maps. If there are n_0 input elements and n_L output elements, the network defines a continuous mapping $NN : \mathbb{R}^{n_0} \to \mathbb{R}^{n_L}$. To enable this map to be surjective (onto), we will choose the output layer to be linear.

Two facts make the networks defined above powerful tools for approximating functions.

Multilayer feedforward neural networks are universal approximators:

It was proved by Cybenko [Cyb89] and Hornik et al. [HSW89] that any continuous mapping over a compact domain can be approximated as accurately as necessary by a feedforward neural network with one hidden layer. This implies that given any $\epsilon > 0$ a neural network with a sufficiently large number of nodes can be determined such that

$$\|f(u) - NN(u)\| < \epsilon \text{ for all } u \in \mathcal{D},$$

where f is the function to be approximated and $\mathcal{D}$ is a compact domain of a finite dimensional normed vector space.

6. Identification of Nonlinear Dynamical Systems

The backpropagation algorithm:

This algorithm, [MRtPRG86], which performs stochastic gradient descent, provides an effective method to train a feedforward neural network to approximate a given continuous function over a compact domain $\mathcal{D}$.

Let $u \in \mathcal{D}$ be a given input. The network approximation error for this input is given by

$$e(u) = f(u) - NN(u).$$

Training $NN(\cdot)$ to closely approximate f over $\mathcal{D}$ is equivalent to minimizing

$$I = \int_{\mathcal{D}} \|e(u)\| du.$$

The training procedure for the network is carried out as follows: The network is presented with a sequence of training data (input–output pairs). Let θ denote a generic parameter (or weight) of the network. Following each training example, the weights of the network are adjusted according to

$$\theta(k+1) = \theta(k) - \eta(k) \frac{\partial I}{\partial \theta}|_{\theta=\theta(k)}.$$

Stochastic approximation theory [Lju77] guarantees that if the step size $\eta(k)$ satisfies certain conditions, I will converge to a local minimum with probability 1. If the performance hypersurface is unimodal, this implies that the global minimum is achieved.

Recurrent Networks

By interconnecting several such feedforward blocks using feedback connections into a recurrent structure, the network's behavior can no longer be described in terms of a static mapping from the input to the output space. Rather, its output will exhibit complex temporal behavior that depends on the current states of the neurons as well as the inputs.

In the same manner that a feedforward layered network can be trained to emulate a static mapping, a training algorithm named *dynamic backpropagation*[2] [WZ89, NP90, NP91] has been proposed to train a recurrent network to follow a temporal sequence. The *dynamic backpropagation algorithm* is based on the fact that the dependence of the output of a dynamical system on a parameter is itself described by a recursive equation. The latter in turn contains terms that depend both explicitly and implicitly on the parameter [NP91], and hence the gradient of the error with respect to a parameter can be described as an output of a linear system.

---

[2] In this chapter we use the name coined by Narendra and Parthasarathy.

The dynamic backpropagation algorithm:

A natural performance criterion for the recurrent network would be the summation of the square of the error between the sequence we want the network to follow, denoted by the vector process $y(k)$, and the outputs of the network denoted by $\hat{y}(k)$:

$$I(k) = \sum_k \|y(k) - \hat{y}(k)\|^2 \triangleq \sum_k \|e(k)\|^2.$$

By its definition, a recurrent network can refer to its inputs $u(k)$, states $x(k)$, and outputs $\hat{y}(k)$. The algorithm presented will make use of these notions.

Let θ denote a generic parameter of the network. The gradient of I with respect to θ is computed as follows:

$$\frac{dI(k)}{d\theta} = -2 \sum_k [y(k) - \hat{y}(k)]^T \frac{d\hat{y}(k)}{d\theta}, \tag{3}$$

$$\frac{d\hat{y}(k)}{d\theta} = \sum_j \frac{\partial \hat{y}(k)}{\partial x_j(k)} \frac{dx_j(k)}{d\theta}, \tag{4}$$

$$\frac{dx_j(k)}{d\theta} = \sum_l \frac{\partial x_j(k)}{\partial x_l(k-1)} \frac{dx_l(k-1)}{d\theta} + \frac{\partial x_j(k)}{\partial \theta}. \tag{5}$$

Thus the gradient of the output with respect to θ is given by the output of the linear system

$$\begin{aligned}\frac{dx(k+1)}{d\theta} &= A\frac{dx(k)}{d\theta} + b\frac{\partial x(k)}{\partial \theta}, \\ \frac{d\hat{y}(k)}{d\theta} &= c^T \frac{dx(k)}{d\theta},\end{aligned} \tag{6}$$

where $\frac{dx(k)}{d\theta}$ is the state vector, $\frac{\partial x(k)}{\partial \theta}$ is the input, and A, b, c are time-varying parameters defined by $a_{ij} \triangleq \frac{\partial x_i(k+1)}{\partial x_j(k)}$, $b_i \triangleq 1$, and $c_i \triangleq \frac{\partial \hat{y}(k)}{\partial x_i(k)}$. Initial conditions for the states are set to zero. This linear system is referred to in the control literature as the *sensitivity network* for θ [JC73, NP90].

2.2 Observability

One of the fundamental concepts of systems theory, which concerns the ability to determine the states of a dynamical system from the observations of its inputs and outputs, is *observability*.

Definition 1 A dynamical system is said to be *observable* if for any two states x_1 and x_2 there exists an input sequence of finite length l, $U_l = (u(0), u(1), \ldots, u(l-1))$, such that $Y_l(x_1, U_l) \neq Y_l(x_2, U_l)$, where Y_l is the output sequence.

6. Identification of Nonlinear Dynamical Systems

The ability to effectively estimate the state of a system, or to identify it based on input–output observations, is determined by the observability properties of the system. However, the definition of observability as given above is too broad to guarantee the existence of efficient methods to perform these tasks. Thus, in the following we will present two specific observability notions: *strong observability* and *generic observability*, based on which practical algorithms can be derived.

Linear systems

Observability has been extensively studied in the context of linear systems and is now part of the standard control literature. A general linear time invariant system is described by the set of equations

$$\begin{aligned} x(k+1) &= Ax(k) + Bu(k), \\ y(k) &= Cx(k), \end{aligned} \quad (7)$$

where $x(k) \in \mathbb{R}^n$, $u(k) \in \mathbb{R}^r$, $y(k) \in \mathbb{R}^m$, and A, B, and C are respectively $n \times n$, $n \times r$, and $m \times n$ matrices. If $r = m = 1$ the system is referred to as *single-input/single-output* (SISO). If $r, m > 1$, it is called *multi-input/multi-output* (MIMO).

Definition 2 (Observability of Linear Systems) A linear time invariant system of order n is said to be *observable* if the state at any instant can be determined by observing the output y over a finite interval of time.

A basic result in linear control theory states that the system (8) will be observable if and only if the $(nr \times n)$ matrix

$$M_o = \begin{bmatrix} C \\ CA \\ \cdots \\ CA^{n-1} \end{bmatrix}$$

is of rank n. For a SISO system this implies that M_o is nonsingular. M_o is called the observability matrix.

Observability of a linear system is a system-theoretic property and remains unchanged even when inputs are present, provided they are known. For a linear observable system of order n, any input sequence of length n will distinguish any state from any other state. If two states are not distinguishable by this randomly chosen input, they cannot be distinguished by any other input sequence. In that case, the input–output behavior of the system can be realized by an observable system of lower dimension, where each state in the new system represents an equivalence class that corresponds to a set of states that could not be distinguished in the original one.

Whereas a single definition (2) is found to be adequate for linear time-invariant systems, the concept of observability is considerably more involved for nonlinear systems [Fit72] (a detailed discussion on different notions of observability is given in [Son79a]). As defined, observability guarantees the existence of an input sequence that can distinguish between any two states. This input sequence may, however, depend on those states. Further, in some cases, the determination of the state of a system, may require the resetting of the system and reexploring it with different inputs, as shown in Example 1.

Example 1 Given the second-order system

$$\begin{aligned} x_1(k+1) &= x_2(k), \\ x_2(k+1) &= \sin[x_1(k)u(k)], \\ y(k) &= x_2(k), \end{aligned}$$

if the sequence of inputs is $U = \{c, u(1), u(2) \ldots\}$, then all states of the form $(\frac{2\pi}{c}, x_2(0))$ cannot be distinguished from $(0, x_2(0))$. However, if the system is reset to the initial state and run with $U' = \{c', u(1), u(2) \ldots\}$ $(c \neq c')$, the initial state can be uniquely determined. △

For observable systems, to assure that a state can be determined by a single input sequence of finite length (*single experiment observability*), we will require that the system be *state invertible*:

Definition 3 We will call the system (1) *state invertible* if for a given u, f defines a diffeomorphism on x.

State invertible systems arise naturally when continuous-time systems are sampled or when an Euler approximation is used to discretize a differential equation [JS90]. For a given input sequence, the invertibility of a system guarantees that the future as well as the past of a state is unique. Whenever necessary, we shall make the assumption that the system is state invertible.

While single-experiment observability concerns the existence of an input such that the state can be determined by applying this input to the system, the input required may still depend upon the state. Hence, to be able to determine the state in a practical context, a stronger form of observability is needed. A desirable situation would be if any input sequence of length l will suffice to determine the state uniquely for some integer l. This form of observability will be referred to as *strong observability*. It readily follows from Definition 2 that any observable linear system is strongly observable with $l = n$, n being the order of the linear system.

As will be shown in Section 4.1, conditions for strong observability can be derived locally around an equilibrium point. Unfortunately, unlike the linear case, global strong observability is too stringent a requirement and may not hold for most nonlinear systems of the form (1). However, practical determination of the state can still be achieved if there exists an integer

l such that almost any input sequence (generic) of length greater than or equal to l will uniquely determine the state. This will be termed *generic observability*.

Example 2 (Generic Observability) Let
$$x(k+1) = x(k) + u(k),$$
$$y(k) = x^2(k).$$

The outputs are given by
$$y(k) = x^2(k),$$
$$y(k+1) = x^2(k) + u^2(k) + 2x(k)u(k)$$
$$= y(k) + u^2(k) + 2x(k)u(k).$$

From the above two equations we have
$$x(k) = \frac{y(k+1) - y(k) - u^2(k)}{2u(k)},$$

and if $u(k) \neq 0$, $x(k)$ can be uniquely determined. Hence, the system is generically observable. △

In the rest of the chapter, only strongly or generically observable systems will be discussed. The notion of generic observability is considered in detail in Section 4.2. That discussion should also help clarify the difference between these two concepts.

2.3 Transversality

The discussion on generic observability will rely on some concepts and results from differential topology — most notably *Transversality*. It will be shown how observability can be described as a transversal intersection between maps. Based on this, the genericness of transversal intersections will be used to prove the genericness of generically observable systems. Our aim in this section is to present these results for the sake of easy reference. The reader may, if he wishes, skip this section on first reading and return to it later, after going through Section 4.2. For an excellent and extensive introduction, the reader is referred to [GP74].

Transversality is a notion that classifies the manner in which smooth manifolds intersect:

Definition 4 Let $\mathcal{X}$ and $\mathcal{Y}$ be smooth manifolds and $f : \mathcal{X} \to \mathcal{Y}$ be a smooth mapping. Let $\mathcal{W}$ be a submanifold of $\mathcal{Y}$ and x a point in $\mathcal{X}$. Then f *intersects* $\mathcal{W}$ *transversally at* x (denoted by $f \pitchfork \mathcal{W}$ at x) if either one of the following hold:

1. $f(x) \notin \mathcal{W}$.

2. $f(x) \in \mathcal{W}$ and $T_{f(x)}\mathcal{Y} = T_{f(x)}\mathcal{W} + (df)_x(T_x\mathcal{X})$
   ($T_a\mathcal{B}$ denoting the tangent space to $\mathcal{B}$ at a).

If $\mathcal{V}$ is a subset of $\mathcal{X}$, then f *intersects $\mathcal{W}$ transversally on $\mathcal{V}$* (denoted by $f \pitchfork \mathcal{W}$ on $\mathcal{V}$) if $f \pitchfork \mathcal{W}$ at x for all $x \in \mathcal{V}$. Finally, f *intersects $\mathcal{W}$ transversally* (denoted by $f \pitchfork \mathcal{W}$) if $f \pitchfork \mathcal{W}$ on $\mathcal{X}$.

Example 3 Let $\mathcal{W}$ be a plane in $\mathbb{R}^3$. Let $f : \mathbb{R} \to \mathbb{R}^3$ be a linear function, i.e., f defines a line in $\mathbb{R}^3$. Now $f \pitchfork \mathcal{W}$ unless $f(x)$ lies inside $\mathcal{W}$. △

An important consequence of the property that a mapping is transversal is given by the following proposition [GG73].

Proposition 1 *Let $\mathcal{X}$ and $\mathcal{Y}$ be smooth manifolds and $\mathcal{W}$ a submanifold of $\mathcal{Y}$. Suppose $\dim \mathcal{W} + \dim \mathcal{X} < \dim \mathcal{Y}$. Let $f : \mathcal{X} \to \mathcal{Y}$ be a smooth mapping, and suppose that $f \pitchfork \mathcal{W}$. Then $f(\mathcal{X}) \cap \mathcal{W} = \emptyset$.*

Thus, in the last example, if $\mathcal{W}$ represented a line in $\mathbb{R}^3$, transversality implies that $f(x)$ and $\mathcal{W}$ do not intersect, i.e., if two lines are picked at random in a three-dimensional space, they will not intersect (which agrees well with our intuition).

The key to transversality is families of mappings. Suppose $f_s : \mathcal{X} \to \mathcal{Y}$ is a family of smooth maps, indexed by a parameter s that ranges over a set $\mathcal{S}$. Consider the map $F : \mathcal{X} \times \mathcal{S} \to \mathcal{Y}$ defined by $F(x, s) = f_s(x)$. We require that the mapping vary smoothly by assuming $\mathcal{S}$ to be a manifold and F to be smooth. The central theorem is:

Theorem 1 (Transversality Theorem) *Suppose $F : \mathcal{X} \times \mathcal{S} \to \mathcal{Y}$ is a smooth map of manifolds and let $\mathcal{W}$ be a submanifold of $\mathcal{Y}$. If $F \pitchfork \mathcal{W}$, then for almost every $s \in \mathcal{S}$ (i.e. generic s) f_s is transversal to $\mathcal{W}$.*

From the transversality theorem it follows that transversality is a generic property of maps:

Theorem 2 *Let $\mathcal{X}$ and $\mathcal{Y}$ be smooth manifolds and $\mathcal{W}$ a closed submanifold of $\mathcal{Y}$. Then the set of smooth mappings $f : \mathcal{X} \to \mathcal{Y}$ that intersect $\mathcal{W}$ transversally is open and dense in $\mathcal{C}^\infty$.*

Another typical behavior of functions that we will make use of is the the *Morse* property:

Definition 5 A function h will be called a *Morse function* if it has only nondegenerate (isolated) critical points.

The set of Morse functions is open and dense in $\mathcal{C}^r$ [GG73]. Hence, we may confidently assume that h in (1) is such a function.

3 State space models for identification

Since by our assumption the system is described by a state equation (1), the natural identification model for the system using neural networks also has the same forms. Relying on the approximation capabilities of feedforward neural networks [Cyb89, HSW89], each of these functions can be approximated by a multilayered neural network with appropriate input and output dimensions. The efficiency of the identification procedure then depends upon the prior information that is assumed.

If the state of the system is assumed to be directly measurable, the identification model can be chosen as

$$\Sigma : \begin{array}{rcl} x(k+1) & = & NN_f[x(k), u(k)], \\ y(k) & = & NN_h[x(k)], \end{array} \qquad (8)$$

where NN_h and NN_f are maps realized by feedforward neural networks (for ease of exposition they will be referred to as neural networks). In this case, the states of the plant to be identified are assumed to be directly accessible, and each of the networks NN_f and NN_h can be independently trained using static learning [LN93, Lev92]. Once constructed, the states of the model provide an approximation to the states of the system.

When the state $x(k)$ of the system is not accessible, the problem of identification is substantially more difficult. In such a case, one cannot obtain an estimate $\hat{x}(k)$ of the $x(k)$, and the identification model has the form

$$\begin{array}{rcl} z(k+1) & = & NN_f[z(k), u(k)], \\ \hat{y}(k) & = & NN_h[z(k)], \end{array} \qquad (9)$$

where again NN_h and NN_f denote feedforward neural networks (Figure 1). This model provides an equivalent representation of the system (1), and its state $z(k) \stackrel{\triangle}{=} [z_1(k), z_2(k), \ldots, z_n(k)]$ is related by a diffeomorphism to $x(k)$, the state of the system.

A natural performance criterion for the model would be the sum of the squares of the errors between the system and the model outputs:

$$I(K) = \sum_{k=0}^{K} \|y(k) - \hat{y}(k)\|^2 \stackrel{\triangle}{=} \sum_{k} \|e(k)\|^2.$$

Since $x(k)$ is not accessible and the error can be measured only at the output, the networks cannot be trained separately. Since the model contains a feedback loop, the gradient of the performance criterion with respect to the weights of NN_f varies with time, and thus dynamic backpropagation needs to be used [NP91].

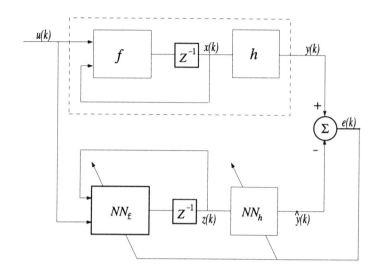

FIGURE 1. State space model for identification.

Let $\theta \in \Theta(NN_f)$ denote a parameter of NN_f. The gradient of I with respect to θ is derived as follows:

$$\frac{dI(K)}{d\theta} = -2\sum_{k=0}^{K}[y(k) - \hat{y}(k)]\frac{d\hat{y}(k)}{d\theta}, \tag{10}$$

$$\frac{d\hat{y}(k)}{d\theta} = \sum_{j=1}^{n} \frac{\partial \hat{y}(k)}{\partial z_j(k)} \frac{dz_j(k)}{d\theta}, \tag{11}$$

$$\frac{dz_j(k)}{d\theta} = \sum_{l=1}^{n} \frac{\partial z_j(k)}{\partial z_l(k-1)} \frac{dz_l(k-1)}{d\theta} + \frac{\partial z_j(k)}{\partial \theta}. \tag{12}$$

Thus the gradient of the output with respect to θ is given by the output of the linear system

$$\begin{aligned}\frac{dz(k+1)}{d\theta} &= A\frac{dz(k)}{d\theta} + b\frac{\partial z(k)}{\partial \theta}, \\ \frac{d\hat{y}(k)}{d\theta} &= c^T\frac{dz(k)}{d\theta},\end{aligned} \tag{13}$$

where $\frac{dz(k)}{d\theta}$ is the state vector, $\frac{\partial z(k)}{\partial \theta}$ is the input, and A, b, c are defined by $a_{ij} \triangleq \frac{\partial z_i(k+1)}{\partial z_j(k)}$, $b_i \triangleq 1$, and $c_i \triangleq \frac{\partial \hat{y}(k)}{\partial z_i(k)}$. The initial conditions of the states are set to zero.

Since NN_h is directly connected to the output, with no feedback loops, the gradients of the error with respect to its parameters are calculated

6. Identification of Nonlinear Dynamical Systems

using static backpropagation. This can be done either on-line or in a batch mode in which the error over a finite number of steps is summed before updating the weights.[3]

As constructed, the model is not unique, and thus the state of the model (after identification is achieved) is given by $z = \phi(x)$ and the neural networks converge to a transform of the system's functions:

$$NN_h(\cdot) \simeq h(\cdot) \circ \phi^{-1},$$
$$NN_f(\cdot, \cdot) \simeq \phi \circ f(\phi^{-1}(\cdot), \cdot),$$

where $\phi : \mathbb{R}^n \to \mathbb{R}^n$ is continuous and invertible.

If the system can be reset at the discretion of the designer to a fixed initial state (which without loss of generality can be assumed to be the origin), the training procedure will be more tractable. The corresponding state for the model can, also without loss of generality, be set to zero, so that each training sequence can start with both the system and the model at the initial state. Thus, in such a framework, the functional relation ϕ between the states of the system and the model will emerge naturally.

On the other hand, if resetting is not possible, the initial state of the model must be treated as an independent parameter. The gradient of the error at time k with respect to the model's initial conditions is given by

$$\frac{dI(K)}{dz(0)} = -2 \sum_{k=0}^{K} [e(k)] \frac{d\hat{y}(k)}{dz(0)}, \tag{14}$$

$$\frac{d\hat{y}(k)}{dz(0)} = \sum_{j=1}^{n} \frac{\partial \hat{y}(k)}{\partial z_j(k)} \frac{dz_j(k)}{dz(0)},$$

$$\frac{dz_j(k)}{dz(0)} = \sum_{l=1}^{n} \frac{\partial z_j(k)}{\partial z_l(k-1)} \frac{dz_l(k-1)}{dz(0)}. \tag{15}$$

This can be described as the output of a homogeneous time-varying linear system

$$\frac{dz(k+1)}{dz(0)} = A \frac{dz(k)}{dz(0)},$$
$$\frac{d\hat{y}(k)}{dz(0)} = c^T \frac{dz(k)}{dz(0)}. \tag{16}$$

This is a system of order n^2, where $\frac{dz(k)}{dz(0)}$ is the state vector at time k and A, c are defined by $a_{ij} \triangleq \frac{\partial z_i(k+1)}{\partial z_j(k)}$ and $c_i \triangleq \frac{\partial \hat{y}(k)}{\partial z_i(k)}$. Initial conditions for the states are set to $I_{n \times n}$, the n-dimensional identity matrix.

---

[3] In many cases, the output is known to be a subset of the state, i.e., h is merely a projection matrix. For such systems, the complexity of the algorithm is greatly reduced, since the gradient of the output with respect to the state is known a priori and the error can be calculated at the state level.

Simulation 1 (Identification: State Model)[4] The system is given by

$$x_1(k+1) = x_2(k)[1 + 0.2u(k)],$$
$$x_2(k+1) = -0.2x_1(k) + 0.5x_2 + u(k),$$
$$y(k) = 0.3[x_1(k) + 2x_2(k)]^2.$$

The neural network–based model used to identify the system is given by

$$\hat{x}_1(k+1) = NN_{f1}[\hat{x}_1(k), \hat{x}_2(k), u(k)],$$
$$\hat{x}_2(k+1) = NN_{f2}[\hat{x}_1(k), \hat{x}_2(k), u(k)], \quad (17)$$
$$\hat{y}(k) = NN_h[\hat{x}_1(k), \hat{x}_2(k)].$$

A separate network was used for the estimation of each of the nonlinear functions f_1, f_2, and h. All three networks were of the class $\mathcal{N}_{1,10,5,1}^3$.

For the training of the networks, it is assumed that the system can be initiated at the discretion of the experimenter. Training[5] was done with a random input uniformly distributed in $[-1, 1]$. Training sequences were gradually increased starting with $k = 10$, and after successful learning was achieved, the length of the sequence was gradually increased by units of ten until $k = 100$ was reached. Parameter adjustment was carried out at the end of each sequence using the summed error square as indicated earlier.

Adaptation was halted after 80,000 steps (with time between consecutive weight adjustments varying between 10 and 100 steps), and the identification model was tested with sinusoidal inputs. A particular example is shown in Figure 2. △

4 Identification Using Input–Output Models

It is clear from Section 3 that choosing state space models for identification requires the use of dynamic backpropagation, which is computationally a

---

[4]The use of gradients with respect to initial conditions requires reinitializing the model with its corrected initial conditions and running it forward to the current time step. Such a process is very tedious and practically infeasible in real time. In the simulations given below, it is assumed that the system can be reset periodically at the discretion of the designer.

[5]When running the dynamic backpropagation algorithm, the following procedure was adopted: The network was run for a predetermined number of steps K_{max} and the weights adjusted so that $I(K_{max})$ was minimized. Our experience showed that better results are achieved if training sequences were gradually increased. Thus, starting the training with short sequences of length k_1, the network was trained on longer sequences of length $k_2, k_3, \ldots$, etc. until K_{max} was reached. For each sequence, the total error at the end of the sequence was used to determine the weight adjustment.

6. Identification of Nonlinear Dynamical Systems

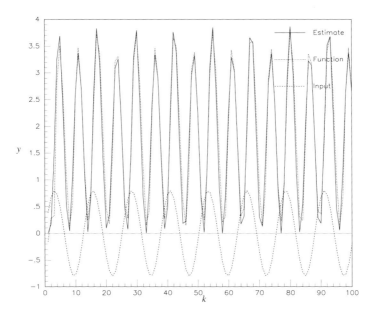

FIGURE 2. Testing of the state space model with sinusoidal input.

very intensive procedure. At the same time, to avoid instabilities while training, one needs to use small gains to adjust the parameters, and this in turn results in long convergence times.

If instead it is possible to determine the future outputs of the system as a function of past observations of the inputs and outputs, i.e., if there exists a number l and a continuous function $\tilde{h} : \mathcal{Y}_l \times \mathcal{U}_l \to \mathcal{Y}$ such that the recursive model

$$y(k+1) = \tilde{h}[Y_l(k-l+1), U_l(k-l+1)] \qquad (18)$$

has the same input–output behavior as the original system (1), then the identification model can be realized by a feedforward neural network with $2l$ inputs and one output. Since both inputs and outputs to the network are directly observable at each instant of time, static backpropagation can be used to train the network (Figure 3).

For linear systems such a model always exists. More specifically, the input–output behavior of any linear system can be realized by a recursive relation of the form

$$y(k) = \sum_{i=1}^{n} a_i y(k-i) + \sum_{i=1}^{n} b_i u(k-i). \qquad (19)$$

Although the use of input–output models for the identification of nonlinear dynamical systems has been suggested in the connectionist literature

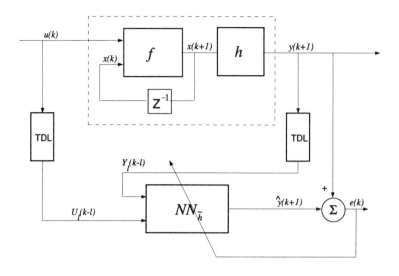

FIGURE 3. Input–output model for identification (TDL represents a tapped delay line).

[Jor86, NP90], it is not at all obvious that such models exist for general systems of the form (1). Actually, the only global results concerning the use of input–output models for the identification of nonlinear dynamical systems are due to Sontag [Son79b], who studied the existence of such realizations for the restricted class of polynomial systems (i.e., systems in which f and h are described by polynomials of finite degree). For this class of systems, he has shown that the input–output realization can be described as a rational function (a ratio of two finite-degree polynomials). In the following we will determine sufficient conditions for the existence of such models for nonlinear systems given by (1). These will be based on the *observability* properties of a system.

4.1 Local Input–Output Models

We first consider the simpler problem of establishing a local input–output model around an equilibrium state of the system (to be referred to as the origin). Intuitively, the problem is stated as follows: Given that the origin is an equilibrium state, does there exist a region Ω_x around the origin such that as long as $x(k) \in \Omega_x$, the output of the system at time k is uniquely determined as a function of a finite number of previous input and output observations? As will be shown here, this can be achieved if the system is locally strongly observable over Ω_x.

6. Identification of Nonlinear Dynamical Systems

Formal Derivation

Sufficient conditions for strong local observability of a system Σ around the origin can be derived from the observability properties of its linearization at the origin:

$$\Sigma_L : \begin{array}{rl} \delta x(k+1) & = f_x|_{0,0}\delta x(k) + f_u|_{0,0}\delta u(k) = A\delta x(k) + b\delta u(k), \\ \delta y(k) & = h_x|_0 \delta x(k) = c^T \delta x(k), \end{array} \quad (20)$$

where $A \stackrel{\triangle}{=} f_x|_{0,0}$, $b \stackrel{\triangle}{=} f_u|_{0,0}$ $c^T \stackrel{\triangle}{=} h_x|_0$.
This is summarized by the following theorem.

Theorem 3 *Let Σ be the nonlinear system (1) and Σ_L its linearization around the equilibrium. If Σ_L is observable, then Σ is locally strongly observable. Furthermore, locally, Σ can be realized by an input–output model.*

Proof: The outputs of Σ given by $Y_n(k) = (y(k), y(k+1), \ldots, y(k+n-1))$ can also be expressed as a function of the initial state and inputs:

$$Y_n(k) = H_n[x(k), U_{n-1}(k)]. \quad (21)$$

The Jacobian of $Y_n(k)$ with respect to $x(k)$ ($\stackrel{\triangle}{=} D_x Y_n(k)$) at the origin is the observability matrix of Σ_l given by

$$M_o = [c^T | c^T A | \ldots | c^T A^{n-1}]^T.$$

Let $\tilde{H} : \mathcal{U}_{n-1} \times \mathcal{X} \to \mathcal{U}_{n-1} \times \mathcal{Y}_n$ be defined by

$$(U_{n-1}(k), Y_n(k)) \stackrel{\triangle}{=} \tilde{H}[U_{n-1}(k), x(k)].$$

The Jacobian matrix of $\tilde{H}(\cdot, \cdot)$ at $(0,0)$ is given by

$$D\tilde{H}|_{(0,0)} = \begin{bmatrix} I & 0 \\ D_{U_{n-1}} Y_n(k) & D_x Y_n(k) \end{bmatrix}.$$

Because of its special form, the determinant of the Jacobian equals $\det[D_x Y_n(k)|_{(0,0)}]$ ($= M_o$). Thus if M_o is of full rank (i.e., Σ_l is observable), $D_{0,0}\tilde{H}$ is of full rank. Now, using the inverse mapping theorem, if M_o is of full rank, there exists a neighborhood $\mathcal{V} \subset \mathcal{X} \times \mathcal{U}_{n-1}$ of $(0,0)$ on which $\tilde{H}$ is invertible. Let $\tilde{\Phi} : \mathcal{Y}_n \times \mathcal{U}_{n-1} \to \mathcal{X} \times \mathcal{U}_{n-1}$ denote the inverse of $\tilde{H}$ and let Φ be the projection on the first n components of $\tilde{\Phi}$. Then locally we have

$$x(k) = \Phi[U_{n-1}(k), Y_n(k)]. \quad (22)$$

The second part follows readily since $y(k+n)$ can be written as a function of $x(k), u(k), \ldots, u(k+n-1)$, and thus after rearranging indices we get

$$y(k+1) = \tilde{h}[Y_n(k-n+1), U_n(k-n+1)]. \quad (23)$$

$\square$

The essence of the previous result is that the existence of a local input–output model for the nonlinear system can be determined by simply testing the observability properties of the underlying linearized system. This is demonstrated by the following example.

Example 4 Let

$$x(k+1) = x(k) + u(k),$$
$$y(k) = x(k) + x^2(k),$$

$$\frac{\partial y(k)}{\partial x(k)} = 2x + 1, \quad \frac{\partial y(k)}{\partial x(k)}\bigg|_{(0,0)} = 1.$$

Hence the linearized system at the origin ($x = 0, u = 0$) is observable, and around the origin there is an input–output representation for the above equation given by

$$\begin{aligned} y(k+1) &= x(k+1) + x^2(k+1) \\ &= x(k) + u(k) + x^2(k) + u^2(k) + 2x(k)u(k) \\ &= y(k) + u^2(k) + 2u(k)\sqrt{1 + 4y(k)}. \end{aligned}$$

△

Sufficient conditions concerning the existence of local input–output realizations have also been established in [LB85]. The derivation there was based on calculating the Hankel matrix of a system. The above result, relying on the properties of the underlying linearized system, is much simpler to derive.

Neural Network Implementation

If strong observability conditions are known (or assumed) to be satisfied in the system's region of operation, then the identification procedure using a feedforward neural network is quite straightforward. At each instant of time, the inputs to the network (not to be confused with the inputs to the system) consisting of the system's past n input values and past n output values (all together $2n$), are fed into the neural network.[6] The network's output is compared with the next observation of the system's output to yield the error

$$e(k+1) = y(k+1) - NN[Y_n(k-n+1), U_n(k-n+1)].$$

The weights of the network are then adjusted using static back propagation to minimize the sum of the squared error.

Once identification is achieved, two modes of operation are possible:

---

[6]It is assumed that the order n of the system is known. If, however, only an upper bound $\bar{n}$ on the order is known, all algorithms have to be modified accordingly, using $\bar{n}$ in place of n.

- **Series Parallel mode:** In this mode, the outputs of the actual system are used as inputs to the model. This scheme can be used only in conjunction with the system, and it can generate only one-step-ahead prediction. The architecture is identical to the one used for identification (Figure 3).

- **Parallel mode:** If more then one-step-ahead prediction is required, the independent mode must be used. In this scheme, the output of the network is fed back into the network (as shown in Figure 4), i.e., the outputs of the network itself are used to generate future predictions. While one cannot expect the identification model to be perfect, this mode of operation provides a viable way to make short-term prediction (> 1). Further, in many cases the objective is not to make specific predictions concerning a system but rather to train the network to generate complex temporal trajectories. In this case, if identification is accurate, the model will exhibit the same type of behavior (in the topological sense) as the original system.

Simulation 2 (Local Identification: An Input–Output Model) The system to be identified is given by

$$\begin{aligned}
x_1(k+1) &= 0.5x_2(k) + 0.2x_1(k)x_2(k), \\
x_2(k+1) &= -0.3x_1(k) + 0.8x_2 + u(k), \\
y(k) &= x_1(k) + [x_2(k)]^2.
\end{aligned}$$

The linearized system around the equilibrium is

$$\begin{aligned}
\delta x_1(k+1) &= 0.5\delta x_2(k), \\
\delta x_2(k+1) &= -0.3\delta x_1(k) + 0.8\delta x_2 + \delta u(k), \\
\delta y(k) &= \delta x_1(k),
\end{aligned}$$

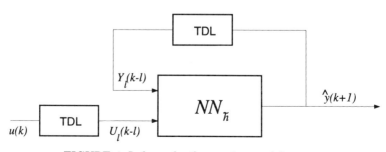

FIGURE 4. Independently running model.

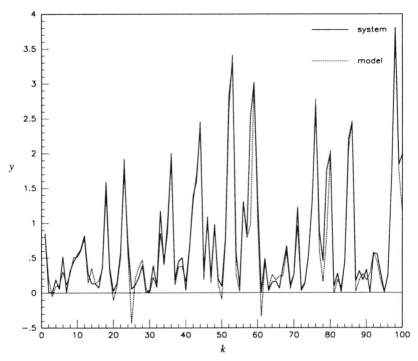

FIGURE 5. Local identification with input–output model.

and its observability matrix

$$M_o = [c|cA] = \begin{bmatrix} 1 & 0 \\ 0 & 0.5 \end{bmatrix}$$

is of full rank. Thus the system can be realized by an input–output model of order 2. A neural network $NN_{\bar{h}} \in \mathcal{N}^3_{4,12,6,1}$ was trained to implement the model. The system was driven with random input $u(k) \in [-1, 1]$. The inputs to the network at each instant of time consisted of $y(k), y(k-1), u(k), u(k-1)$, and the output of the network $\hat{y}(k+1)$ was compared to the output of the system $y(k+1)$. The error $e(k+1) = y(k+1) - \hat{y}(k+1)$ was used as the performance criterion for the network, and the weights were adjusted using static backpropagation along the negative gradient.

Figure 5 shows the performance of the network after 20,000 training steps. The system is driven with a random input, and prediction of the network at the next step is compared to the actual output. △

4.2 Global Input–Output Models

The input–output results presented so far are local in nature, and one cannot be certain that the conditions upon which these results rest are actually satisfied in the system's domain of operation. While strong global observability is desirable, the conditions under which it can be achieved are too restrictive to be satisfied by most systems. Also, even though the existence of a region over which the system is strongly observable can be determined by examining the observability properties of the linearized system, determining the actual size of that region can be extremely cumbersome [Fit72]. Hence, practical use of the result assumes that the conditions for strong observability are satisfied over the system's domain of operation.

Once we relax the observability requirement to generic observability (i.e., almost any input of sufficient length will make the states observable), global results can be attained. As will be shown, almost all observable systems are globally generically observable. Hence, with no need for further testing, one can assume that the particular system under consideration is generically observable. This in turn can be used to derive a global input–output identification model for the system.

In addition to the knowledge of the order of the system, the ensuing development will rely on the following two assumptions.

Assumption 1 *f and h are smooth functions.*

Assumption 2 *the system is state invertible (as defined in Section 2.2).*

Formal Derivation

The central idea of this section is to show how observability can be described as a transversal intersection between maps. Through that, the genericness of transversal intersections will be used to prove the genericness of generically observable systems. On the other hand, we prove that a generically observable system can be realized by an input–output model. Bringing the two together, we conclude that generic systems of the form (1) can be identified using a recursive model of the form (18).

For *continuous time homogeneous* dynamical systems described by

$$\dot{x} = f(x), \ y = h(x), \tag{24}$$

the question of the genericness of observability has been investigated by Aeyels [Aey81]. By expressing the observability property in terms of transversality conditions, he has shown that almost any such system will be observable if at least $2n + 1$ measurements of the output are taken.

Following similar reasoning, we first wish to extend this result to nonhomogeneous systems of the form (1). In order to express the observability of Σ in terms of transversality conditions, we need the notion of the *diagonal*:

Definition 6 *Let $\mathcal{X}$ be a smooth manifold and let $x \in \mathcal{X}$. The diagonal $\Delta(\mathcal{X} \times \mathcal{X})$ is the set of points of the form (x, x).*

Recalling the definition of observability, a system is observable if for a given input the mapping from the state space to the output is injective, i.e., $Y(x_1, U) = Y(x_2, U)$ if and only if $x_1 = x_2$. This is equivalent to saying that for any $x_1 \neq x_2$, $Y_l(x_1, U_l), Y_l(x_2, U_l) \notin \Delta(\mathcal{Y}_l \times \mathcal{Y}_l)$. Now, from Proposition 1, transversality implies empty intersection if

$$\dim \Delta(\mathcal{Y}_l \times \mathcal{Y}_l) + 2 \dim \mathcal{X} < 2 \dim \mathcal{Y}_l,$$

and since $\dim \Delta(\mathcal{Y}_l \times \mathcal{Y}_l) = \dim \mathcal{Y}_l \geq l$ and $\dim \mathcal{X} = n$, observability can be expressed in terms of the transversality condition if

$$l \geq 2n + 1.$$

With this in mind the following result, which is the equivalent of Aeyels's result for discrete systems, can be stated:

Lemma 1 *Let $h : \mathcal{X} \to \mathcal{Y}$ be a Morse function with distinct critical points. Let $U^*_{2n+1} \in \mathcal{U}_{2n+1}$ be a given input sequence. Then the set of smooth functions $f \in \mathcal{C}^\infty$ for which the system*

$$\begin{aligned} x(k+1) &= f[x(k), u^*(k)], \\ y(k) &= h[x(k)] \end{aligned}$$

is observable is open and dense in $\mathcal{C}^\infty$.

The proof is long, and since it is not pertinent to the ensuing development, it is given in the Appendix.

Using Lemma 1 we can deduce that $(2n+1)$-step generic observability is a natural assumption for nonhomogeneous discrete time systems described by (1), i.e., it holds for almost all systems. More precisely we have the following theorem.

Theorem 4 *Let $h : \mathcal{X} \to \mathcal{Y}$ be a Morse function. Then the set of functions $f \in \mathcal{C}^\infty$ for which the system (1) is $(2n+1)$-step generically observable is open and dense in $\mathcal{C}^\infty$.*

Proof: Let $\mathcal{F} \subset \mathcal{C}^\infty$ and $\mathcal{V} \subset \mathcal{U}_{2n+1}$ be compact and let $\mathcal{A} = \mathcal{F} \times \mathcal{V}$.

Open: Assume for a given $v^* \in \mathcal{V}$ and a given $f^* \in \mathcal{F}$ the system (1) is observable. Observability means that the map $H^*_l(v^*) : \mathcal{X} \to \mathcal{Y}_{2n+1}$ is injective (the definition of H_l was given in Section 1). Injectivity is a stable property, thus there exists a neighborhood $\mathcal{B} \subset \mathcal{A}$ such that for all $(v, f) \in \mathcal{B}$ the system is observable.

Dense: For any neighborhood $\mathcal{B}$ of a given v^* and a given f^* there exist $\mathcal{W} \subset \mathcal{V}$ and $\mathcal{G} \subset \mathcal{F}$ such that $\mathcal{W} \times \mathcal{G} \subset \mathcal{B}$. From Lemma 1, for a given v^* there exists $\tilde{f} \in \mathcal{G}$ for which the triplet $\tilde{f}, h, v^*$ is observable. Thus $(\tilde{f}, v^*) \in \mathcal{B}$. □

To understand the importance of the result, the following short discussion may prove useful. In the real world of perceptions and measurements, no

6. Identification of Nonlinear Dynamical Systems

continuous quantity or functional relationship is ever perfectly determined. The only physically meaningful properties of a mapping, consequently, are those that remain valid when the map is slightly deformed. Such properties are *stable* properties, and the collection of maps that possess a particular stable property may be referred to as a *stable class* of maps. A property is *generic* if it is stable and dense, that is, if any function may be deformed by an arbitrarily small amount into a map that possesses that property. Physically, only stable maps can be observed, but if a property is generic, all observed maps will possess it. Hence, the above theorem states that in practice only generically observable systems will ever be observed.

For a generically observable system[7] we wish to show that an observer can be realized (by a neural network) that for almost all values of u will give the state as a function of the observed inputs and outputs. The above theorem suggests that this set is generic. To build an input–output model we will also need to assume that the complement of this set (i.e., the set of input sequences for which the system is not observable) is of measure zero. More formally:

Assumption 3 *In the systems under consideration, the complement of the generic input set for which the system is observable is of measure zero.*

With this preamble, the following result can be stated:

Theorem 5 *Let Σ be a generically observable system(1). Let $\mathcal{K} \subset \mathcal{X}$ and $\mathcal{C} \subset \mathcal{U}_{2n+1}$ be compact. Let $\mathcal{A}^s \subset \mathcal{C}$ denote the set of input sequences for which the system is not observable. If Assumption 3 holds, then for all $\epsilon > 0$ there exists an open set $\mathcal{A}^\epsilon \supset \mathcal{A}^s$ such that:*

1. *$\mu(\mathcal{A}^\epsilon) < \epsilon$ (μ denoting the measure).*

2. *There exists a continuous function $\Phi : \mathbb{R}^{2(2n+1)} \to \mathbb{R}^n$ such that for all $x(k) \in \mathcal{K}$ and all $U_{2n+1}(k) \in \mathcal{A}^{1-\epsilon}$ (denoting the complement of $\mathcal{A}^\epsilon$ in $\mathcal{C}$) we have*

$$x(k) = \Phi[Y_{2n+1}(k), U_{2n+1}(k)]. \tag{25}$$

3. *There exists a feedforward neural network NN_Φ such that for all $x(k) \in \mathcal{K}$ and all $U_{2n+1}(k) \in \mathcal{A}^{1-\epsilon}$ we have*

$$\|x(k) - NN_\Phi[Y_{2n+1}(k), U_{2n+1}(k)]\| < \epsilon. \tag{26}$$

Proof: Since $\mathcal{A}^s$ is of measure zero, for any ϵ there exists an open set $\mathcal{A}^\epsilon$ such that $\mathcal{A}^s \subset \mathcal{A}^\epsilon$ and $\mu(\mathcal{A}^\epsilon) < \epsilon$.

---

[7]Since generic observability requires $2n + 1$ measurements, from now on by *generic observability* we will mean $(2n + 1)$-*step generic observability*.

To prove part 2, consider the mapping $\tilde{H} : \mathcal{K} \times \mathcal{A}^{1-\epsilon} \to \mathcal{B} \times \mathcal{A}^{1-\epsilon}$ defined by
$$(Y_{2n+1}(k), U_{2n+1}(k)) = \tilde{H}[x(k), U_{2n+1}(k)],$$
where $\mathcal{B}$ denotes the image of this map in the $\mathcal{Y}_{2n+1}$ space. $\tilde{H}$ is continuous and bijective on the compact set $\mathcal{K} \times \mathcal{A}^{1-\epsilon}$; hence $\mathcal{B}$ is compact and there exists a continuous inverse $\tilde{\Phi} : \mathcal{B} \times \mathcal{A}^{1-\epsilon} \to \mathcal{K} \times \mathcal{A}^{1-\epsilon}$ such that
$$[x(k), U_{2n+1}(k)] = \tilde{\Phi}[Y_{2n+1}(k), U_{2n+1}(k)].$$

Since this map is continuous on the compact set $\mathcal{B} \times \mathcal{A}^{1-\epsilon}$, by the Tietze extension theorem [RS80], it can be extended to all of $\mathcal{Y}_{2n+1} \times \mathcal{C}$, and if we denote its first n components by Φ we get (25).

The last part follows immediately from the approximation properties [Cyb89, HSW89] of feedforward neural networks. □

Finally, combining Theorems 4 and 5, the existence of an input–output model can be established.

Theorem 6 *Let Σ be defined by (1). Then for generic f and h and for every $\epsilon > 0$, there exists a set $\mathcal{A}^\epsilon$ such that $\mu(\mathcal{A}^\epsilon) < \epsilon$, a continuous function $\tilde{h} : \mathbb{R}^{2n+1} \times \mathbb{R}^{2n+1} \to \mathbb{R}$, and a multilayer feedforward neural network $NN_{\tilde{h}}$ such that:*

1. *For all input sequences $U_{2n+1}(k-2n) \notin \mathcal{A}^\epsilon$,*

$$y(k+1) = \tilde{h}[Y_{2n-1}(k-2n), U_{2n+1}(k-2n)]. \tag{27}$$

2. *For all input sequences $U_{2n+1}(k-2n)$,*

$$\|\tilde{h}[Y_{2n-1}(k-2n), U_{2n+1}(k-2n)] - NN_{\tilde{h}}[Y_{2n-1}(k-2n), U_{2n+1}(k-2n)]\| < \epsilon. \tag{28}$$

Proof: From Theorem 4 we have that for generic f and h, Σ is generically observable. Hence, from Theorem 5, for any $\epsilon > 0$, for all input sequences not contained in a set $\mathcal{A}^\epsilon$, $x(k-2n)$ can be written as a function of $Y_{2n+1}(k-2n)$ and $U_{2n+1}(k-2n)$ (n denoting the order of the system). Now, $y(k+1)$ can be written as a continuous function of $x(k-2n), u(k-2n), \ldots, u(k)$, and thus there exists a continuous function $\tilde{h}$ such that

$$\begin{aligned} y(k+1) &= \tilde{h}[y(k), \ldots, y(k-2n), u(k), \ldots, u(k-2n)] \\ &= \tilde{h}[Y_{2n+1}(k-2n), U_{2n+1}(k-2n)] \end{aligned} \tag{29}$$

for all $U_{2n+1}(k-2n) \notin \mathcal{A}^\epsilon$.

The second part follows immediately from the approximation properties of feedforward neural networks [Cyb89, HSW89]. □

Hence, generically, input–output models can be used to identify systems whose underlying behavior is given by (1). Thus the result implies that

6. Identification of Nonlinear Dynamical Systems

practically all systems can be identified using input–output models. Further, even though the algorithm presented relied on the knowledge of the system's order (which may not be available), we are guaranteed that even without this information a finite number of past observations suffices to predict the future (as opposed to the Volterra or Wiener series [Rug81]).

Neural Network Implementation

The input–output model based on the assumption of generic observability is similar to the one introduced for the local input–output model with a few modifications. First, a minimum of $2n+1$ observations of the system's inputs and outputs need to be fed into the network at each time instant. Further, for a generic $2n+1$ sequence of inputs, for any $x_1 \neq x_2$ we have

$$Y_{2n+1}(x_1, U_{2n+1}) \neq Y_{2n+1}(x_2, U_{2n+1}),$$

but there is no lower bound on the distance between the two values. This may cause the inverse map (25), upon which the recursive model is based, to be very steep. In theory, a neural network should be able to approximate any continuous function. However, the more rugged the function to be approximated, the more difficult is the task. Thus, practically, it might prove advantageous to use even longer sequences as inputs to the neural networks, which can only increase the distance between the image of any two points, thus resulting in a smoother inverse map to be approximated and thus easier to identify.

Simulation 3 (Identification: A Generically Observable System)
The system to be identified is given by

$$\begin{aligned}
x_1(k+1) &= -0.7x_2(k) + x_3(k), \\
x_2(k+1) &= \tanh[0.3x_1(k) + x_3(k) + (1 + 0.3x_2(k))u(k)], \\
x_3(k+1) &= \tanh[-0.8x_1(k) + 0.6x_2(k) + 0.2x_2(k)x_3(k)], \\
y(k) &= [x_1(k)]^2.
\end{aligned}$$

Since $c = \frac{\partial y}{\partial x}|_0 = 0$, the linearized system is unobservable. From the above result we have that a third-order system can be realized by an input–output model of order $7 = (2 \cdot 3 + 1)$, i.e, the prediction relies on 7 past observations of the inputs and outputs (a total of 14). To test the relevance of this number, we tried to identify the system's different input–output models, with the recursion varying between $l = 1$ and $l = 10$. The models were implemented using a feedforward network of size $NN_{\tilde{h}} \in \mathcal{N}^3_{2l,12,6,1}$. Thus, for a given l the input–output model is given by

$$\hat{y}(k+1) = NN_{\tilde{h}}[y(k), \ldots y(k-l+1), u(k), \ldots u(k-l+1)].$$

Training was done by driving the system and the model using a random input signal $u(k)$ uniformly distributed in the interval $[-1, 1]$. At each instant of time, the prediction error is given by $e(k) = y(k) - \hat{y}(k)$, and using

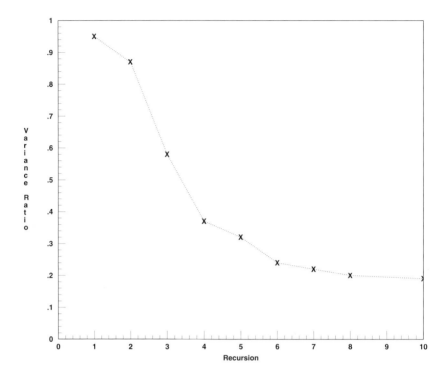

FIGURE 6. Identification error as a function of the number of past observations used for the identification model.

the backpropagation algorithm, the weights of $NN_{\tilde{h}}$ are adjusted along the negative gradient of the squared error. The comparative performance of the different models after 50,000 training iterations is shown in Figure 6. As a figure of merit for the identification error we chose the ratio between the variance of the error and the variance of the output of the system. It is seen that initially the error drops rapidly and reaches a plateau approximately around $l = 7$. To have an intuitive appreciation as to what this error means, Figure 7 compares the next step prediction of the system and the model with $l = 7$, when both are driven with a random input signal. As can be seen, the model approximates the input–output behavior of the system quite accurately. △

5 Conclusion

The identification of nonlinear dynamical systems by neural networks is treated in this chapter for both state space and input–output models. It

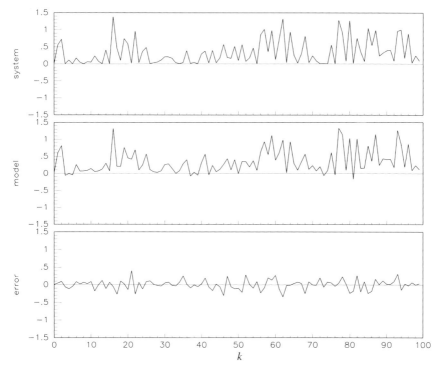

FIGURE 7. Identification of a generically observable system using a seventh-order recursive input–output model.

is shown how prior assumptions concerning the properties of the system influence the type of architectures that can be used.

The state space model offers a more compact representation. However, learning such a model involves the use of dynamic backpropagation, which is a very slow and computationally intensive algorithm. Furthermore, practical use of such models requires the ability to reset the system periodically. Both these disadvantages are overcome when input–output models are used. Thus, the latter offers a much more viable solution to the identification of real-world systems.

The most important result presented in this chapter is the demonstration of the existence of a global input–output model based on generic observability. The fact that generic observability is a generic property of systems implies that almost all systems can be identified using input–output models and hence realized by feedforward networks. The algorithm presented is based on the knowledge of an upper bound on the system's order. While the latter may not always be available, this does not detract from the utility of the proposed method. In such a case the number of past observations used for the identification process can be increased to achieve a good prediction. The result guarantees that this procedure will converge, since a

finite number of past observations suffices to predict the future.
Further work related to this chapter has been presented in [LN96].

Acknowledgment

The first author wishes to thank Felipe Pait for many stimulating discussions and Eduardo Sontag for insightful suggestions concerning the issue of generic observability. This work was supported by NSF grant ECS-8912397.

6 Appendix: Proof of Lemma 1

First the following lemma is necessary:

Lemma 2 *Let Σ be the system (1). Let h in Σ be a Morse function with distinct critical points. The set of functions f that satisfy the conditions:*

1. *No two trajectories with period $\leq 2n+1$ belong to the same level surface of h.*

2. *No trajectory with period $\leq 2n+1$ coincides with a critical point of h.*

3. *No integral trajectory contains two or more critical points of h.*

4. *No integral trajectory (except equilibrium points) belongs to a single level surface of h.*

is open and dense in $\mathcal{C}^\infty$.

Proof: The proof of the lemma is an immediate consequence of transversality theory. Violation of any of the above conditions involves the intersection of manifolds whose sum of dimensions is less then n, i.e., manifolds that do not intersect transversally. Since transversal intersections are generic, the conditions follow. □

Proof of Lemma 1: Let $f_i(x) \triangleq f(x, u_i)$, where u_i denotes the input at time i. For a given f, Σ will be observable if the mapping $\phi : \Delta(\mathcal{F} \times \mathcal{F}) \times \mathcal{X} \times \mathcal{X} \to \mathbb{R}^{2n+1} \times \mathbb{R}^{2n+1}$ defined by

$$\phi(f, f, x, z, u^*) = \left(\begin{bmatrix} h \circ f_1(x) \\ h \circ f_2 \circ f_1(x) \\ \vdots \\ h \circ f_{2n+1} \ldots f_1(x) \end{bmatrix}, \begin{bmatrix} h \circ f_1(z) \\ h \circ f_2 \circ f_1(z) \\ \vdots \\ h \circ f_{2n+1} \ldots f_1(z) \end{bmatrix} \right) \quad (30)$$

is transversal to $\mathcal{W} = \Delta(\mathbb{R}^{2n+1} \times \mathbb{R}^{2n+1})$.

6. Identification of Nonlinear Dynamical Systems

To prove that this is true for a generic f, we will consider the family of maps $F(x,s) = f(x) + sg(x)$, where s is a parameter and g is a smooth function. In the same manner that ϕ was defined, we can define $\Phi(f, f, x, z, u^*, s)$ by replacing $f_i(x)$ in (30) with $F_i(x,s)$.

Now, from the transversality theorem, if $\Phi \pitchfork \mathcal{W}$ then for a generic f, $\phi \pitchfork \mathcal{W}$, i.e., the system is observable. By definition, $\Phi \pitchfork \mathcal{W}$ if for each $x \neq z$ either $\phi(f, f, x, z) \notin \mathcal{W}$ or $\frac{\partial \Phi}{\partial s}|_{s=0}$ spans $\mathcal{W}^c$ (the complement of $\mathcal{W}$).

Since all elements of $\mathcal{W}$ are of the form (w, w), then if we can find g such that whenever $\phi(f, f, x, z) \in \mathcal{W}$, $\frac{\partial \Phi}{\partial s}|_{s=0}$ is of the form

$$\begin{bmatrix} a_1 & 0 & \cdots & 0 & b_1 & 0 & \cdots & 0 \\ * & a_2 & \cdots & 0 & * & b_2 & \cdots & 0 \\ \vdots & & & & \vdots & & & \\ * & * & \cdots & a_{2n+1} & * & * & \cdots & b_{2n+1} \end{bmatrix}, \quad (31)$$

where $a_i \neq b_i$ for all i, then $\frac{\partial \Phi}{\partial s}|_{s=0}$ will span $\mathcal{W}^c$ and thus $\phi \pitchfork \mathcal{W}$.

Four possible cases need to be considered:

Case I: Neither x nor z is periodic with period $\leq 2n + 1$.

The trajectories of both x and z consist of at least $2n + 1$ distinct points. If $\phi(f, f, x, z) \notin \mathcal{W}$, the mapping is transversal, else we need to show that $\frac{\partial \Phi}{\partial s}|_{s=0}$ spans $\mathcal{W}^c$. Let $N = 2n + 1$. Then

$$\frac{\partial \Phi}{\partial s}|_{s=0} =$$

$$\begin{bmatrix} \frac{\partial y_1}{\partial x_1}g(x_1) & 0 & \cdots & 0 & \frac{\partial y_1}{\partial z_1}g(z_1) & 0 & \cdots & 0 \\ \frac{\partial y_2}{\partial x_1}g(x_1) & \frac{\partial y_2}{\partial x_2}g(x_2) & \cdots & 0 & \frac{\partial y_2}{\partial z_1}g(z_1) & \frac{\partial y_2}{\partial x_2}g(z_2) & \cdots & 0 \\ \vdots & & & & \vdots & & & \\ \frac{\partial y_N}{\partial x_1}g(x_1) & \frac{\partial y_N}{\partial x_2}g(x_2) & \cdots & \frac{\partial y_N}{\partial x_N}g(x_N) & \frac{\partial y_N}{\partial z_1}g(z_1) & \frac{\partial y_N}{\partial z_2}g(z_2) & \cdots & \frac{\partial y_N}{\partial z_N}g(z_N) \end{bmatrix}.$$

If for all i

$$\frac{\partial y_i}{\partial x_i}g(x_i) \neq \frac{\partial y_i}{\partial z_1}g(z_i), \quad (32)$$

then (32) is of the form (31), and hence $\frac{\partial \Phi}{\partial s}|_{s=0}$ spans $\mathcal{W}^c$. From condition 1 in Lemma 2, $\frac{\partial h}{\partial x_i}$ and $\frac{\partial h}{\partial z_i}$ cannot be zero simultaneously; thus, g can always be chosen such that (32) holds.

Case II: Either x or z is periodic with period $\leq N$.

Without loss of generality let x be periodic. By condition 2 of Lemma 2, $\frac{\partial h}{\partial z_i}$ can be zero for at most a single value of $i(= m)$. For all $i \neq m$, $g(z_i)$ can be chosen such that $\frac{\partial y_i}{\partial x_i}g(x_i) \neq \frac{\partial y_i}{\partial z_i}g(z_i)$. Now, from condition 2 of Lemma 2, no periodic trajectory with period $\leq N$ coincides with a critical point of h; thus $\frac{\partial y_m}{\partial x_m} \neq 0$ and $g(x_m)$ can be selected such that $\frac{\partial y_m}{\partial x_m}g(x_m) \neq \frac{\partial y_m}{\partial z_m}g(z_m)$.

Case III: Both x and z are periodic with period $\leq N$.

By condition 1 of Lemma 2, no two orbits with period $\leq N$ belong to the same level surface of h; thus $\phi(f, f, x, z) \notin \mathcal{W}$.

Case IV: x and z are on the same trajectory.

From condition 4 in Lemma 2, no integral trajectory belongs to a single level surface of h. Thus for some i, $y_i(x) \neq y_i(z)$, and thus $\phi(f, f, x, z) \notin \mathcal{W}$.

Since the family of systems parameterized by s is transversal to $\mathcal{W}$, it follows from the transversality theorem that transversality will hold for almost all s, both in the sense that it is satisfied on an open and dense set as well as in the sense that the set of parameters for which the system is unobservable is of measure zero. □

7 REFERENCES

[Aey81] D. Aeyels. Generic observability of differentiable systems. *SIAM Journal of Control and Optimization*, 19:595–603, 5 1981.

[Cyb89] G. Cybenko. Approximation by superpositions of a sigmoidal function. *Mathematics of Control, Signals, and Systems*, 2:303–314, 1989.

[Fit72] J. M. Fitts. On the observability of non-linear systems with applications to non-linear regression analysis. *Information Sciences*, 4:129–156, 1972.

[GG73] M. Golubitsky and V. Guillemin. *Stable Mappings and Their Singularities*. Springer-Verlag, 1973.

[GP74] V. Guillemin and A. Pollack. *Differential Topology*. Prentice-Hall, Englewood Cliffs, New Jersey, 1974.

[HSW89] K. Hornik, M. Stinchcombe, and H. White. Multilayer feedforward networks are universal approximators. *Neural Networks*, 2:359–366, 1989.

[JC73] J. B. Cruz, Jr., editor. *System Sensitivity Analysis*. Dowden, Hutchinson and Ross Inc., Stroudsburg, Pennsylvania, 1973.

[Jor86] M. I. Jordan. Attractor dynamics and parallelism in a connectionist sequential machine. In *Proceedings of the Eighth Annual Conference of the Cognitive Science Society, Amherst, Connecticut, 1986*, pages 531–546. Lawrence Erlbaum, Hillsdale, New Jersey, 1986.

[JS90] B. Jakubczyk and E. D. Sontag. Controllability of nonlinear discrete-time systems: A lie algebraic approach. *SIAM Journal of Control and Optimization*, 28:1–33, January 1990.

[LB85] I. J. Leontaritis and S. A. Billings. Input-output parametric models for non-linear systems, part I: Deterministic nonlinear systems. *International Journal of Control*, 41:303–328, 1985.

[Lev92] A. U. Levin. *Neural Networks in Dynamical Systems: a System Theoretic Approach*. Ph.D. Thesis, Yale University, New Haven, Connecticut, November 1992.

[Lju77] L. Ljung. Analysis of recursive stochastic algorithms. *IEEE Transactions on Automatic Control*, 22:551–575, 1977.

[LL91] L. Ljung. Issues in system identification. *IEEE Control Systems Magazine*, 11:25–29, 1991.

[LN93] A. U. Levin and K. S. Narendra. Control of non-linear dynamical systems using neural networks. Controllability and stabilization. *IEEE Transactions on Neural Networks*, 4:192–206, March 1993.

[LN96] A. U. Levin and K. S. Narendra. Control of non-linear dynamical systems using neural networks– Part II: Observability, identification and control. *IEEE Transactions on Neural Networks*, 7:30–42, January 1996.

[MRtPRG86] J. L. McClelland, D. E. Rumelhart, and the PDP Research Group. *Parallel Distributed Processing: Explorations in the Microstructure of Cognition*, volume 2. MIT Press, Cambridge, Massachusetts, 1986.

[NP90] K. S. Narendra and K. Parthasarathy. Identification and control of dynamical systems using neural networks. *IEEE Transactions on Neural Networks*, 1:4–27, March 1990.

[NP91] K. S. Narendra and K. Parthasarathy. Gradient methods for the optimization of dynamical systems containing neural networks. *IEEE Transactions on Neural Networks*, 2:252–261, March 1991.

[RS80] M. Reed and B. Simon. *Methods of Modern Mathematical Physics I: Functional Analysis*. Academic Press, New York, 1980.

[Rug81] W. J. Rugh. *Nonlinear System Theory: the Volterra/Wiener Approach*. The Johns Hopkins Univ. Press, Baltimore, Maryland, 1981.

[Son79a] E. D. Sontag. On the observability of polynomial systems, I: Finite time problems. *SIAM Journal of Control and Optimization*, 17:139–150, 1979.

[Son79b] E. D. Sontag. *Polynomial Response Maps*. Springer-Verlag, Berlin, 1979.

[WZ89] R. J. Williams and D. Zipser. A learning algorithm for continually running fully recurrent neural networks. *Neural Computation*, 1:270–280, 1989.

Chapter 7

Neural Network Control of Robot Arms and Nonlinear Systems

F. L. Lewis
S. Jagannathan
A. Yeşildirek

> ABSTRACT Neural network (NN) controllers are designed that give guaranteed closed-loop performance in terms of small tracking errors and bounded controls. Applications are given to rigid-link robot arms and a class of nonlinear systems. Both continuous-time and discrete-time NN tuning algorithms are given. New NN properties such as strict passivity avoid the need for persistence of excitation. New NN controller structures avoid the need for preliminary off-line learning, so that the NN weights are easily initialized and the NN learns on-line in real-time. No regression matrix need be found, in contrast to adaptive control. No certainty equivalence assumption is needed, as Lyapunov proofs guarantee simultaneously that both tracking errors and weight estimation errors are bounded.

1 Introduction

Neural networks (NN) can be used for classification and decision-making or for controls applications. Some background on NN is given in [MSW91, MB92, Pao89, PG89, RHW86, Wer74, Wer89]. In classification and decision-making NN have by now achieved common usage and are very effective in solving certain types of problems, so that their use is commonplace in image and signal processing and elsewhere. A major reason for this is the existence of a mathematical framework for selecting the NN weights using proofs based on the notion of energy function, or of algorithms that effectively tune the weights on-line.

1.1 Neural Networks for Control

In controls there have been many applications of NN, but few rigorous justifications or guarantees of performance. The use of ad hoc controller structures and tuning strategies has resulted in uncertainty on how to select the initial NN weights, so that a so-called "learning phase" is often needed that can last up to 50,000 iterations. Although preliminary NN offline training may appear to have a mystique due to its anthropomorphic connotations, it is not a suitable strategy for controls purposes.

There are two sorts of controls applications for NN: identification and control. Some background on robotics and controls applications of NN is given in [CS92, CS93, HHA92, IST91, MSW91, Nar91, NA87, NP90, YY92]. In identification the problems associated with implementation are easier to solve, and there has been good success (see references). Since the system being identified is usually stable, it is only necessary to guarantee that the weights remain bounded. This can generally be accomplished using standard tuning techniques such as the delta rule with, for instance, backpropagation of error. In identification, it is generally not a problem to have a learning phase.

Unfortunately, in closed-loop control using NN the issues are much more complicated, so approaches that are suitable for NN classification applications are of questionable use. A long learning phase is detrimental to closed-loop applications. Uncertainty on how to initialize the NN weights to give initial stability means that during the learning phase the NN controller cannot be switched on line. Most importantly, in closed-loop control applications one must guarantee two things — boundedness of the NN weights and boundedness of the regulation or tracking errors, with the latter being the prime concern of the engineer. This is difficult using approaches to NN that are suitable for classification applications. Some work that successfully uses NN rigorously for control appears in [CK92, LC93, PI91, PI92, RC95, Sad91, SS91], though most of these papers that contain proofs are for 2-layer (linear-in-the-parameters) NN.

The background work for this chapter appears in [JL96, LLY95, LYL96, YL95]. To guarantee performance and stability in closed-loop control applications using multilayer (nonlinear) NN, it is found herein that the standard delta rule does not suffice. Indeed, we see that the tuning rules must be modified with extra terms. In this chapter we give new controller structures that make it easy to initialize the NN weights and still guarantee stability. No off-line learning phase is needed, and tuning to small errors occurs in real time in fractions of a second. New NN properties such as passivity and robustness make the controller robust to unmodeled dynamics and bounded disturbances.

Our primary application is NN for control of rigid robotic manipulators, though a section on nonlinear system control shows how the technique can be generalized to other classes of systems in a straightforward manner. Our work provides continuous-time update algorithms for the NN weights; a section is added to show how to use the same approach to derive discrete-time weight tuning algorithms, which are directly applicable in digital control.

1.2 Relation to Adaptive Control

One will notice, of course, the close connection between NN control and adaptive control [Cra88, Goo91, KC91, SB89]; in fact, from this chapter one may infer that NN comprise a special class of nonlinear adaptive controllers with very important properties. Thus, this chapter considerably extends the capabilities of linear-in-the-parameters adaptive control. In indirect adaptive control, especially in discrete time, one makes a certainty equivalence assumption that allows one to decouple the controller design from the adaptive identification phase. This is akin to current approaches to NN control. This chapter shows how to perform direct NN control, even in the discrete-time case, so that the certainty equivalence assumption is not needed. The importance of this is that closed-loop performance in terms of small tracking errors and bounded controls is *guaranteed*.

In adaptive control it is often necessary to make assumptions, like those of Erzberger or the model-matching conditions, on approximation capabilities of the system, which may not hold. By contrast, the NN approximation capabilities employed in this chapter always hold. In the NN controllers of this chapter, no persistence of excitation condition is needed. Finally, a major debility of adaptive control is the need to find a "regression matrix," which often entails determining the full dynamics of the system. In NN control, no regression matrix is needed; the NN learns in real time the dynamical structure of the unknown system.

2 Background in Neural Networks, Stability, and Passivity

Some fairly standard notation is needed prior to beginning. Let $\mathbb{R}$ denote the real numbers, $\mathbb{R}^n$ the real n-vectors, $\mathbb{R}^{m \times n}$ the real $m \times n$ matrices. Let $\mathbf{S}$ be a compact, simply connected set of $\mathbb{R}^n$. With map $f : \mathbf{S} \to \mathbb{R}^m$, define $C^m(\mathbf{S})$ as the space such that f is continuous. We denote by $\|.\|$ any suitable vector norm; when it is required to be specific, we denote the p-norm by $\|.\|_p$. The *supremum norm* of $f(x)$ (over $\mathbf{S}$) is defined as [Bar64]

$$\sup_{x \in \mathbf{S}} \|f(x)\|, \quad f : \mathbf{S} \to \mathbb{R}^m.$$

Given $A = [a_{ij}], B \in \mathbb{R}^{m \times n}$, the *Frobenius norm* is defined by

$$\|A\|_F^2 = \operatorname{tr}(A^T A) = \sum_{i,j} a_{ij}^2,$$

with tr() the trace. The associated inner product is $< A, B >_F = tr(A^T B)$. The Frobenius norm is nothing but the vector 2-norm over the space defined by stacking the matrix columns into a vector. As such, it cannot be defined as the induced matrix norm for any vector norm, but it is *compatible* with the 2-norm, so that $\|Ax\|_2 \leq \|A\|_F \|x\|_2$, with $A \in \mathbb{R}^{m \times n}$ and $x \in \mathbb{R}^n$.

When $x(t) \in \mathbb{R}^n$ is a function of time, we may refer to the standard L_p norms [LAD93] denoted $\|x(.)\|_p$. We say vector $x(t)$ is *bounded* if the L_∞ norm is bounded. We say matrix $A(t) \in \mathbb{R}^{m \times n}$ is *bounded* if its induced matrix ∞-norm is bounded.

2.1 Neural Networks

Given $x = [x_1 \; x_2 \; \ldots x_{N_1}]^T \in \mathbb{R}^{N_1}$, a three-layer neural net (NN) (Figure 1) has a net output given by

$$y_i = \sum_{j=1}^{N_2} w_{ij} \left[\sigma \left(\sum_{k=1}^{N_1} v_{jk} x_k + \theta_{vj} \right) + \theta_{wi} \right]; \quad i = 1, \ldots, N_3, \qquad (1)$$

with $\sigma(.)$ the activation function, v_{jk} the first-to-second layer interconnection weights, and w_{ij} the second-to-third layer interconnection weights. The θ_{vj}, θ_{wi}, $i, j = 1, 2, \ldots$, are threshold offsets, and the number of neurons in layer l is N_l, with N_2 the number of hidden-layer neurons. In the NN we should like to adapt the weights and thresholds on-line in real time to provide suitable performance of the net. That is, the NN should exhibit "learning behavior."

Typical selections for the activation functions $\sigma(.)$ include, with $z \in \mathbb{R}$,

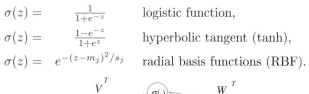

$$\sigma(z) = \frac{1}{1+e^{-z}} \quad \text{logistic function,}$$
$$\sigma(z) = \frac{1-e^{-z}}{1+e^z} \quad \text{hyperbolic tangent (tanh),}$$
$$\sigma(z) = e^{-(z-m_j)^2/s_j} \quad \text{radial basis functions (RBF).}$$

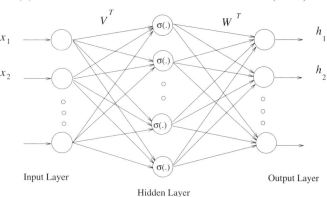

FIGURE 1. Three-layer neural net structure.

Matrix Formulation

The NN equation may be conveniently expressed in matrix format by redefining $x = [x_0\ x_1\ x_2\ \ldots x_{N1}]^T$ and defining $y = [y_1\ y_2\ \ldots y_{N3}]^T$ and weight matrices $W^T = [w_{ij}], V^T = [v_{jk}]$. Including $x_0 \triangleq 1$ in x allows one to include the threshold vector $[\theta_{v1}\ \theta_{v2}\ \ldots\ \theta_{vN2}]^T$ as the first column of V^T, so that V^T contains both the weights and thresholds of the first-to-second layer connections. Then,

$$y = W^T \sigma(V^T x), \qquad (2)$$

where if $z = [z_1\ z_2\ \ldots]^T$ is a vector, we define the activation function componentwise as $\sigma(z) = [\sigma(z_1)\ \sigma(z_2)\ \ldots]^T$. Including 1 as a first term in the vector $\sigma(V^T x)$ (i.e., prior to $\sigma(z_1)$) allows one to incorporate the thresholds θ_{wi} as the first column of W^T. Any tuning of W and V then includes tuning of the thresholds as well.

Although to account for nonzero thresholds the vector x may be augmented by $x_0 = 1$ and the vector $\sigma(V^T x)$ by the constant first entry of 1, we loosely say that $x \in \mathbb{R}^{N1}$ and $\sigma: \mathbb{R}^{N2} \to \mathbb{R}^{N2}$.

Approximation Property of NN

With $x \in \mathbb{R}^n$, A general function $f(x) \in C^m(\mathbf{S})$ can be written as

$$f(x) = W^T \sigma(V^T x) + \varepsilon(x), \qquad (3)$$

with $N_1 = n,\ N_3 = m$, and $\varepsilon(x)$ an *NN functional reconstruction error* vector. If there exist N_2 and constant "ideal" weights W and V such that $\varepsilon = 0$ for all $x \in \mathbf{S}$, we say $f(x)$ is *in the functional range of the NN*. In general, given a real number $\varepsilon_N > 0$, we say $f(x)$ is *within ε_N of the NN range* if there exist N_2 and constant weights such that for all $x \in \mathbb{R}^n$, (3) holds with $\|\varepsilon\| < \varepsilon_N$. Unless the net is "minimal," the weights minimizing may not be unique [AS92, Sus92].

Various well-known results for various activation functions $\sigma(.)$, based, e.g., on the Stone-Weierstrass theorem, say that any sufficiently smooth function can be approximated by a suitably large net [Cyb89, HSW89, PS91, SS91]. The functional range of NN(2) is said to be *dense* in $C^m(\mathbf{S})$ if for any $f \in C^m(\mathbf{S})$ and $\varepsilon_N > 0$ there exist finite N_2, and W and V, such that (3) holds with $\varepsilon < \varepsilon_N$, $N_1 = n, N_3 = m$. Typical results are like the following, for the case of $\sigma(.)$ any "squashing function" (a bounded, measurable, nondecreasing function from the real numbers onto $(0, 1)$), for instance the logistic function.

Theorem 2.1 *Set $N_1 = n$, $N_3 = m$, and let $\sigma(.)$ be any squashing function. Then the functional range of NN(2) is dense in $C^m(\mathbf{S})$.*

□

In this result, the metric defining denseness is the supremum norm. Moreover, the last-layer thresholds θ_{wi} are not needed for this result. The engineering design issues of selecting $\sigma(.)$, and of choosing N_2 for a specified $\mathbf{S} \subset \mathbb{R}^n$ and ε_N are current topics of research (see, e.g., [PS91]).

2.2 Stability and Passivity of Dynamical Systems

Some stability notions are now needed [LAD93]. Consider the nonlinear system

$$\dot{x} = f(x, u, t), y = h(x, t). \tag{4}$$

We say the solution is *uniformly ultimately bounded (UUB)* if there exists a compact set $\mathbf{U} \subset \mathbb{R}^n$ such that for all $x(t_0) = x_0 \in \mathbf{U}$ there exists an $\varepsilon > 0$ and a number $T(\varepsilon, x_0)$ such that $x(t) < \varepsilon$ for all $t \geq t_0 + T$. UUB is a notion of stability in a practical sense that is good enough for suitable tracking performance of robotic manipulators if, of course, the bound is small enough.

Passive systems are important in robust control, where bounded disturbances or unmodeled dynamics are present. Since we intend to define some new passivity properties of NN, some aspects of passivity will subsequently be important [GS84, Lan79, LAD93, SL91]. A system with input $u(t)$ and output $y(t)$ is said to be *passive* if it satisfies an equality of the so-called "power form"

$$\dot{L}(t) = y^T u - g(t), \tag{5}$$

with $L(t)$ lower bounded and $g(t) \geq 0$. That is,

$$\int_0^T y^T(\tau) u(t) \, d\tau \geq \int_0^T g(t) \, dt - \gamma^2 \tag{6}$$

for all $T \geq 0$ and some $\gamma \geq 0$.

We say the system is *dissipative* if it is passive and in addition

$$\int_o^\infty y^T(\tau) u(\tau) \, dt \neq 0 \text{ implies } \int_o^\infty g(\tau) \, d\tau > 0. \tag{7}$$

A special sort of dissipativity occurs if $g(t)$ is a monic quadratic function of x with bounded coefficients, where $x(t)$ is the internal state of the system. We call this *state strict passivity*, and are not aware of its use previously in the literature (although cf. [GS84]). Then the L_2 norm of the state is bounded above in terms of the L_2 inner product of output and input (i.e. the power delivered to the system). This we use to advantage to conclude some internal boundedness properties of the system without the usual assumptions of observability (e.g., persistence of excitation), stability, etc.

3 Dynamics of Rigid Robot Arms

In some sense the application of NN controllers to rigid robot arms turns out to be very natural. A main reason is that the robot dynamics satisfy some important properties, including passivity, that are very easy to preserve in closed loop by considering the corresponding properties on the NN. Thus, one is motivated in robotics applications to discover new properties of NN. The dynamics of robotic manipulators and some of their properties are now discussed.

3.1 Robot Dynamics and Properties

The dynamics of an n-link rigid (i.e., no flexible links or joints) robotic manipulator may be expressed in the Lagrange form [Cra88, LAD93]

$$M(q)\ddot{q} + V_m(q,\dot{q})\dot{q} + G(q) + F(\dot{q}) + t_d = t, \qquad (8)$$

with $q(t) \in \mathbb{R}^n$ the joint variable vector, $M(q)$ the inertia matrix, $V_m(q,\dot{q})$ the Coriolis/centripetal matrix, $G(q)$ the gravity vector, and $F(\dot{q})$ the friction. Bounded unknown disturbances (including, e.g., unstructured unmodeled dynamics) are denoted by t_d, and the control input torque is $\tau(t)$.

The following standard properties of the robot dynamics are required [LAD93].

Property 1: $M(q)$ is a positive definite symmetric matrix bounded by $m_1 I \leq M(q) \leq m_2 I$ with m_1, m_2 known positive constants.

Property 2: $V_m(q,\dot{q})$ is bounded by $v_b(q)\|\dot{q}\|$, with $v_b(q) \in C^1(\mathbf{S})$.

Property 3: The matrix $\dot{M} - 2V_m$ is skew-symmetric.

Property 4: The unknown disturbance satisfies $\|\tau_d\| < b_d$, with b_d a known positive constant.

3.2 Tracking a Desired Trajectory and the Error Dynamics

An important application in robot arm control is for the manipulator to follow a prescribed trajectory.

Error Dynamics

Given a desired arm trajectory $q_d(t) \mathbb{R}^n$, the *tracking error* is

$$e(t) = q_d(t) - q(t). \qquad (9)$$

It is typical in robotics to define a so-called *filtered tracking error* as

$$r = \dot{e} + \Lambda e, \qquad (10)$$

where $\Lambda = \Lambda^T > 0$ is a design parameter matrix, usually selected to be diagonal. Differentiating $r(t)$ and using (8), the arm dynamics may be written in terms of the filtered tracking error as

$$M\dot{r} = -V_m r - \tau + f + \tau_d, \qquad (11)$$

where the important *nonlinear robot function* is

$$f(x) = M(q)(\ddot{q}_d + \dot{e}) + V_m(q,\dot{q})(\dot{q}_d + e) + G(q) + F(\dot{q}) \qquad (12)$$

and we may define, for instance,

$$x \triangleq [e^T \ \dot{e}^T \ q_d^T \ \dot{q}_d^T \ \ddot{q}_d^T]^T. \qquad (13)$$

A suitable control input for trajectory following is given by

$$\tau_o = \hat{f} + K_v r, \qquad (14)$$

$K_v = K_v^T > 0$ a gain matrix and $\hat{f}(x)$ an estimate of $f(x)$ found by some means not yet discussed. Using this control, the closed-loop system becomes

$$M\dot{r} = -(K_v + V_m)r + \tilde{f} + \tau_d \triangleq -(K_v + V_m)r + \zeta_o, \qquad (15)$$

where the *functional estimation error* is given by

$$\tilde{f} = f - \hat{f}. \qquad (16)$$

This is an *error system* wherein the filtered tracking error is driven by the functional estimation error. The control τ_0 incorporates a proportional-plus-derivative (PD) term in $K_v r = K_v(\dot{e} + \Lambda e)$.

The Control Problem

In the remainder of the chapter we shall use (15) to focus on selecting NN tuning algorithms that guarantee the stability of the filtered tracking error $r(t)$. Then, since (10), with the input considered as $r(t)$ and the output as $e(t)$ describes a stable system, standard techniques [LL92, SL91] guarantee that $e(t)$ exhibits stable behavior. In fact, one may show using the notion of "operator gain" that $\|e\|_2 \leq \|r\|_2/\sigma_{min}(\Lambda)$, $\|\dot{e}\|_2 \leq \|r\|_2$, with $\sigma_{min}(\Lambda)$ the minimum singular value of Λ. Generally, Λ is diagonal, so that $\sigma_{min}(\Lambda)$ is the smallest element of Λ.

Therefore, the control design problem is to complete the definition of the controller so that both the error $r(t)$ *and* the control signals are bounded. It is important to note that the latter conclusion hinges on showing that the estimate $\hat{f}(x)$ is bounded. Moreover, for good performance, the bounds on $r(t)$ should be in some sense "small enough."

Passivity Property

The next property is important in the design of robust NN controllers.

Property 5: The dynamics (15) from $\zeta_o(t)$ to $r(t)$ are a state strict passive system.

Proof of Property 5:
Take the nonnegative function $L = \frac{1}{2} r^T M r$ so that using (15)

$$\dot{L} = r^T M \dot{r} + \tfrac{1}{2} r^T \dot{M} r = -r^T K_v r + r^T (\dot{M} - 2V_m) r + r^T \zeta_o,$$

whence skew-symmetry yields the power form $\dot{L} = r^T \zeta_o - r^T K_v r$. □

4 NN Controller for Robot Arms

In this section we derive an NN controller for the robot dynamics in Section 3. This controller implements the control strategy developed in that section, where the robot function estimate $\hat{f}(x)$ is now provided by an NN. Since we must demonstrate boundedness of both the NN weights and the tracking error, it will be found that the standard delta rule does not suffice in tuning this NN, but extra terms must be added.

4.1 Some Assumptions and Facts

Some required mild assumptions are now stated. The assumptions will be true in every practical situation and are standard in the existing literature.

Assumption 1 The nonlinear robot function (12) is given by a neural net as in (3) for some constant "target" NN weights W and V, where the net reconstruction error $\varepsilon(x)$ is bounded by a known constant ε_N. □

Unless the net is "minimal," suitable target weights may not be unique [AS92, Sus92]. The "best" weights may then be defined as those that minimize the supremum norm over **S** of $\varepsilon(x)$. This issue is not of major concern here, as *we only need to know that such target weights exist; their actual values are not required.* According to the discussion in Section 2, results on the approximation properties of NN guarantee that this assumption does in fact hold. This is in direct contrast to the situation often arising in adaptive control, where assumptions (e.g., Erzberger, model-matching) on the plant structure often do not hold in practical applications.

For notational convenience define the matrix of all the weights as

$$Z = \begin{bmatrix} W & 0 \\ 0 & V \end{bmatrix}. \tag{17}$$

Assumption 2 The target weights are bounded by known positive values so that $\|V\|_F \leq V_M$, $\|W\|_F \leq W_M$, or

$$\|Z\|_F \leq Z_M, \tag{18}$$

with Z_M known. □

Assumption 3 The desired trajectory is bounded in the sense, for instance, that

$$\left\| \begin{array}{c} q_d \\ \dot{q}_d \\ \ddot{q} \end{array} \right\| \leq Q_d, \tag{19}$$

where $Q_d \in \mathbb{R}$ is a known constant. □

The next fact follows directly from the assumptions and previous definitions.

Fact 4 For each time t, $x(t)$ in (13) is bounded by

$$\|x\| \leq c_1 Q_d + c_2 \|r\| \tag{20}$$

for computable positive constants c_i (c_2 decreases as Λ increases.) □

4.2 A Property of the Hidden-Layer Output Error

The next discussion is of major importance in this chapter (cf. [PI92]). It shows a key structural property of the hidden-layer output error that plays a major role in the upcoming closed-loop stability proof. It is in effect the step that allows one to progress to *nonlinear* adaptive control as opposed to linear-in-the-parameters control. The analysis introduces some novel terms that will appear directly in the NN weight-tuning algorithms, effectively adding additional terms to the standard delta rule weight updates.

With $\hat{V}$, $\hat{W}$ some estimates of the target weight values, define the weight deviations, or *weight estimation errors*, as

$$\tilde{V} = V - \hat{V}, \quad \tilde{W} = W - \hat{W}, \quad \tilde{Z} = Z - \hat{Z}. \tag{21}$$

In applications, the weight estimates are provided by the NN weight-tuning rules. Define the *hidden-layer output error* for a given x as

$$\tilde{\sigma} = \sigma - \hat{\sigma} \triangleq \sigma(V^T x) - \sigma(\hat{V}^T x). \tag{22}$$

The Taylor series expansion for a given x may be written as

$$\sigma(V^T x) = \sigma(\hat{V}^T x) + \sigma'(\hat{V}^T x)\tilde{V}^T x + O(\tilde{V}^T x)^2, \tag{23}$$

with

$$\sigma'(\hat{z}) \triangleq \frac{d\sigma(z)}{dz}\Big|_{z=\hat{z}},$$

and $O(z)^2$ denoting terms of order two. Setting $\hat{\sigma}' = \sigma'(\hat{V}^T x)$, we have

$$\tilde{\sigma} = \sigma'(\hat{V}^T x)\tilde{V}^T x + O(\tilde{V}^T x)^2 = \hat{\sigma}'\tilde{V}^T x + O(\tilde{V}^T x)^2. \quad (24)$$

Different bounds may be put on the Taylor series' higher-order terms, depending on the choice for $\sigma(.)$. Noting that

$$O(\tilde{V}^T x)^2 = [\sigma(V^T x) - \sigma(\hat{V}^T x)] - \sigma'(\hat{V}^T x)\tilde{V}^T x,$$

we take note of the following:

Fact 5 For logistic, RBF, and tanh activation functions the higher-order terms in the Taylor series are bounded by

$$\|O(\tilde{V}^T x)^2\| \leq c_3 + c_4 Q_d \|\tilde{V}\|_F + c_5 \|\tilde{V}\|_F \|r\|,$$

where c_i are computable positive constants. □

Fact 5 is direct to show using (20), some standard norm inequalities, and the fact that $\sigma(.)$ and its derivative are bounded by constants for RBF, logistic, and tanh functions.

The extension of these ideas to nets with more than three layers is not difficult and leads to composite function terms in the Taylor series (giving rise to backpropagation filtered error terms for the multilayer net case (see Theorem 4.1).

4.3 Controller Structure and Error System Dynamics

The NN controller structure will now be defined; it appears in Figure 2, where $\underline{q} \triangleq [q^T \ \dot{q}^T]^T$, $\underline{e} \triangleq [e^T \ \dot{e}^T]^T$. It is important that the NN controller structure is not ad hoc, but follows directly from a proper treatment of the robot error dynamics and its properties; it is not open to question.

NN Controller

Define the *NN functional estimate* of (12) by

$$\hat{f}(x) = \hat{W}^T \sigma(\hat{V}^T x), \quad (25)$$

with $\hat{V}$, $\hat{W}$ the current (estimated) values of the target NN weights V, W. These estimates will be provided by the weight-tuning algorithms. With $\tau_o(t)$ defined in (14), select the control input

$$\tau = \tau_o - v = \hat{W}^T \sigma(\hat{V}^T x) + K_v r - v, \quad (26)$$

with $v(t)$ a function to be detailed subsequently that provides robustness in the face of higher-order terms in the Taylor series.

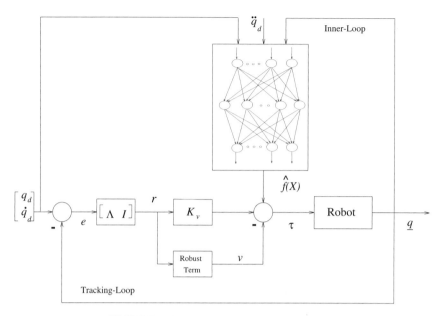

FIGURE 2. Neural net controller structure.

Closed-Loop Error Dynamics and Disturbance Bounds

Using this controller, the closed-loop filtered error dynamics become

$$M\dot{r} = -(K_v + V_m)r + W^T\sigma(V^Tx) - \hat{W}^T\sigma(\hat{V}^Tx) + (\varepsilon + \tau_d) + v.$$

Adding and subtracting $W^T\hat{\sigma}$ yields

$$M\dot{r} = -(K_v + V_m)r + \tilde{W}^T\hat{\sigma} + W^T\tilde{\sigma} + (\varepsilon + \tau_d) + v, \qquad (27)$$

with $\hat{\sigma}$ and $\tilde{\sigma}$ defined in (22). Adding and subtracting now $\hat{W}^T\tilde{\sigma}$ yields

$$M\dot{r} = -(K_v + V_m)r + \tilde{W}^T\hat{\sigma} + \hat{W}^T\tilde{\sigma} + \tilde{W}^T\tilde{\sigma} + (\varepsilon + \tau_d) + v. \qquad (28)$$

A key step is the use now of the Taylor series approximation (24) for $\tilde{\sigma}$, according to which the closed-loop error system is

$$M\dot{r} = -(K_v + V_m)r + \tilde{W}^T\hat{\sigma} + \hat{W}^T\hat{\sigma}'\tilde{V}^Tx + w_1 + v, \qquad (29)$$

where the disturbance terms are

$$w_1(t) = \tilde{W}^T\hat{\sigma}'\tilde{V}^Tx + W^TO(\tilde{V}^Tx)^2 + (\varepsilon + \tau_d). \qquad (30)$$

Unfortunately, using this error system does not yield a compact set outside which a certain Lyapunov function derivative is negative; this makes the upcoming stability proof extremely difficult. Therefore, write finally the error system

$$M\dot{r} = -(K_v + V_m)r + \tilde{W}^T(\hat{\sigma} - \hat{\sigma}'\hat{V}^T x) + \hat{W}^T\hat{\sigma}'\tilde{V}^T x + w + v$$
$$\triangleq -(K_v + V_m)r + \zeta_1, \qquad (31)$$

where the disturbance terms are

$$w(t) = \tilde{W}^T\hat{\sigma}'V^T x + W^T O(\tilde{V}^T x)^2 + (\varepsilon + \tau_d). \qquad (32)$$

It is important to note that the NN reconstruction error $\varepsilon(x)$, the robot disturbances τ_d, and the higher-order terms in the Taylor series expansion of $f(x)$ all have exactly the same influence as disturbances in the error system. The next key bound is required. Its importance is in allowing one to bound the unknown disturbance $w(t)$ at each time by a *known computable function*; it follows from Fact 5 and some standard norm inequalities.

Fact 6
The disturbance term (32) is bounded according to

$$\|w(t)\| \leq (\varepsilon_N + b_d + c_3 Z_M) + c_6 Z_M \|\tilde{Z}\|_F + c_7 Z_M \|\tilde{Z}\|_F r$$
or
$$w(t) \leq C_0 + C_1 \|\tilde{Z}\|_F + C_2 \|\tilde{Z}\|_F \|r\|, \qquad (33)$$

with C_i known positive constants. □

4.4 NN Weight Updates for Guaranteed Tracking Performance

We give here an NN weight tuning algorithm that guarantees the performance of the closed-loop system. To confront the stability and tracking performance of the closed-loop NN robot arm controller we require: (1) the modification of the delta rule weight-tuning algorithm, and (2) the addition of a robustifying term $v(t)$. The problem in the closed-loop control case is that though it is not difficult to conclude that the error $r(t)$ is bounded, it is very hard without these modifications to show that the NN weights are bounded in general. Boundedness of the weights is needed to verify that the control input $\tau(t)$ remains bounded.

The next main theorem relies on an extension to Lyapunov theory. The disturbance τ_d, the NN reconstruction error ε, and the nonlinearity of $f(x)$ make it impossible to show that the Lyapunov derivative $\dot{L}$ is nonpositive for all $r(t)$ and weight values. In fact, it is only possible to show that $\dot{L}$ is negative outside a *compact set in the state space*. This, however, allows one to conclude boundedness of the tracking error and the neural net weights. In fact, explicit bounds are discovered during the proof.

Theorem 4.1 *Let the desired trajectory be bounded by (19). Take the control input for the robot (8) as (26) with robustifying term*

$$v(t) = -K_Z(\|\hat{Z}\|_F + Z_M)r \quad (34)$$

and gain
$$K_Z > C_2, \quad (35)$$

with C_2 the known constant in (33). Let NN weight tuning be provided by

$$\dot{\hat{W}} = F\hat{\sigma}r^T - F\hat{\sigma}'\hat{V}^T xr^T - \kappa F\|r\|\hat{W}, \quad (36)$$
$$\dot{\hat{V}} = Gx(\hat{\sigma}'^T \hat{W}r)^T - \kappa G\|r\|\hat{V}, \quad (37)$$

with any constant matrices $F = F^T > 0$, $G = G^T > 0$, and scalar design parameter $\kappa > 0$. Then, for large enough control gain K_v, the filtered tracking error $r(t)$ and NN weight estimates $\hat{V}, \hat{W}$ are UUB, with practical bounds given specifically by the right-hand sides of (39) and (40). Moreover, the tracking error may be kept as small as desired by increasing the gains K_v in (26).

Proof: Let the approximation property (3) hold for $f(x)$ in (12) with a given accuracy ε_N for all x in the compact set $U_x \triangleq \{x | \|x\| \leq b_x\}$, with $b_x > c_1 Q_d$ in (20). Define $U_r = \{r | \|r\| \leq (b_x - c_1 Q_d)/c_2\}$. Let $r(0) \in U_r$. Then the approximation property holds.

Define the Lyapunov function candidate

$$L = r^T Mr + \operatorname{tr}(\tilde{W}^T F^{-1}\tilde{W}) + \operatorname{tr}(\tilde{V}^T G^{-1}\tilde{V}). \quad (38)$$

Differentiating yields

$$\dot{L} = \frac{1}{2}r^T \dot{M}r + r^T M\dot{r} + \operatorname{tr}(\tilde{W}^T F^{-1}\dot{\tilde{W}}) + \operatorname{tr}(\tilde{V}^T G^{-1}\dot{\tilde{V}}).$$

Substituting now from the error system (31) yields

$$\dot{L} = -r^T K_v r + \frac{1}{2}r^T(\dot{M} - 2V_m)r + \operatorname{tr}\tilde{W}^T(F^{-1}\dot{\tilde{W}} + \hat{\sigma}r^T - \hat{\sigma}'\hat{V}^T xr^T)$$
$$+ \operatorname{tr}\tilde{V}^T(G^{-1}\dot{\tilde{V}} + xr^T \hat{W}^T \hat{\sigma}') + r^T(w + v).$$

The tuning rules give

$$\dot{L} = -r^T K_v r + \kappa\|r\|\operatorname{tr}\tilde{W}^T(W - \tilde{W}) + \kappa\|r\|\operatorname{tr}\tilde{V}^T(V - \tilde{V}) + r^T(w + v)$$
$$= -r^T K_v r + \kappa\|r\|\operatorname{tr}\tilde{Z}^T(Z - \tilde{Z}) + r^T(w + v).$$

Since

$$\operatorname{tr}\tilde{Z}^T(Z - \tilde{Z}) = <\tilde{Z}, Z>_F - \|\tilde{Z}\|_F^2 \leq \|\tilde{Z}\|_F \|Z\|_F - \|\tilde{Z}\|_F^2,$$

there results

$$\begin{aligned}\dot{L} &\leq -K_{vmin}\|r\|^2 + \kappa\|r\|\|\tilde{Z}\|_F(Z_M - \|\tilde{Z}\|_F) - K_Z(\|\hat{Z}\|_F + Z_M)\|r\|^2 \\ &\quad + \|r\|\|w\| \\ &\leq -K_{vmin}\|r\|^2 + \kappa\|r\|\|\tilde{Z}\|_F(Z_M - \|\tilde{Z}\|_F) - K_Z(\|\hat{Z}\|_F + Z_M)\|r\|^2 \\ &\quad + \|r\|[C_0 + C_1\|\tilde{Z}\|_F + C_2\|\tilde{Z}\|_F\|r\|] \\ &\leq -\|r\|[K_{vmin}\|r\| + \kappa\|\tilde{Z}\|_F(\|\tilde{Z}\|_F - Z_M) - C_0 - C_1\|\tilde{Z}\|_F],\end{aligned}$$

where K_{vmin} is the minimum singular value of K_v and the last inequality holds due to (35). Thus, $\dot{L}$ is negative as long as the term in braces is positive. We show next that this occurs outside a compact set in the $(\|r\|, \|\tilde{Z}\|_F)$ plane.

Defining $C_3 = Z_M + C_1/\kappa$ and completing the square yields

$$\begin{aligned}&K_{vmin}\|r\| + \kappa\|\tilde{Z}\|_F(\|\tilde{Z}\|_F - C_3) - C_0 \\ &= \kappa(\|\tilde{Z}\|_F - C_3/2)^2 - \kappa C_3^2/4 + K_{vmin}\|r\| - C_0,\end{aligned}$$

that is guaranteed positive as long as either

$$\|r\| > \frac{\kappa C_3^2/4 + C_0}{K_{vmin}} \triangleq b_r \tag{39}$$

or

$$\|\tilde{Z}\|_F > C_3/2 + \sqrt{C_3^2/4 + C_0/\kappa} \triangleq b_Z, \tag{40}$$

where

$$C_3 = Z_M + C_1/\kappa. \tag{41}$$

Thus, $\dot{L}$ is negative outside a compact set. The form of the right-hand side of (39) shows that the control gain K_v can be selected large enough so that $b_r < (b_x - c_1 Q_d)/c_2$. Then, any trajectory $r(t)$ beginning in U_r evolves completely within U_r. According to a standard Lyapunov theorem extension [LAD93, NA87], this demonstrates the UUB of both $\|r\|$ and $\|\tilde{Z}\|_F$. □

The complete NN controller is given in Table 1 and illustrated in Figure 2. It is important to note that this is a novel control structure with an inner NN loop and an *outer robust tracking loop* that has important ramifications as delineated below. Some discussion of these results is now given.

Bounded Errors and Controls

The dynamical behavior induced by this controller is as follows. Due to the presence of the disturbance terms, it is not possible to use Lyapunov's theorem directly as it cannot be demonstrated that $\dot{L}$ is always negative; instead an extension to Lyapunov's theorem is used (cf. [NA87] and Theorem 1.5-6 in [LAD93]). In this extension, it is shown that $\dot{L}$ is negative if

TABLE 1. Neural net robotic controller.

NN Controller:
$$\tau = \hat{W}^T \sigma(\hat{V}^T x) + K_v r - v \quad (42)$$

Robustifying term:
$$v(t) = -K_Z(\|\hat{Z}\|_F + Z_M)r \quad (43)$$

NN weight tuning:
$$\dot{\hat{W}} = F\hat{\sigma}r^T - F\hat{\sigma}'\hat{V}^T x r^T - \kappa F\|r\|\hat{W} \quad (44)$$

$$\dot{\hat{V}} = Gx(\hat{\sigma}'^T \hat{W} r)^T - \kappa G\|r\|\hat{V} \quad (45)$$

Signals:
$$e = q(t) - q_d(t), \text{ tracking error} \quad (46)$$
$$r(t) = \dot{e}(t) + \Lambda e(t), \text{ filtered tracking error} \quad (47)$$

with Λ a symmetric positive definite matrix

$$x \triangleq [e^T \ \dot{e}^T \ q_d^T \ \dot{q}_d^T \ \ddot{q}_d^T]^T, \text{NN input signal vector} \quad (48)$$

Design parameters:

Gains K_v, K_Z symmetric and positive definite.
Z_M a bound on the unknown target weight norms.
Tuning matrices F, G symmetric and positive definite.
Scalar $\kappa > 0$.

---

either $\|r\|$ or $\|\tilde{Z}\|$ is above some specific bounds. Therefore, if either norm increases too much, L decreases, so that both norms decrease as well. If both norms are small, nothing may be said about $\dot{L}$ except that it is probably positive, so that L increases. This has the effect of making *the boundary of a compact set* an attractive region for the closed-loop system. Thus the errors are guaranteed bounded, but in all probability nonzero.

In applications, therefore, the right-hand sides of (39) and (40) may be taken as *practical bounds* on the norms of the error $r(t)$ and the weight errors $\tilde{Z}(t)$. Since the target weights Z are bounded, it follows that the NN weights $\hat{W}(t)$ and $\hat{V}(t)$ provided by the tuning algorithms are bounded; hence the control input is bounded.

In fact, it is important to note that according to (39), *arbitrarily small tracking error bounds may be achieved* by selecting large control gains K_v. (If K_v is taken as a diagonal matrix, K_{vmin} is simply the smallest gain element.) On the other hand, (40) reveals that *the NN weight errors are fundamentally bounded by Z_M* (through C_3). The parameter κ offers a design tradeoff between the relative eventual magnitudes of $\|r\|$ and $\|\tilde{Z}\|$.

An alternative to guaranteeing the boundedness of the NN weights for the 2-layer case $V = I$ (i.e., linear in the parameters) is presented in [PI91, RC95], where a projection algorithm is used for tuning $\hat{W}$.

Initializing the NN Weights and Real-Time Learning

Note that the problem of *net weight initialization* occurring in other approaches in the literature does not arise. In fact, selecting the initial weights $\hat{W}(0), \hat{V}(0)$ as zero takes the NN out of the circuit and leaves only the outer tracking loop in Figure 2. It is well known that the PD term $K_v r$ in (42) can then stabilize the robot arm on an interim basis until the NN begins to learn. A formal proof reveals that K_v should be large enough and the initial filtered error $r(0)$ small enough. The exact value of K_v needed for initial stabilization is given in [DQLD90], though for practical purposes it is only necessary to select K_v large.

This means that *there is no off-line learning phase* for this NN controller. Results in a simulation example soon to be presented show that convergence of the tracking error occurs in real time in a fraction of a second.

Extension of Delta Rule with Error Backpropagation

The first terms of (44), (45) are nothing but continuous-time versions of the standard backpropagation algorithm. In fact, the first terms are

$$\dot{\hat{W}} = F\hat{\sigma}r^T, \qquad (49)$$
$$\dot{\hat{V}} = Gx(\hat{\sigma}'^T \hat{W} r)^T. \qquad (50)$$

In the scalar case the logistic function satisfies

$$\sigma'(z) = \sigma(z)(1 - \sigma(z)),$$

so that

$$\hat{\sigma}'^T \hat{W} r = \text{diag}\{\sigma(\hat{V}^T x)\}[I - \text{diag}\{\sigma(\hat{V}^T x)\}]\hat{W} r,$$

which is the filtered error weighted by the current estimate $\hat{W}$ and multiplied by the usual product involving the hidden-layer outputs.

The last terms in (44), (45) correspond to the ϵ-modification [NA87] in standard use in adaptive control to guarantee bounded parameter estimates. They are needed due to the presence of the NN reconstruction error ε and the robot unmodeled disturbances $\tau_d(t)$.

The second term in (44) is a novel one and bears discussion. The standard backprop terms can be thought of as backward propagating signals in a

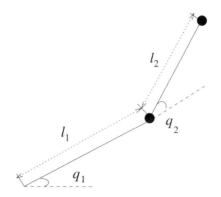

FIGURE 3. 2-link planar elbow arm.

nonlinear "backprop" network [NP90] that contains multipliers. The second term in (44) corresponds to a *forward traveling wave in the backprop net* that provides a second-order correction to the weight tuning for $\hat{W}$. This term is needed to bound certain of the higher-order terms in the Taylor series expansion of $\tilde{\sigma}$, and arises from the extension of adaptive control from the linear-in-the-parameters case to the nonlinear case.

Design Freedom in NN Complexity

Note that there is *design freedom in the degree of complexity (e.g., size) of the NN*. For a more complex NN (e.g., more hidden units), the bounding constants will decrease, resulting in smaller tracking errors. On the other hand, a simplified NN with fewer hidden units will result in larger error bounds; this degradation can be compensated for, as long as bound ε_N is known, by selecting a larger value for K_z in the robustifying signal $v(t)$ or for Λ in (47).

Example 4.1: NN Control of 2-Link Robot Arm

A planar 2-link arm used extensively in the literature for illustration purposes appears in Figure 3. The dynamics are given in, for instance, [LAD93]; no friction term was used in this example. The joint variable is $q = [q_1 \; q_2]^T$. We should like to illustrate the NN control scheme derived herein, which will require no knowledge of the dynamics, not even their structure, which is needed for adaptive control.

Adaptive Controller: Baseline Design

For comparison, a standard adaptive controller is given by [SL88]:

$$\tau = Y\hat{\psi} + K_v r, \tag{51}$$
$$\dot{\hat{\psi}} = F Y^T r, \tag{52}$$

with $F = F^T > 0$ a design parameter matrix and $Y(e, \dot{e}, q_d, \dot{q}_d, \ddot{q}_d)$ the *regression matrix*, a fairly complicated matrix of robot functions *that must be explicitly derived from the dynamics for each arm*. The is the vector of unknown parameters, in this case simply the link masses m_1, m_2.

We took the arm parameters as $l_1 = l_2 = 1$ meter, $m_1 = 0.8$ kg, $m_2 = 2.3$ kg, and selected $q_{1d}(t) = \sin(t)$, $q_{2d}(t) = \cos(t)$, $K_v = \text{diag}\{20, 20\}$, $F = \text{diag}\{10, 10\}$, $\Lambda = \text{diag}\{5, 5\}$. The response with this controller when $q(0) = 0$, $\dot{q}(0) = 0$, $\hat{m}_1(0) = 0$, $\hat{m}_2(0) = 0$ is shown in Figure 4. Note the good behavior, which obtains since there are only two unknown parameters, so that the single mode (e.g., 2 poles) of $q_d(t)$ guarantees persistence of excitation [GS84].

The (1, 1) entry of the robot function matrix Y is $l_1^2(\ddot{q}_{d1} + \lambda_1 \dot{e}_1) + l_1 g \cos(q_1)$ (with $\Lambda = \text{diag}\{\lambda_1, \lambda_2\}$). To demonstrate the deleterious effects of unmodeled dynamics in adaptive control, the term $l_1 g \cos(q_1)$ was now dropped in the controller. The result appears in Figure 5 and is unsatisfactory. This demonstrates conclusively the fact that the adaptive controller cannot deal with *unmodeled dynamics*. It is now emphasized that in the NN controller *all the dynamics are unmodeled*.

NN Controller

Some preprocessing of signals yields a more advantageous choice for $x(t)$ than (12), one that already contains some of the nonlinearities inherent to robot arm dynamics. Since the only occurrences of the revolute joint variables are as sines and cosines, the vector x can be taken for a general n-link revolute robot arm as (componentwise)

$$x = [\ \zeta_1^T \quad \zeta_2^T \quad \cos(q)^T \quad \sin(q)^T \quad \dot{q}^T \quad \text{sgn}(\dot{q})^T]^T, \tag{53}$$

where $\zeta_1 = \ddot{q}_d + \dot{e}$, $\zeta_2 = \dot{q}_d + \Lambda e$, and the signum function is needed in the friction terms (not used in this example). The NN controller appears in Figure 2.

The logistic function activation functions were used, and 10 hidden-layer neurons. The values for $q_d(t)$, Λ, F, K_v were the same as before, and we selected $G = \text{diag}\{10, 10\}$. The response of the controller with the weight tuning in Theorem 4.1 appears in Figure 6, where we took $\kappa = 0.1$. The comparison with the performance of the standard adaptive controller in Figure 4 is impressive, even though *the dynamics of the arm were not required to implement the NN controller*. That is, no regression matrix was needed.

No initial NN training or learning phase was needed. The NN weights were simply initialized at zero in this figure.

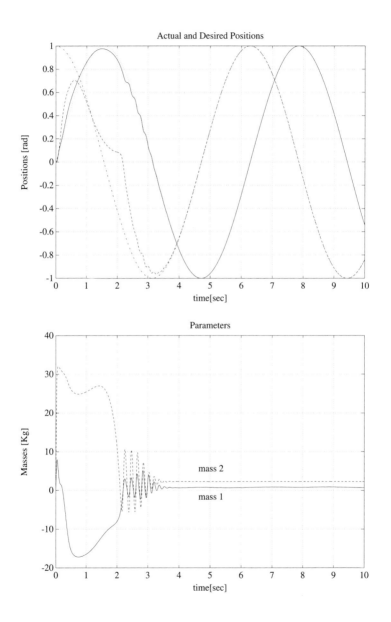

FIGURE 4. Response of adaptive controller. (a) Actual and desired joint angles. (b) Parameter estimates.

7. Neural Network Control of Robot Arms 181

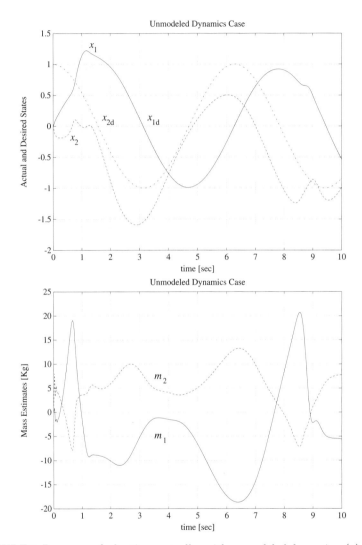

FIGURE 5. Response of adaptive controller with unmodeled dynamics. (a) Actual and desired joint angles. (b) Representative weight estimates.

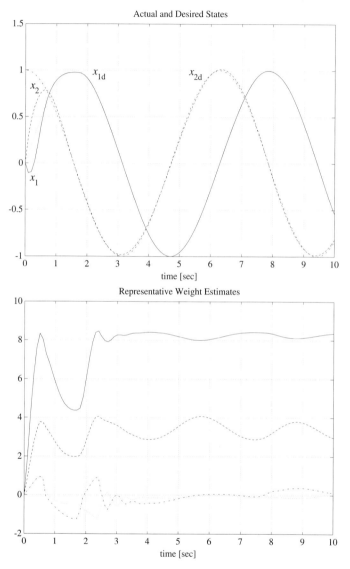

FIGURE 6. Response of NN controller. (a) Actual and joint angles. (b) Representative weight estimates.

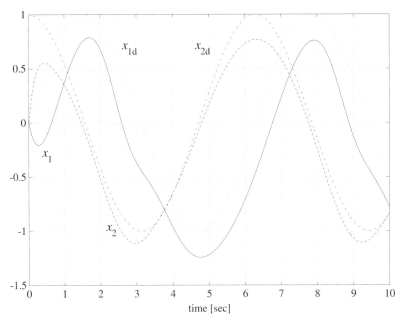

FIGURE 7. Response of controller in Figure 2 without NN. Actual and desired joint angles.

To study the contribution of the NN, Figure 7 shows the response with the controller $\tau = K_v r$, that is, with no neural net. Standard results in the robotics literature indicate that such a PD controller should give bounded errors if K_v is large enough. This is observed in the figure. However, it is very clear that the addition of the NN makes a very significant improvement in the tracking performance.

5 Passivity and Structure Properties of the NN

A major advantage of the NN controller is that it has some important passivity properties that result in robust closed-loop performance, as well as some structure properties that make it easier to design and implement.

5.1 Neural Network Passivity and Robustness

The *closed-loop error system* appears in Figure 8, with signal ζ_2 defined as

$$\zeta_2(t) \triangleq -\tilde{W}^T(\hat{\sigma} - \hat{\sigma}'\hat{V}^T x). \tag{54}$$

Note the role of the NN, which is decomposed into two effective blocks appearing in a typical feedback configuration, in contrast to the role of the NN in the controller in Figure 2.

Passivity is important in a closed-loop system as it guarantees the boundedness of signals, and hence suitable performance, even in the presence of

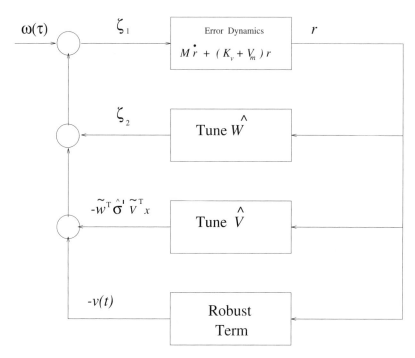

FIGURE 8. Neural net closed-loop error system.

additional unforeseen disturbances as long as they are bounded. In general, an NN cannot be guaranteed to be passive. The next results show, however, that the weight tuning algorithm given here does in fact guarantee desirable passivity properties of the NN and hence of the closed-loop system.

Theorem 5.1 *The weight-tuning algorithms (44), (45) make the map from $r(t)$ to $-\tilde{W}^T(\hat{\sigma} - \hat{\sigma}'\hat{V}^T x)$ and the map from $r(t)$ to $-\hat{W}^T \hat{\sigma}' \tilde{V}^T x$ both state strict passive (SSP) maps.*

Proof:
The dynamics relative to $\tilde{W}$, $\tilde{V}$ are given by

$$\dot{\tilde{W}} = -F\hat{\sigma}r^T + F\hat{\sigma}'\hat{V}^T x r^T + \kappa F\|r\|\hat{W}, \tag{55}$$
$$\dot{\tilde{V}} = -Gx(\hat{\sigma}'^T \hat{W} r)^T + \kappa Gr\hat{V}. \tag{56}$$

1. Selecting the nonnegative function

$$L = \frac{1}{2}\mathrm{tr}(\tilde{W}^T F^{-1} \tilde{W})$$

and evaluating $\dot{L}$ yields

$$\dot{L} = \mathrm{tr}(\tilde{W}^T F^{-1} \dot{\tilde{W}}) = \mathrm{tr}\{[-\tilde{W}^T(\hat{\sigma} - \hat{\sigma}'\hat{V}^T x)]r^T + \kappa\|r\|\tilde{W}^T \hat{W}\}.$$

Since
$$\operatorname{tr}(\tilde{W}^T(W-\tilde{W})) = <\tilde{W},W>_F - \|\tilde{W}\|_F^2 \leq \|\tilde{W}\|_F\|W\|_F - \|\tilde{W}\|_F^2,$$
there results
$$\begin{aligned}\dot{L} &\leq r^T[-\tilde{W}^T(\hat{\sigma}-\hat{\sigma}'\hat{V}^Tx)] - \kappa\|r\|(\|\tilde{W}\|_F^2 - \|\tilde{W}\|_F\|\|W\|_F) \\ &\leq r^T[-\tilde{W}^T(\hat{\sigma}-\hat{\sigma}'\hat{V}^Tx)] - \kappa\|r\|(\|\tilde{W}\|_F^2 - W_M\|\tilde{W}\|_F), \end{aligned} \quad (57)$$
which is in power form, with the last function quadratic in $\tilde{W}_F$.
2. Selecting the nonnegative function
$$L = \frac{1}{2}\operatorname{tr}(\tilde{V}^T G^{-1}\tilde{V})$$
and evaluating $\dot{L}$ yields
$$\begin{aligned}\dot{L} &= \operatorname{tr}(\tilde{V}^T G^{-1}\dot{\tilde{V}}) & (58) \\ &= r^T(-\hat{W}^T\hat{\sigma}'\tilde{V}^T x) - \kappa\|r\|(\|\tilde{V}\|_F^2 - <\tilde{V},V>_F) & (59) \\ &\leq r^T(-\hat{W}^T\hat{\sigma}'\tilde{V}^T x) - \kappa\|r\|(\|\tilde{V}\|_F^2 - V_M\tilde{V}_F), & (60)\end{aligned}$$
which is in power form, with the last function quadratic in $\|\tilde{V}_F\|$. □

Thus, the robot error system in Figure 8 is state strict passive (SSP) and the weight error blocks are SSP; this guarantees the SSP of the closed-loop system (cf. [SL91]). Using the passivity theorem, one may now conclude that the input/output signals of each block are bounded as long as the external inputs are bounded. Now, the state-strictness of the passivity guarantees that all signals *internal* to the blocks are bounded as well. This means specifically that the tracking error $r(t)$ and the weight estimates $\hat{W}(t)$, $\hat{V}(t)$ are bounded (since $\tilde{W}, W, \tilde{V}, V$ are all bounded).

We define an NN as *robust* if in the error formulation it guarantees the SSP of the weight tuning subsystems. Then the weights are bounded if the power into the system is bounded. Note that: (1) SSP of the open-loop plant error system is needed in addition for tracking stability, and (2) the NN passivity properties are *dependent on the weight-tuning algorithm used*. It can be shown, for instance, that using only the first (backprop) terms in weight tuning as in (49), (50), the weight-tuning blocks are only passive, so that no bounds on the weights can be concluded without extra (persistence of excitation) conditions.

5.2 Partitioned Neural Nets and Preprocessing of Inputs

A major advantage of the NN approach is that it allows one to partition the controller in terms of *partitioned NN or neural subnets*. This (1) simplifies the design, (2) gives added controller structure, and (3) makes for faster weight-tuning algorithms.

Partitioned Neural Nets

In [OSF+91] an NN scheme was presented for robot arms that used separate NNs for the inertia and Coriolis terms in (12). We now give a rigorous approach to this simplified NN structure.

The nonlinear robot function (12) is

$$f(x) = M(q)\zeta_1(t) + V_m(q,\dot{q})\zeta_2(t) + G(q) + F(\dot{q}), \qquad (61)$$

where for control purposes, $\zeta_1(t) = \ddot{q}_d + \Lambda\dot{e}$, $\zeta_2(t) = \dot{q}_d + \Lambda e$.

Let $q \in \mathbb{R}^n$. Taking the four terms in $f(x)$ one at a time, use separate NNs to reconstruct each one so that

$$\begin{aligned}
M(q)\zeta_1(t) &= W_M^T \sigma_M(V_M^T x_M), \\
V_m(q,\dot{q})\zeta_2(t) &= W_V^T \sigma_V(V_V^T x_V), \\
G(q) &= W_G^T \sigma_G(V_G^T x_G), \\
F(\dot{q}) &= W_F^T \sigma_F(V_F^T x_F).
\end{aligned} \qquad (62)$$

Now, write $f(x)$ as

$$f(x) = [W_M^T \ W_V^T \ W_G^T \ W_F^T] \begin{bmatrix} \sigma_M \\ \sigma_V \\ \sigma_G \\ \sigma_F \end{bmatrix}, \qquad (63)$$

so that $\sigma(\cdot)$ is a diagonal function composed of the activation function vectors $\sigma_M, \sigma_V, \sigma_G, \sigma_F$ of the separate partitioned NNs. Formulation (63) reveals that the theory developed herein for stability analysis applies when individual NNs are designed for each of the terms in $f(x)$.

This procedure results in four neural subnets, which we term a *structured NN*, as shown in Figure 9. It is straightforward to show that the individual partitioned NNs can be separately tuned, making for a faster weight update procedure. That is, each of the NNs in (62) can be tuned individually using the rules in Theorem 4.1.

Preprocessing of Neural Net Inputs

The selection of a suitable $x(t)$ for computation remains to be addressed; some preprocessing of signals, as used in Example 4.1, yields a more advantageous choice than (48) since it already contains some of the nonlinearities inherent to robot arm dynamics. Let an n-link robot have n_r revolute joints with joint variables q_r, and n_p prismatic joints with joint variables q_p. Define $n = n_r + n_p$. Since the only occurrences of the revolute joint variables are as sines and cosines, transform $q = [q_r^T \ q_p^T]^T$ by preprocessing to $[\cos(q_r)^T \ \sin(q_r)^T \ q_p^T]^T$ to be used as arguments for the basis functions. Then the vector x can be taken as

$$x = [\zeta_1^T \ \zeta_2^T \ \cos(q_r)^T \ \sin(q_r)^T \ q_p^T \ \dot{q}^T \ \text{sgn}(\dot{q})^T]^T,$$

where the signum function is needed in the friction terms.

6 Neural Networks for Control of Nonlinear Systems

In this section, for a class of continuous-time systems we will give a design procedure for a multilayer NN controller. That is, a stable NN adaptation rules and feedback structures will be derived so that systems of interest perform a desired behavior while all the generated signals remain bounded.

6.1 The Class of Nonlinear Systems

When the input/output representation of a plant is in "affine form," the problem of control is significantly simplified. Consequently, there has been considerable interest in studying those systems. Consider a single-input single-output (SISO) system having a state space representation in the Brunovsky canonical form

$$
\begin{aligned}
\dot{x}_1 &= x_2, \\
\dot{x}_2 &= x_3, \\
&\vdots \\
\dot{x}_n &= f(\mathbf{x}) + u + d, \\
y &= x_1,
\end{aligned}
\tag{64}
$$

with a state vector $\mathbf{x} = [x_1, x_2, \ldots, x_n]^T$, bounded unknown disturbances $d(t)$, which is bounded by a known constant b_d, and an unknown smooth

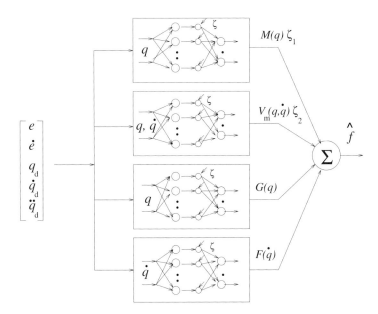

FIGURE 9. Structured neural net.

function $f: \mathbb{R}^n \to \mathbb{R}$.

6.2 Tracking Problem

Control action will be used for output tracking, which can be described as, Given a desired output $y_d(t)$, find a bounded control action $u(t)$ such that the plant follows the desired trajectory with an acceptable accuracy (i.e., bounded-error tracking), while all the states remain bounded.

For this purpose we will make some mild assumptions that are widely used. First, define a vector

$$\mathbf{x}_d(t) = \begin{bmatrix} y_d \\ \dot{y}_d \\ \vdots \\ y_d^{(n-1)} \end{bmatrix}.$$

The desired trajectory $\mathbf{x}_d(t)$ is assumed to be continuous, available for measurement, and have a bounded norm,

$$\|\mathbf{x}_d(t)\| \leq Q, \tag{65}$$

with Q a known bound.

6.3 Error Dynamics

Define a state error vector as

$$\mathbf{e} = \mathbf{x} - \mathbf{x}_d \tag{66}$$

and a filtered error as

$$r = \Lambda^T \mathbf{e}, \tag{67}$$

where $\Lambda = [\bar{\Lambda}\ 1]^T$ with $\bar{\Lambda} = [\lambda_1, \lambda_2, \cdots, \lambda_{n-1}]$ is an appropriately chosen coefficient vector such that the state error vector $\mathbf{e}(t)$ exponentially goes to $\mathbf{0}$ as the filtered error $r(t)$ tends to 0, i.e., $s^{n-1} + \lambda_{n-1}s^{n-2} + \cdots + \lambda_2 s + \lambda_1$ is Hurwitz. Then the time derivative of the filtered error can be written as

$$\dot{r} = f(\mathbf{x}) + u + Y_d + d, \tag{68}$$

with

$$Y_d = -\mathbf{x}_d^{(n)} + \sum_{i=1}^{n-1} \lambda_i e_{i+1}.$$

Next we will construct an NN controller to regulate the error system dynamics (68), which guarantees that the desired tracking performance is achieved.

6.4 Neural Network Controller

If we knew the exact form of the nonlinear function $f(\mathbf{x})$, then the control action
$$u = -f(\mathbf{x}) - K_v r - Y_d$$
would bring $r(t)$ to zero exponentially for any $K_v > 0$ when there was no disturbance $d(t)$. Since in general, $f(\mathbf{x})$ is not known to us exactly, we will choose a control signal as
$$u_c = -\hat{f}(\mathbf{x}) - K_v r - Y_d + v, \qquad (69)$$
where the estimate of $f(\mathbf{x})$ is $\hat{f}(\mathbf{x})$ and the auxiliary robustifying term $v(t)$ that provides robustness will be revealed later. Hence, the filtered error dynamics (68) becomes
$$\dot{r} = -K_v r + \tilde{f} + d + v. \qquad (70)$$

As shown in Theorem 2.1, multilayer neural networks that have linear activation in the input and output layers and a nonlinear activation function in the hidden layer can approximate any continuous function uniformly on a compact set arbitrarily well, provided that enough neurons are used. Let $f(\mathbf{x})$ be a continuous function; then there exists a best set of weights W and V such that the equation
$$f(\mathbf{x}) = W^T \sigma(V^T \mathbf{x}) + \varepsilon \qquad (71)$$
holds for any $\varepsilon > 0$. Therefore, $f(\mathbf{x})$ may be constructed by a multilayer neural network as
$$\hat{f}(\mathbf{x}) = \hat{W}^T \sigma(\hat{V}^T \mathbf{x}). \qquad (72)$$
Using steps similar to Section 4.3, we can write the functional approximation error by using the Taylor series expansion of $\sigma(V^T \mathbf{x})$ as
$$\tilde{f}(\mathbf{x}) = \tilde{W}^T(\hat{\sigma} - \hat{\sigma}'\hat{V}^T \mathbf{x}) + \hat{W}^T \hat{\sigma}' \tilde{V}^T \mathbf{x} + w \qquad (73)$$
with (cf. (33))
$$|w(t)| \leq C_0 + C_1 \|\tilde{Z}\|_F + C_2 |r| \|\tilde{Z}\|_F, \qquad (74)$$
where the C_i are some computable constants and the generalized weight matrix Z is defined in (18). In the sequel, $\|\cdot\|$ will indicate the Frobenius norm, unless otherwise mentioned. Also, recall that the Frobenius norm of a vector is equivalent to its 2-norm, i.e., these norms are compatible.

6.5 Stable NN Control System

In order to give theoretical justification for the proposed controller structure, which is shown in Figure 10, we will choose the NN weight update rules as
$$\begin{aligned} \dot{\hat{W}} &= M(\hat{\sigma} - \hat{\sigma}'\hat{V}^T \mathbf{x})r - \kappa |r| M \hat{W}, \\ \dot{\hat{V}} &= N \mathbf{x} r \hat{W}^T \hat{\sigma}' - \kappa |r| N \hat{V}. \end{aligned} \qquad (75)$$
Now we can reveal the stability properties of the system (64) by the following theorem.

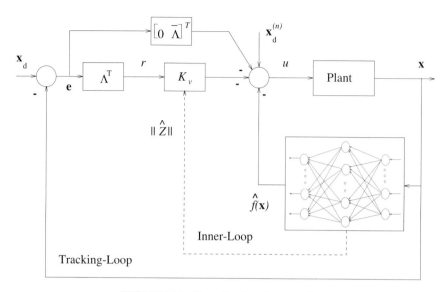

FIGURE 10. Neural network controller.

Theorem 6.1 *Assume that the system has a representation in the reachability form as in (64) and the control input is given by (69) with the auxiliary control signal*

$$v = -K_z(\|\hat{Z}\| + Z_m), \tag{76}$$

with $K_z \geq C_2 > 0$. Let the neural net weight update law be provided by (75). Then the filtered tracking error $r(t)$ and neural net weight error $\tilde{Z}$ are UUB with specific bounds given by (80).

Proof: Since $f(\mathbf{x})$ is continuous in $\mathbf{x}$, then the NN approximation property holds in any compact subset of $\mathbb{R}^n$. Given $\mathbf{x}_d(t) \in U_d$, define a bound b_x such that $U = \{\mathbf{x} \mid \|\mathbf{x}\| \leq b_x\}$ and $U_d \subset U$. Let $|r(0)| \leq b_r$, with b_r defined in (80).

Substitution of the functional approximation error as shown in (73) into the error system dynamics for $\tilde{f}$ yields

$$\dot{r} = -K_v r + \tilde{W}^T(\hat{\sigma} - \hat{\sigma}'\hat{V}^T \mathbf{x}) + \hat{W}^T \hat{\sigma}' \tilde{V}^T \mathbf{x} + d + w. \tag{77}$$

Let the Lyapunov function candidate be

$$L = \frac{1}{2}r^2 + \frac{1}{2}\mathrm{tr}\left\{\tilde{W}^T M^{-1} \tilde{W}\right\} + \frac{1}{2}\mathrm{tr}\left\{\tilde{V}^T N^{-1} \tilde{V}\right\}. \tag{78}$$

Now substitute (77) into the time derivative of (78) and perform a simple manipulation, (i.e., using the equality

$$\mathbf{x}^T \mathbf{y} = \mathrm{tr}\left\{\mathbf{x}^T \mathbf{y}\right\} = \mathrm{tr}\left\{\mathbf{y}\mathbf{x}^T\right\},$$

one can place weight matrices inside a trace operator) to obtain

$$\dot{L} = -K_v r^2 + \mathrm{tr}\left\{\tilde{W}^T(\hat{\sigma} - \hat{\sigma}'\hat{V}^T\mathbf{x})r + M^{-1}\dot{\tilde{W}}\right\}$$
$$+\mathrm{tr}\left\{\tilde{V}^T(\mathbf{x}r\hat{W}^T\hat{\sigma}' + N^{-1}\dot{\tilde{V}})\right\} + r(d+w).$$

With the update rules given in (75) one has

$$\dot{L} = -K_v r^2 + r(d+w+v) + \kappa|r|\mathrm{tr}\{\tilde{Z}^T\hat{Z}\}.$$

From the inequality

$$\mathrm{tr}\left\{\tilde{Z}^T\hat{Z}\right\} =<\tilde{Z}^T, Z> -\mathrm{tr}\left\{\tilde{Z}^T\tilde{Z}\right\} \leq \|\tilde{Z}\|(Z_m - \|\tilde{Z}\|),$$

it follows that

$$\dot{L} \leq -K_v r^2 + r(d+w+v) + \kappa|r|\|\tilde{Z}\|(Z_m - \|\tilde{Z}\|).$$

Substitute the upper bound of w according to (74), b_d for disturbances and v from (76) to yield

$$\dot{L} \leq -K_v r^2 - K_z(\|\hat{Z}\| + Z_m)r^2 + \kappa|r|\|\tilde{Z}\|(Z_m - \|\tilde{Z}\|)$$
$$+ \left[C_2\|\tilde{Z}\||r| + C_1\|\tilde{Z}\| + (b_d + C_0)\right]|r|.$$

Picking $K_z > C_2$ and completing the squares yields

$$\dot{L} \leq -|r|\left\{K_v|r| + \kappa(\|\tilde{Z}\| - C_3/2)^2 - D_1\right\}, \tag{79}$$

where

$$D_1 = b_d + C_0 + \frac{\kappa}{4}C_3^2$$

and

$$C_3 = Z_m + C_1/\kappa.$$

Observe that the terms in braces in (79) define a compact set around the origin of the error space ($|r|, \|\tilde{Z}\|$) outside of which $\dot{L} \leq 0$. We can, therefore, deduce from (79) that if either $|r| > b_r$ or $\|\tilde{Z}\| > b_f$ then $\dot{L} \leq 0$, where

$$b_r = \frac{D_1}{K_v}, \quad b_f = \frac{C_3}{2} + \sqrt{\frac{D_1}{\kappa}}. \tag{80}$$

Note that b_r can be kept small by adjusting the design parameter K_v, which ensures that $\mathbf{x}(t)$ stays in the compact set U. Thus, the NN approximation property remains valid. According to a standard Lyapunov theorem extension (cf. Theorem 4.1), this demonstrates the UUB of both $|r|$ and $\|\tilde{Z}\|$. This concludes the proof. □

TABLE 2. Neural Net Controller.

NN Controller:
$$u = -\hat{W}^T \sigma(\hat{V}^T \mathbf{x}) - K_v r + v \qquad (81)$$

Robustifying Term:
$$v = -K_z(\|\hat{Z}\| + Z_m) \qquad (82)$$

NN Weight Tuning:
$$\begin{aligned} \dot{\hat{W}} &= M(\hat{\sigma} - \hat{\sigma}'\hat{V}^T\mathbf{x})r - \kappa|r|M\hat{W} \\ \dot{\hat{V}} &= N\mathbf{x}r\hat{W}^T\hat{\sigma}' - \kappa|r|N\hat{V}. \end{aligned} \qquad (83)$$

Signals:
$$\mathbf{e}(t) = \mathbf{x}(t) - \mathbf{x}_d(t), \text{Tracking error} \qquad (84)$$
$$r(t) = \Lambda^T \mathbf{e}(t), \text{Filtered tracking error} \qquad (85)$$
$$\mathbf{x}(t) = [x_1, x_2, \cdots, x_n]^T, \text{NN input signal vector} \qquad (86)$$

Design parameters:

Gains K_v, K_z positive
Λ, a coefficient vector of a Hurwitz function.
Z_m, a bound on the unknown target weight norms.
Tuning matrices M, N symmetric and positive definite.
Scalar $\kappa > 0$.

The NN functional construction error ε, the bounded disturbances, the norm of the desired performance and the neural network size are all contained in the constants C_j, and they increase the bounds on error signals. Nevertheless, the bound on the tracking error may be kept arbitrarily small by increasing the gain K_v. Therefore, for the class of systems, stability of the closed-loop system is shown in the sense of Lyapunov without making any assumptions on the initial weight values. We may simply select $\hat{Z}(0) = \mathbf{0}$. (See Table 2 for a summary of the design rules.)

Example 6.1

Let us illustrate the stable NN controller design on a van der Pol system
$$\begin{aligned} \dot{x}_1 &= x_2, \\ \dot{x}_2 &= (1 - x_1^2)x_2 - x_1 + u, \end{aligned} \qquad (87)$$
which is in the Brunovsky canonical form. Note that (87) has an unstable equilibrium point at the origin $\mathbf{x} = (0,0)$ and a stable limit cycle. A typical trajectory for this system is illustrated in Figure 11.

The neural net that is used for estimation of $f(x_1, x_2) = (1 - x_1^2)x_2 - x_1$

consists of 10 neurons. Design parameters are set to $K_v = 20$, $\Lambda = 5$, $K_z = 10$, $Z_m = 1$, $M = N = 20$, and $\kappa = 1$. Initial conditions are $\hat{Z}(0) = 0$ and $x_1 = x_2 = 1$. The desired trajectory is defined as $y_d(t) = \sin t$. Actual and desired outputs are shown in Figures 12 and 13. Recall that the dynamic model (87) has not been used to implement the NN-based control of Theorem 6.1. The control input is illustrated in Figure 14.

7 Neural Network Control with Discrete-Time Tuning

In Section 4 we designed a robot arm neural net controller and in Section 6 an NN controller for a fairly general class of nonlinear systems. We gave algorithms for tuning the NN weights in continuous time; the algorithms in those sections are virtually identical. However, it is often more convenient to implement control systems in discrete time. Therefore, in this section we present discrete-time NN weight-tuning algorithms for digital control purposes. This will also provide a connection to the usual form of tuning algorithms based on the delta rule as used by the NN community.

The notation is similar to that in previous sections, but variables are now functions of the time index k. Though the development follows that in Section 6, the derivation and proofs for the control algorithm are more complex, as is usual for discrete-time analysis.

The approach in this section is unusual even from the point of view of linear adaptive control for discrete-time systems. This is because for

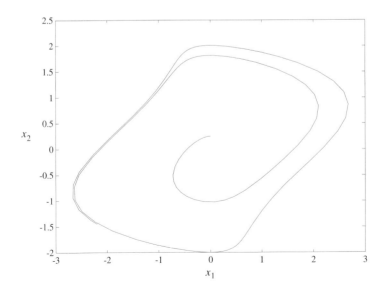

FIGURE 11. State trajectory of the van der Pol system.

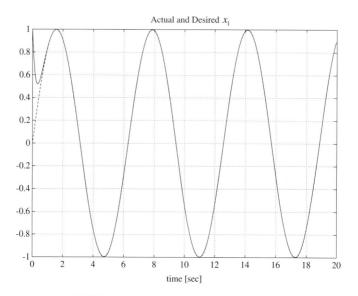

FIGURE 12. Actual and desired state x_1.

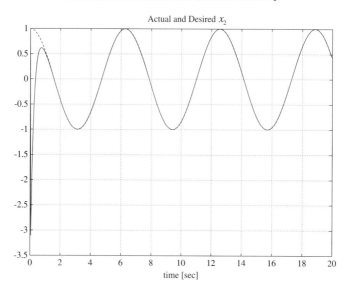

FIGURE 13. Actual and desired state x_2.

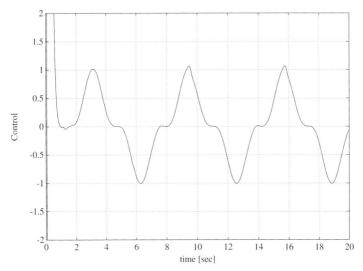

FIGURE 14. Control Input.

adaptive control of discrete-time systems, it is usual to design a controller that requires an estimate of some unknown function. Then one makes two assumptions: "linearity in the parameters" and "certainty equivalence." According to the former, a parameter vector is extracted from the functional estimate that is tuned using a derived algorithm. According to the latter, one uses the resulting estimate for the function in the control law. A third assumption of "persistence of excitation" is needed to show the boundedness of the parameter estimation errors.

Unfortunately, a great deal of extra analysis is needed to show that both the tracking error and the parameter estimation error are bounded (e.g., the so called "averaging methods"). In contrast, our approach selects a Lyapunov function containing both the tracking error and the functional estimation error, so that closed-loop performance is guaranteed from the start. It is a key factor that our work requires none of the usual assumptions of linearity in the parameters, certainty equivalence, or persistence of excitation. As such, this NN controller may be considered as a nonlinear adaptive controller for discrete-time systems.

7.1 A Class of Discrete-Time Nonlinear Systems

Consider an mnth order multi-input and multi-output discrete-time nonlinear system to be controlled, given in the form

$$x_1(k+1) = x_k,$$
$$\vdots$$
$$x_{n-1}(k+1) = x_{n(k)},$$

$$x_n(k+1) = f(x(k)) + u(k) + d(k), \tag{88}$$

with state $x(k) = [x_1(k) \ \ldots \ x_n(k)]^T$, with each $x_i(k) \in \mathbb{R}^m$, $i = 1, \ldots, n$; control $u(k) \in \mathbb{R}^m$; $d(k) \in \mathbb{R}^m$ an unknown disturbance vector acting on the system at time instant k with a known constant upper bound d_M (i.e., $\|d(k)\| \leq d_M$) and $f(x(k))$ an unknown smooth function.

7.2 Tracking Problem

Given a desired trajectory and its delayed values, define the *tracking error* as
$$e_n(k) = x_n(k) - x_{nd}(k)$$
and the *filtered tracking error*, $r(k) \in \mathbb{R}^m$,
$$r(k) = e_n(k) + \lambda_1 e_{n-1}(k) + \ldots + \lambda_{n-1} e_1(k), \tag{89}$$
where $e_{n-1}(k), \ldots, e_1(k)$ are the delayed values of the error, and $\lambda_1 \ldots \lambda_{n-1}$ are constant matrices selected such that $\det(z^{n-1} + \lambda_1 z^{n-2} + \ldots + \lambda_{n-1})$ is stable. Equation (89) can be further expressed as
$$r(k+1) = e_n(k+1) + \lambda_1 e_{n-1}(k+1) + \ldots + \lambda_{n-1} e_1(k+1). \tag{90}$$

Using (88) in (90), the dynamics of the MIMO system can be written in terms of the filtered tracking error as

$$r(k+1) = f(x(k)) - x_{nd}(k+1) + \lambda_1 e_n(k) + \ldots + \lambda_{n-1} e_2(k) + u(k) + d(k). \tag{91}$$

Define the control input $u(k)$ as

$$u(k) = k_v r(k) - \hat{f}(x(k)) + x_{nd}(k+1) - \lambda_1 e_n(k) - \ldots - \lambda_{n-1} e_2(k), \tag{92}$$

with the diagonal gain matrix $k_v > 0$ and $\hat{f}(x(k))$ an estimate of $f(x(k))$. Then the closed-loop error system becomes

$$r(k+1) = k_v r(k) + \tilde{f}(x(k)) + d(k), \tag{93}$$

where the *functional estimation error* is given by

$$\tilde{f}(x(k)) = f(x(k)) - \hat{f}(x(k)).$$

This is an error system wherein the filtered tracking error is driven by the functional estimation error.

In the remainder of this chapter, Equation (93) is used to focus on selecting NN tuning algorithms that guarantee the stability of the filtered tracking error $r(k)$. Since (89) (with the input considered as $r(k)$ and the output as $e(k)$) describes a stable system, standard techniques [SB89] guarantee that $e(k)$ exhibits stable behavior.

7.3 Neural Net Controller Design

Approaches such as σ-modification [PS91] or ϵ-modification [Nar91] are available for the robust adaptive control of continuous systems wherein a persistency of excitation condition is not needed. However, modification of the standard weight-tuning mechanisms in discrete-time to avoid a PE-like condition is, to our knowledge, yet to be investigated. In this section an approach similar to σ- or ϵ-modification is derived for discrete-time adaptive control of dynamical systems. Then it is applied to nonlinear NN tuning.

Assume that there exist some constant ideal weights W and V for a 3-layer NN (Figure 1), so that the nonlinear function in (88) can be written as

$$f(x(k)) = W^T \phi(\tilde{V}^T \phi(x(k)) + \varepsilon,$$

where the NN reconstruction error $\varepsilon(k)$ satisfies $\|\varepsilon(k)\| \leq \varepsilon_N$, with the bounding constant ε_N known. *It is needed to know only the existence of such ideal weights; their actual values are not required.* For notational convenience define the matrix of all the ideal weights as

$$Z = \begin{bmatrix} W & 0 \\ 0 & V \end{bmatrix}.$$

The bounding assumption provided in Section 4.1 is needed on the ideal weights, with the bound on $\|Z\|$ denoted in this section as Z_M.

Structure of the NN Controller and Error System Dynamics

Now suppose the estimate for $f(x(k))$ is provided by an NN, so that the NN functional estimate is

$$\hat{f}(x(k)) = \hat{W}^T(k) \phi(\hat{V}^T(k) \phi(x(k))),$$

with $\hat{W}$ and $\hat{V}$ the current values of the weights given by the tuning algorithms. The vector of input layer activation functions is given by $\hat{\phi}_1(k) \triangleq \phi_1(k) = \phi(x(k))$. Then the vector of activation functions of the hidden layer with the actual weights at the instant k is denoted by

$$\hat{\phi}_2(k) \triangleq \phi(\hat{V}^T(k) \phi(x(k))).$$

Fact 7 The usual activation functions, such as tanh, RBF, and logistic functions, are bounded by known positive values, so that

$$\|\phi_1(k)\| \leq \phi_{1\,\max} \text{ and } \|\phi_2(k)\| \leq \phi_{2\,\max}.$$

□

The error in the weights, or *weight estimation errors*, are defined by

$$\tilde{W}(k) = W - \hat{W}(k), \quad \tilde{V}(k) = V - \hat{V}(k), \quad \tilde{Z}(k) = \hat{Z}(k),$$

where

$$\hat{Z}(k) = \begin{bmatrix} \hat{W} & 0 \\ 0 & \hat{V} \end{bmatrix}, \qquad (94)$$

and the *hidden-layer output errors* are defined as

$$\tilde{\phi}_2(k) = \phi_2(k) - \hat{\phi}_2(k).$$

Now the control input (92) is

$$u(k) = x_{nd}(k+1) - \hat{W}^T(k)\hat{\phi}_2(k) - \lambda_1 e_n(k) - \ldots - \lambda_{n-1} e_2(k) + k_v r(k).$$

The closed-loop filtered error dynamics become

$$r(k+1) = k_v r(k) + \bar{e}_i(k) + W^T(k)\tilde{\phi}_2(k) + \varepsilon(k) + d(k), \qquad (95)$$

where the *identification error* is defined by

$$\bar{e}_i(k) = \tilde{W}^T(k)\hat{\phi}_2(k).$$

The proposed NN controller structure is shown in Figure 15. The output of the plant is processed through a series of delays in order to obtain the past values of the output and fed into the NN so that the nonlinear function in (88) can be suitably approximated. Thus, the NN controller derived in a straightforward manner using a filtered error notion naturally provides a dynamical NN control structure. Note that neither the input $u(k)$ nor its past values are needed by the NN. The next step is to determine the weight-tuning updates so that the tracking performance of the closed-loop filtered error dynamics is guaranteed.

7.4 Weight Updates for Guaranteed Tracking Performance

A novel NN weight-tuning paradigm that guarantees the stability of the closed-loop system (95) is presented in this section. It is required to demonstrate that the tracking error $r(k)$ is suitably small and that the NN weights $\hat{W}$ and $\hat{V}$ remain bounded, for then the control $u(k)$ is bounded. The upcoming theorem shows a tuning algorithm that guarantees the performance in this case of a multilayer NN.

The theorem relies on the extension to Lyapunov theory for dynamical systems given as Theorem 1.5-6 in [LAD93]. The nonlinearity $f(x)$, the bounded disturbance $d(k)$, and the NN reconstruction error $\varepsilon(k)$ make it impossible to show that the first difference for a Lyapunov function is nonpositive for all values of $r(k)$ and weight values. In fact, it is only possible to show that the first difference is negative outside a compact set in the state space, that is, if either $\|r(k)\|$ or $\|\tilde{Z}(k)\|$ is above some specific bound. Therefore, if either norm increases too much, the Lyapunov function decreases, so that both norms decrease as well. If both norms are small, nothing may be said about the first difference of the Lyapunov function except that it is probably positive, so that the Lyapunov function increases. This has the effect of making the boundary of a compact set an attractive region for the closed-loop system. This, however allows one to conclude the boundedness of the output tracking error and the neural net weights.

7. Neural Network Control of Robot Arms

Theorem 7.1 Let the reference input $r(k)$ be bounded and the NN functional reconstruction error and the disturbance bounds, ε_N, d_M, respectively, be known constants. Let the weight tuning for the input and hidden layers be provided as

$$\hat{V}(k+1) = \hat{V}(k) - \alpha_1 \hat{\phi}_1(k)[\hat{y}_1(k) + B_1 k_v r(k)]^T - \Gamma \|I - \alpha \hat{\phi}_1(k)\hat{\phi}_1^T(k)\|\hat{V}^T(k), \quad (96)$$

where $\hat{y}_1(k) = \hat{V}^T(k)\hat{\phi}_1(k)$ and B_1 is a constant design-parameter matrix. Let the weight tuning for the output layer be given by

$$\hat{W}(k+1) = \hat{W}(k) - \alpha_2 \hat{\phi}_2(k) r^T(k+1) - \Gamma \|I - \alpha_2 \hat{\phi}_2(k)\hat{\phi}_2^T(k)\|\hat{W}(k). \quad (97)$$

In both of these $\Gamma > 0$ is a design parameter. Then the tracking error $r(k)$ and the NN weight estimates $\hat{W}$ and $\hat{V}$ are uniformly ultimately bounded provided the following conditions hold:

(1) $\quad \alpha_1 \phi_{1\,\text{max}}^2 < 2,$
$\quad\quad\;\; \alpha_2 \phi_{2\,\text{max}}^2 < 1,$ (98)

(2) $\quad 0 < \Gamma < 1,$ (99)

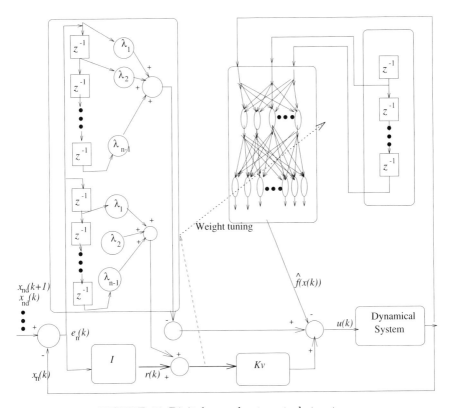

FIGURE 15. Digital neural net control structure.

$$(3) \quad k_{v\,\max} < \frac{1}{\sqrt{\bar{\sigma}}}, \qquad (100)$$

where $\bar{\sigma}$ is given by

$$\bar{\sigma} = \beta_1 + \kappa_1^2 \beta_2,$$

where κ_1 is a bound on B_1 such that $\|B_1\| \le \kappa_1$ and

$$\beta_1 = 1 + \alpha_2 \phi_{2\,\max}^2 + \frac{[-\alpha_2 \phi_{2\,\max}^2 + \Gamma(1 - \alpha_2 \phi_{2\,\max}^2)]^2}{1 - \alpha_2 \phi_{2\,\max}^2}, \qquad (101a)$$

$$\beta_2 = 1 + \alpha_1 \phi_{1\,\max}^2 + \frac{[\alpha_1 \phi_{1\,\max}^2 + \Gamma(1 - \alpha_1 \phi_{1\,\max}^2)]^2}{2 - \alpha_1 \phi_{1\,\max}^2}. \qquad (101b)$$

Proof: Select the Lyapunov function candidate

$$J = r^T(k)r(k) + \frac{1}{\alpha_1}\mathrm{tr}[\tilde{V}^T(k)\tilde{V}(k)] + \frac{1}{\alpha_2}\mathrm{tr}[\tilde{W}^T(k)\tilde{W}(k)], \qquad (102a)$$

whose first difference is given by

$$\Delta J = \Delta J_1 + \Delta J_2, \qquad (102b)$$

where

$$\Delta J_1 = r^T(k+1)r(k+1) - r^T(k)r(k) \qquad (102c)$$

and

$$\begin{aligned}\Delta J_2 &= \frac{1}{\alpha_2}\mathrm{tr}[\tilde{W}^T(k+1)\tilde{W}(k+1) - \tilde{W}^T(k)\tilde{W}(k)] \\ &\quad + \frac{1}{\alpha_1}\mathrm{tr}[\tilde{V}^T(k+1)\tilde{V}(k+1) - \tilde{V}^T(k)\tilde{V}(k)]. \end{aligned} \qquad (102d)$$

Using the tracking error dynamics (95), the term ΔJ_1 in (102c) is obtained as

$$\begin{aligned}\Delta J_1 &= \\ &-r^T(k)[1 - k_v^T k_v] + 2(k_v r(k))^T\big(\bar{e}_i(k) + W^T \tilde{\phi}_2(k) + \varepsilon(k) + d(k)\big) \\ &\times \bar{e}_i^T(k)\big(\bar{e}_i(k) + 2W^T \tilde{\phi}_2(k) + 2(\varepsilon(k) + d(k))\big) + \big(W^T \tilde{\phi}_2(k)\big)^T \\ &\times \big(W^T \tilde{\phi}_2(k) + 2(\varepsilon(k) + d(k))\big)(\varepsilon(k) + d(k))^T(\varepsilon(k) + d(k)). \end{aligned} \qquad (102e)$$

Considering the input and hidden (96), output (97) layer weight updates, using these in (102d) and combining with (102c), one may obtain

$$\Delta J \leq -(1 - \bar{\sigma}k_{v\,\text{max}}^2)\|r(k)\|^2 + 2\gamma k_{v\,\text{max}}\|r(k)\| + \rho$$
$$- \left\| \tilde{V}^T(k)\hat{\phi}_1(k) \frac{(1 - \alpha_1\hat{\phi}_1^T(k)\hat{\phi}_1(k)) - \Gamma\|I - \alpha_1\hat{\phi}_1(k)\hat{\phi}_1^T(k)\|}{2 - \alpha_1\hat{\phi}_1^T(k)\hat{\phi}_1(k)} \right.$$
$$\left. \times \left(V^T\hat{\phi}_1(k) + B_1 k_v r(k)\right) \right\|^2 [2 - \alpha_1\hat{\phi}_1^T(k)\hat{\phi}_1(k)]$$
$$- \left\| \bar{e}_i(k) - \frac{\alpha_2\hat{\phi}_2^T(k)\hat{\phi}_2(k) + \Gamma\|I - \alpha_2\hat{\phi}_2(k)\hat{\phi}_2^T(k)\|}{1 - \alpha_2\hat{\phi}_2^T(k)\hat{\phi}_2(k)} \right.$$
$$\left. \times \left(k_v r(k) + W^T\hat{\phi}_2(k) + \varepsilon(k) + d(k)\right) \right\|^2 [1 - \alpha_2\hat{\phi}_2^T(k)\hat{\phi}_2(k)]$$
$$+ \frac{1}{\alpha_1}\|I - \alpha_1\hat{\phi}_1^T(k)\hat{\phi}_1(k)\|^2 \,\text{tr}[\Gamma^2 \hat{V}^T(k)\hat{V}(k) + 2\Gamma \hat{V}^T \hat{V}(k)]$$
$$+ \frac{1}{\alpha_2}\|I - \alpha_2\hat{\phi}_2^T(k)\hat{\phi}_2(k)\|^2 \,\text{tr}[\Gamma^2 \hat{W}^T(k)\hat{W}(k) + 2\Gamma \hat{W}^T \hat{W}(k)], \tag{103}$$

where

$$\gamma = \beta_1(W_{\text{max}}\tilde{\phi}_{2\,\text{max}} + \varepsilon_N + d_M + \Gamma(1 - \alpha_2)\phi_{2\,\text{max}}^2)\phi_{2\,\text{max}}W_{\text{max}}$$
$$+ \kappa_1(\beta_2 + \Gamma(1 - \alpha_1\phi_{1\,\text{max}}^2)\phi_{1\,\text{max}}V_{\text{max}}$$

and

$$\rho = [\beta_1(W_{\text{max}}\tilde{\phi}_{2\,\text{max}} + \varepsilon_N + d_M) + 2\Gamma(1 - \alpha_2\phi_{2\,\text{max}}^2)\|\phi_{2\,\text{max}}W_{\text{max}}]$$
$$\times (W_{\text{max}}\tilde{\phi}_{2\,\text{max}} + \varepsilon_N + d_M)[(\beta_2 + \Gamma(1 - \alpha_1\phi_{1\,\text{max}}^2)\phi_{1\,\text{max}}^2 V_{\text{max}}^2].$$

Completing the squares for $\|\tilde{Z}\|(k)$ in (103) yields $\Delta J \leq 0$ as long as the conditions in (98) through (100) are satisfied and the tracking error given is larger than

$$\|r(k)\| > \frac{1}{(1 - \bar{\sigma}k_{v\,\text{max}}^2)} \left(\gamma k_{v\,\text{max}} + \sqrt{\gamma^2 k_{v\,\text{max}}^2 + (\rho + \frac{\Gamma}{(2 - \Gamma)}Z_M^2)(1 - \bar{\sigma}k_{v\,\text{max}}^2)} \right). \tag{104}$$

On the other hand, completing the squares for $\|r(k)\|$ in (103) results in $\Delta J < 0$ as long as the conditions (98) through (100) are satisfied and

$$\|\tilde{Z}(k)\| > \frac{1}{\Gamma(2 - \Gamma)} \left(\Gamma(1 - \Gamma)Z_M + \sqrt{\Gamma^2(1 - \Gamma)^2 Z_M^2 + \Gamma(2 - \Gamma)\theta} \right), \tag{105}$$

where

$$\theta = \Gamma^2 Z_M^2 + \frac{\Gamma^2 k_{v\,\text{max}}^2}{1 - \bar{\sigma}k_{v\,\text{max}}^2} + \rho.$$

Therefore $\Delta J \leq 0$ as long as (98) through (100) are satisfied outside a compact set (i.e., either (104) or (105) holds). In other words, if the right-hand sides of (104) and (105) are denoted by two constants δ_1 and δ_2 respectively, then $\Delta J \leq 0$ whenever $\|r(k)\| > \delta_1$ or $\|\tilde{Z}(k)\| > \delta_2$. Let us represent $(\|r(k)\|, \|\tilde{Z}(k)\|)$ in a new coordinate system $(\vartheta_1, \vartheta_2)$. Define the region
$$D : \{\vartheta | \vartheta_1 < \delta_1, \vartheta_2 < \delta_2\}.$$
Then there exists an open set
$$\Omega : \{\vartheta | \vartheta_1 < \bar{\delta}_1, \vartheta_2 < \bar{\delta}_2\},$$
where $\bar{\delta}_i > \delta_i$ implies that $D \subset \Omega$. This further implies that the Lyapunov function J will stay in the region Ω, which is an invariant set. Therefore, from (104) or (105), it can be concluded that the Lyapunov function decreases outside a compact set, so that the tracking error $r(k)$ and the error in weight estimates are UUB. □

Remarks: For practical purposes, (104) (105) can be considered as bounds for $\|r(k)\|$ and $\|\tilde{Z}(k)\|$.

The complete discrete-time NN controller is given in Table 3 and shown in Figure 15. The NN reconstruction error bound ε_N and the bounded disturbances d_M affect the bounds on $\|r(k)\|$ and $\|\tilde{Z}(k)\|$ in a very interesting way. Note that small tracking error bounds may be achieved by placing the closed-loop poles inside the unit circle and near the origin through the selection of the largest eigenvalue, $k_{v\,\max}$, of k_v. On the other hand, the NN weight error estimates are fundamentally bounded by Z_M, the known bound on the ideal weights Z. The parameter Γ offers a design tradeoff between the relative eventual magnitudes of $\|r(k)\|$ and $\|\tilde{Z}(k)\|$; a smaller Γ yields a smaller $\|r(k)\|$ and a larger $\|\tilde{Z}(k)\|$, and vice versa.

It is important to note that the *problem of initializing the net weights* (referred to as symmetry breaking [RC95]) occurring in other techniques in the literature does not arise, since when $W(0)$ and $V(0)$ are taken as zero, the PD term (i.e., the outer loop in Figure 15) stabilizes the plant, on an interim basis, for instance in a certain restricted class of nonlinear systems. Thus, the NN controller requires *no off-line learning phase.* In other words, this algorithm exhibits a learning-while-functioning feature instead of learning-then-control.

The tuning algorithms (96) and (97) are similar to the delta rule, but with a "forgetting term" added. It can be shown [JL96] that if the forgetting term is omitted, the performance of the delta rule cannot be guaranteed without a stringent persistency of excitation condition.

7.5 Projection Algorithm

The theorem reveals that the NN tuning mechanisms given have a major drawback, shared by delta-rule tuning algorithms in the literature. Since

TABLE 3. Discrete-time neural net controller.

NN controller:

$$u(k) = x_{nd}(k+1) - \tilde{W}^T(k)\hat{\phi}_2(k) - \lambda_1 e_n(k) - \ldots - \lambda_{n-1}e_2(k) + k_v r(k)$$

NN weight tuning:

$$\hat{V}(k+1)$$
$$= \hat{V}(k) - \alpha_1 \hat{\phi}_1(k)[\hat{y}_1(k) + B_1 k_v r(k)]^T - \Gamma\|I - \alpha_1 \hat{\phi}_1(k)\hat{\phi}_1^T(k)\|\hat{V}(k)$$
$$\hat{W}(k+1)$$
$$= \hat{W}(k) - \alpha_2 \hat{\phi}_2(k) r^T(k+1) - \Gamma\|I - \alpha_2 \hat{\phi}_2(k)\hat{\phi}_2^T(k)\|\hat{W}(k)$$

Signals:

$e_n(k) = x_n(k) - x_{nd}(k)$, tracking error
$r(k) = e_n(k) + \lambda_1 e_{n-1}(k) + \ldots + \lambda_{n-1}e_1(k)$, filtered tracking error

with $\lambda_1, \ldots, \lambda_{n-1}$ constant matrices
selected so that $\det(z^{n-1} + \lambda_1 z^{n-2} + \ldots + \lambda_{n-1})$ is stable.

Design parameters:

k_v, gain matrix, positive definite.
B_1, a known parameter matrix.
Z_M, a bound on the unknown target weight norms.
α_1 and α_2, scalar tuning rates.
$0 < \Gamma < 1$, a scalar.

$\hat{\phi}_i(k) \in \mathbb{R}^{N_{pi}}$, with N_p the number of hidden-layer neurons in the ith layer and the maximum value of each hidden-node output in the ith layer taken as unity (as for the logistic function), then the bounds on the adaptation gain in order to assure stability of the closed-loop system are in effect given by

$$0 < \alpha_1 < \frac{2}{N_p},$$
$$0 < \alpha_2 < \frac{1}{N_p}.$$

In other words, the upper bound on the adaptation gain at each layer decreases with an increase in the number of hidden-layer nodes in that particular layer, so that learning must slow down as the NN gets larger for guaranteed performance. This behavior has been noted in the literature [MSW91] using the delta rule for tuning but has never to our knowledge been explained.

This major drawback can be easily overcome by modifying the update rule at each layer using a projection algorithm [GS84]. To wit, replace the constant adaptation gain at each layer by

$$\alpha_i = \frac{\xi_i}{\zeta_i + \|\hat{\phi}(k)\|^2}, \quad i = 1, 2, \tag{106}$$

where $\zeta_i > 0$, $i = 1, 2$, and $0 < \xi_i < 2$, $i = 1, 2$, are constants. Note that the ξ_i are now the new adaptation gains at each layer, and it is always true that

$$\frac{\xi_i}{\zeta_i + \|\hat{\phi}_i(k)\|^2}\|\hat{\phi}(k)\|^2 \begin{cases} < 2, & i = 1, \\ < 1, & i = 2, \end{cases} \tag{107}$$

thus guaranteeing (97) for every N_p at each layer.

From the bounds indicated for the adaptation gains in both (98) and (107), it is interesting to note that the *upper bound on the adaptation gains for the input and hidden layers is 2, whereas for the output layer the upper bound is given by 1*. It appears that the *hidden layers act as pattern extractors* [RC95]. In other words, the *hidden layers of a multilayer NN are employed for the identification of the nonlinear plant, and the output layer is used for controlling the plant*.

The weight-tuning paradigm for the discrete-time NN controller is based on the delta rule but includes a correction term. This discrete-time NN controller offers guaranteed performance without a persistency of excitation condition on internal signals. In addition, it was shown that the adaptation gains in the case of the given tuning mechanisms at each layer must decrease with an increase in the number of hidden-layer neurons in that layer, so that learning must *slow down for large NN*. The constant learning-rate parameters employed in these weight-tuning updates were modified using a projection algorithm, so that the learning rate is independent of the number of hidden-layer neurons.

Example 7.1: NN control of a Discrete-Time Nonlinear System

Consider the first-order multi-input multi-output discrete-time nonlinear system described by

$$\begin{bmatrix} x_1(k+1) \\ x_2(k+1) \end{bmatrix} = \begin{bmatrix} \frac{x_2(k)}{1+x_1^2(k)} \\ \frac{x_1(k)}{1+x_1^2(k)} \end{bmatrix} + \begin{bmatrix} u_1(k) \\ u_2(k) \end{bmatrix}. \tag{108}$$

The objective is to track a periodic step input of magnitude two units with a period of 30 sec.

The elements in the diagonal gain matrix were chosen as

$$k_v = \begin{bmatrix} 0.1 & 0 \\ 0 & 0.1 \end{bmatrix},$$

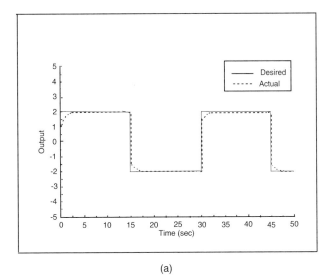

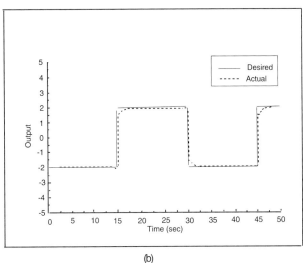

FIGURE 16. Tracking errors using discrete-time NN controller. (a) Output 1. (b) Output 2.

and a sampling interval of T= 10 msec was considered. A three-layer NN was selected with two input, six hidden and two output nodes. Sigmoidal activation functions were employed in all the nodes in the hidden layer. The initial conditions for the plant were chosen to be $[1, -1]^T$. The weights were initialized to zero with an initial threshold value of 3.0. No learning was performed initially to train the network.

The response of the NN controller with weight tuning in (96) and (97) with projection algorithm (106) is illustrated in Figure 16. The design pa-

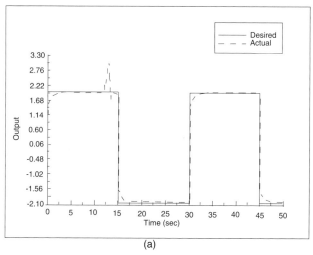

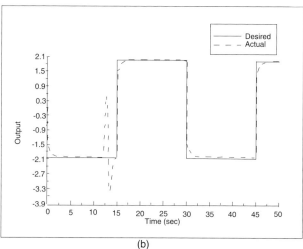

FIGURE 17. Tracking errors with a disturbance input: (a) output 1, (b) output 2.

rameters were selected as $\xi_1 = 1.0$, $\xi_2 = 0.7$ with ζ_1, ζ_2 chosen as 0.001. The parameter Γ was selected as 0.01. All the elements of the design parameter matrix B_1 were chosen to be 0.1. Note that with the weight-tuning algorithms given here, the weights are guaranteed to be bounded without the need of a PE condition.

Let us consider the case when a bounded disturbance given by

$$d(k) = \begin{cases} 0.0, & 0 \leq kT < 12, \\ 0.5, & kT \geq 12 \end{cases} \qquad (109)$$

is acting on the plant at the time instant k. Figure 17 presents the tracking response using the NN controller with the projection algorithm. It can be seen from the figure that the bounded disturbance induces bounded tracking errors at the output of the plant, but the performance is satisfactory. Smaller tracking errors can be achieved by modifying k_v.

8 Conclusion

Continuous-time and digital neural network (NN) controllers were designed that give guaranteed closed-loop performance in terms of small tracking errors and bounded controls. New NN properties such as strict passivity avoided the need for persistence of excitation. New NN controller structures avoided the need for preliminary off-line learning, so that the NN weights are easily initialized and the NN learns on-line in real time. No regression matrix need be found, in contrast to adaptive control. No certainty equivalence assumption is needed, as Lyapunov proofs guarantee simultaneously that both tracking errors and weight estimation errors are bounded.

To guarantee performance and stability in closed-loop control applications using multilayer (nonlinear) NNs it was found that the standard delta rule does not suffice, but that the NN tuning rules must be modified with extra terms.

Our primary application was NNs for control of rigid robotic manipulators, though a section on nonlinear system control shows how the technique can be generalized to other classes of systems in a straightforward manner.

9 REFERENCES

[AS92] F. Albertini and E. D. Sontag. For neural nets, function determines form. In *Proceedings of the 31st IEEE Conference on Decision and Control*, pages 26–31, December 1992.

[Bar64] R. G. Bartle. *The Elements of Real Analysis*. Wiley, New York, 1964.

[CK92] F.-C. Chen and H. K. Khalil. Adaptive control of nonlinear systems using neural networks. *International Journal of Control*, 55(6):1299–1317, 1992.

[Cra88] J. J. Craig. *Adaptive Control of Robot Manipulators*. Addison-Wesley, Reading, Massachusetts, 1988.

[CS92] S. R. Chu and R. Shoureshi. Neural-based adaptive nonlinear system identification. In *Intelligent Control Systems, ASME Winter Annual Meeting*, volume DSC 45, 1992.

[CS93] X. Cui and K. G. Shin. Direct control and coordination using neural networks. *IEEE Transactions on Systems, Man, and Cybernetics*, 23(3), May/June 1993.

[Cyb89] G. Cybenko. Approximation by superpositions of a sigmoidal function. *Mathematics of Control. Signals and Systems*, 2(4):303–314, 1989.

[DQLD90] D. M. Dawson, Z. Qu, F. L. Lewis, and J. F. Dorsey. Robust control for the tracking of robot motion. *International Journal of Control*, 52(3):581–595, 1990.

[Goo91] G. C. Goodwin. Can we identify adaptive control? In *Proceedings of the European Control Conference*, pages 1714–1725, July 1991.

[GS84] G. C. Goodwin and K. S. Sin. *Adaptive Filtering, Prediction, and Control*. Prentice-Hall, Englewood Cliffs, New Jersey, 1984.

[HHA92] B. Horn, D. Hush, and C. Abdallah. The state space recurrent neural network for robot identification. In *Advanced Control Issues for Robot Manipulators, ASME Winter Annual Meeting*, DSC-volume 39, 1992.

[HSW89] K. Hornik, M. Stinchcombe, and H. White. Multilayer feedforward networks are universal approximators. *Neural Networks*, 2:359–366, 1989.

[IST91] Y. Iiguni, H. Sakai, and H. Tokumaru. A nonlinear regulator design in the presence of system uncertainties using multilayer neural networks. *IEEE Transactions on Neural Networks*, 2(4):410–417, July 1991.

[JL96] S. Jagannathan and F. L. Lewis. Multilayer discrete-time neural net controller with guaranteed performance. *IEEE Transactions on Neural Networks*, 7(1):107–130, January 1996.

[KC91] L. G. Kraft and D. P. Campagna. A summary comparison of cmac neural network and traditional adaptive control systems. In W. T. Miller, R. S. Sutton, and P. J. Werbos, editors, *Neural Networks for Control*, pages 143–169. MIT Press, Cambridge, Massachusetts, 1991.

[LAD93] F. L. Lewis, C. T. Abdallah, and D. M. Dawson. *Control of Robot Manipulators*. Macmillan, New York, 1993.

[Lan79] Y. D. Landau. *Adaptive Control: The Model Reference Approach*. Dekker, 1979.

[LC93] C.-C. Liu and F.-C. Chen. Adaptive control of nonlinear continuous-time systems using neural networks— general relative degree and mimo cases. *International Journal of Control*, 58(2):317–335, 1993.

[LL92]   K. Liu and F. L. Lewis. Robust control techniques for general dynamic systems. *Journal of Intelligence and Robotic Systems*, 6:33–49, 1992.

[LLY95]   F. L. Lewis, K. Liu, and A. Yeşildirek. Neural net robot controller with guaranteed tracking performance. *IEEE Transactions on Neural Networks*, 6(3):703–715, 1995.

[LYL96]   F. L. Lewis, A. Yeşildirek, and K. Liu. Multilayer neural net robot controller: Structure and stability proofs. *IEEE Transactions on Neural Networks*, 7(2):1–12, March 1996.

[MB92]   G. J. Mpitsos and R. M. Burton, Jr. Convergence and divergence in neural networks: Processing of chaos and biological analogy. *Neural Networks*, 5:605–625, 1992.

[MSW91]   W. T. Miller, R. S. Sutton, and P. J. Werbos, editors. *Neural Networks for Control*. MIT Press, Cambridge, 1991.

[NA87]   K. S. Narendra and A. M. Annaswamy. A new adaptive law for robust adaptation without persistent excitation. *IEEE Transactions on Automatic Control*, AC-32(2):134–145, February 1987.

[Nar91]   K. S. Narendra. Adaptive control using neural networks. In W. T. Miller, R. S. Sutton, and P. J. Werbos, editors, *Neural Networks for Control*, pages 115–142. MIT Press, Cambridge, Massachusetts, 1991.

[NP90]   K. S. Narendra and K. Parthasarathy. Identification and control of dynamical systems using neural networks. *IEEE Transactions on Neural Networks*, 1:4–27, March 1990.

[OSF+91] T. Ozaki, T. Suzuki, T. Furuhashi, S. Okuma, and Y. Uchikawa. Trajectory control of robotic manipulators. *IEEE Transactions on Industrial Electronics*, 38:195–202, June 1991.

[Pao89]   Y. H. Pao. *Adaptive Pattern Recognition and Neural Networks*. Addison-Wesley, Reading, Massachusetts, 1989.

[PG89]   T. Poggio and F. Girosi. A theory of networks for approximation and learning. Technical Report 1140, Artificial Intelligence Lab, MIT, Cambridge, Massachuetts, 1989.

[PI91]   M. M. Polycarpou and P. A. Ioannu. Identification and control using neural network models: Design and stability analysis. Technical Report 91-09-01, Dept. of Electrical Engineering Systems, University of Southern California, Los Angeles, September 1991.

[PI92] M. M. Polycarpou and P. A. Ioannu. Neural networks as on-line approximators of nonlinear systems. In *Proceedings of the 31st IEEE Conference on Decision and Control*, pages 7–12, Tucson, Arizona, Dec. 1992.

[PS91] J. Park and I. W. Sandberg. Universal approximation using radial-basis-function networks. *Neural Computation*, 3:246–257, 1991.

[RC95] G. A. Rovithakis and M. A. Christodoulou. Adaptive control of unknown plants using dynamical neural networks. *IEEE Transactions on Systems, Man and Cybernetics*, 24(3):400–412.

[RHW86] D. E. Rumelhart, G. E. Hinton, and R. J. Williams. Learning internal representations by error propagation. In D. E. Rumelhart and J. L. McClelland, editors, *Parallel Distributed Processing*. MIT Press, Cambridge, Massachusetts, 1986.

[Sad91] N. Sadegh. Nonlinear identification and control via neural networks. In *Control Systems with Inexact Dynamics Models*, DSC-volume 33, ASME Winter Annual Meeting, 1991. ASME, New York, 1991.

[SB89] S. S. Sastry and M. Bodson. *Adaptive Control*. Prentice-Hall, Englewood Cliffs, New Jersey, 1989.

[SL88] J.-J. E. Slotine and W. Li. Adaptive manipulator control: A case study. *IEEE Transactions on Automatic Control*, 33(11):995–1003, Nov. 1988.

[SL91] J.-J. E. Slotine and W. Li. *Applied Nonlinear Control*. Prentice-Hall, Englewood Cliffs, New Jersey, 1991.

[SS91] R. M. Sanner and J.-J. E. Slotine. Stable adaptive control and recursive identification using radial gaussian networks. In *Proceedings of the 30th IEEE Conference on Decision and Control*, Brighton, England, 1991.

[Sus92] H. J. Sussmann. Uniqueness of the weights for minimal feedforward nets with a given input–output map. *Neural Networks*, 5:589–593, 1992.

[Wer74] P. J. Werbos. *Beyond Regression: New Tools for Prediction and Analysis in the Behavioral Sciences*. Ph.D. Thesis, Committee on Applied Mathematics, Harvard University, Cambridge, Massachusetts,, 1974.

[Wer89]   P. J. Werbos. Back propagation: Past and future. In *Proceedings of the 1988 International Conference on Neural Networks*, volume 1, pages I343–I353. Lawrence Erlbaum, Hillsdale, New Jersey, 1989.

[YL95]   A. Yeşildirek and F. L. Lewis. Feedback linearization using neural networks. *Automatica*, 31(11):1659–1664, November 1995.

[YY92]   T. Yabuta and T. Yamada. Neural network controller characteristics with regard to adaptive control. *IEEE Transactions on Systems, Man and Cybernetics*, 22(1):170–176, January 1992.

Chapter 8

Neural Networks for Intelligent Sensors and Control — Practical Issues and Some Solutions

S. Joe Qin

> ABSTRACT Multilayer neural networks have been successfully applied as intelligent sensors for process modeling and control. In this chapter, a few practical issues are discussed and some solutions are presented. Several biased regression approaches, including ridge regression, PCA, and PLS, are integrated with neural net training to reduce the prediction variance.

1 Introduction

The availability of process control computers and associated data historians makes it easy to generate neural network solutions for process modeling and control. Numerous applications of neural networks in the field of process engineering have been reported in recent annual meetings and technical journals. Neural network solutions are well accepted in process industries since they are cost-effective, easy to understand, nonlinear, and data-driven.

This chapter addresses several practical issues and some solutions regarding to use of neural networks for intelligent sensors and control. The chapter begins with an introduction to neural network applications as intelligent soft sensors to predict process variables that are not measurable online. Then several practical issues in the intelligent sensor applications are presented: (i) outlier detection and missing value treatment; (ii) variable selection; (iii) correlated input data versus network training, prediction, and control; and (iv) neural network training and validation. Approaches to integrating biased statistical methods with neural network training are discussed to handle correlated input data. The integration of neural networks with partial least squares is discussed and illustrated with a real process application. Last, conclusions are given and further issues are recommended for future investigation.

Artificial neural networks have found many applications in process modeling and control. These applications include: (i) building intelligent soft sensors to estimate variables that usually need to be measured through lab tests [M+89, WDM+91]; (ii) dynamic system identification [BMMW90, NP90]; (iii) fault detection and diagnosis [VVY90, UPK90]; (iv) process data screening and analysis [Kra92]; and (v) use of neural nets for control [HA90, HH92].

Among all types of neural networks, *multilayer feedforward networks* (MFN) [RHW86] have primarily been applied in process modeling and control. There are a number of reasons that can explain why MFNs are widely applied to process industries. First of all, multilayer feedforward networks are good nonlinear function approximators. A number of researchers (Hornik, Stinchcombe, and White [HSW89]; Cybenko [Cyb89]) have proved that an MFN can approximate any continuous function sufficiently well. Since many chemical processes are highly nonlinear, the nonlinear capability of neural nets is promising to process engineers. Second, a neural network can be trained to learn a chemical process by using historical process data. With plenty of process data available from distributed control systems in industrial processes, building a neural network based on process data is cost-effective. A third reason is the ease of use for process engineers. Building a neural network model does not necessarily require as much knowledge as the first principles approach or statistical approaches, although fundamental understanding of the process is required.

Owing to the availability of distributed control systems (DCS) and associated database historians in process industries, huge amounts of historical data exist that can be used for neural network applications without additional investment. Typically, the historical data have been poorly utilized in the past, although much useful information could be extracted from the data. This is one of the areas in process industries where neural networks can be applied without further investment in data acquisition. One such application is to predict key process variables that cannot be measured on-line. This type of application is known as soft sensor or intelligent sensor. The hard-to-measure variables are usually quality variables or those directly related to the economic interest of the production. These quality variables are often observed from analyzing product samples off-line in a laboratory. An obvious time delay is incurred in analyzing the test samples, which can be in the range of one to ten hours. Although one could know the product quality after this delay, it might be too late to make timely control adjustment if required. Figure 1 illustrates the intelligent sensor application using neural networks. A neural network is trained on historical data to predict process quality variables so that it can replace the lab-test procedure. An immediate benefit of building intelligent sensors is that the neural network can predict product quality in a timely manner. If the product quality does not meet the requirement and hence a correction is needed, the intelligent sensors allow early control actions to be made, which can avoid continuing manufacturing of poor quality product.

This chapter presents a few practical issues that are often encountered in

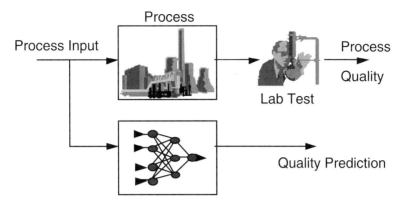

FIGURE 1. Use of neural networks as intelligent sensors to predict process quality variables that can only be measured through lab test.

using neural networks for intelligent soft sensors and control. These issues are: (i) data preprocessing, including missing values and outlier detection; (ii) variable selection; (iii) network training and control issues with correlated variables; and (iv) integrating statistical approaches and neural networks for intelligent sensors.

The organization of rest of the chapter is as follows. Section 2 discusses the typical characteristics of historical process data. Section 3 addresses the issues in data preprocessing, including outlier detection and missing value replacement. Section 4 presents the issue of variable selection in dealing with real applications. Sections 5 analyzes the effect of collinearity on neural network training, generalization, and control. Section 6 presents approaches to integrating neural networks with statistical methods, including partial least squares (PLS), principal component analysis (PCA), and ridge regression. Section 7 uses a real application example to illustrate how the integrated approach may outperform a direct neural network approach. The final section gives concluding remarks.

2 Characteristics of Process Data

Most industrial processes are well equipped with on-line process sensors, such as temperature, flowrate, and pressure sensors, and in some cases analytical sensors. These sensors allow the process computer to acquire on-line process information and make appropriate control to maintain consistent product quality. Most of the control strategies are feedback control, where PID control and model-based control are dominant. A data historian that collects and stores historical data is usually linked to the process computer. Some variables, especially the quality variables, do not have on-line sensors, or the available sensors are not cost-effective or reliable. However, since these variables are of crucial importance, lab tests of the product samples are usually conducted to measure the product quality off-line on a specified interval base.

In the situation where lab tests are conducted, a time delay of one to ten hours is often incurred. In order to detect the quality variables in a timely manner, one can build intelligent sensors to infer the quality variables from other on-line measured process variables. The foundation of building intelligent sensors is that the product quality variables have a functional relationship with other process variables that can be measured on-line, such as compositions of raw materials, temperature, pressure, and residence time during which the product is made. Since the functional relationship between the quality variables and other variables is usually nonlinear, the neural network approach is a convenient choice for modeling the relationship.

The neural network approach to building intelligent sensors is fundamentally an empirical approach based on process data. This approach is efficient when alternative approaches such as the first principles approach are difficult or too expensive to obtain. However, one of the difficulties involved in the neural network approach is that many issues are not well defined and have to be determined based on data analysis. These issues include data acquisition, outliers and missing values, input variable selection, and variable collinearity. Practical experience shows that resolving these issues takes much more time than training a neural network.

There are three categories of data that could be used to build intelligent sensors. The first category is historical data that are collected from on-line sensors during the process operation. Most of the available data are in this category. The nature of the historical data is known as *data-rich and information-poor* [PO91]. Usually, a large portion of the data is collected under a particular operating condition. Relatively little information is available under other operating conditions. In order to collect data that cover various types of operating conditions, one needs to use data that cover a long period of history. In the case where the process is slowly varying over time or where significant change was made in the process, old data may be obsolete and useless for building intelligent sensors for the current process. Therefore, a tradeoff is needed in data acquisition to get the greatest amount of truly representative data.

The second category of data is lab-test data that are collected for hard-to-measure quality variables. These data are often available with much larger time intervals than the historical data. Typically, the lab-test data are not collected at a regular rate. This is due to the fact that the lab test is done by a human being instead of a digital computer. Furthermore, the lab-test data often have missing data points. The sampling rate must be regularized with interpolation or other techniques. When there are missing values in the lab-test data, they should be either replaced or not used for training a neural network. Dayal et al.[DMT$^+$92] discuss a practical treatment of missing values in this regard.

The third category of data are experimental data that can be collected by conducting an experimental design. This category of data is most desirable in terms of the data quality, but it is often not available due to the high cost of obtaining it or because of safety considerations. Sometimes an experimental design may be allowed to collect a few samples as a complementary approach to the historical data. In this chapter, we focus on the case where the experimental data are generally not available for intelligent sensor modeling.

Building intelligent sensors based on historical and lab-test data is a passive approach, since no experimental designs are conducted. This approach has the least intervention to process operation. Therefore, it is generally applicable to all kinds of processes. However, due to the limited information content in the historical data, the resulting intelligent sensors are valid only on the particular region in which the data are collected. Therefore, it is important to identify the valid region associated with the intelligent sensor. It is necessary, though not sufficient, to check the lower and upper bounds for new data when a trained neural network is implemented on-line. When the process operation changes significantly, the intelligent sensor has to be recalibrated on new data in order to give a valid prediction.

3 Data Preprocessing

Before training a neural network, data preprocessing is normally required since most of the historical data and lab-test data are not ready to use for training. The first step in preprocessing is to identify and remove outliers, since industrial data bases typically have outliers. Outliers are treated in statistics as samples that carry high leverage [MN89] yet are too few as to be statistically meaningful. Outliers can result from sensor failure, misreading from lab tests, and other unknown upsets to the process. Some outliers can be normal data and represent important information, but additional knowledge is needed to distinguish beetween these and bad outliers. A distinctive feature of outliers is that they normally have extremely large influence on the model, since neural network training employs a least-squares type of objective function with individual training error

$$E_i = \sum_{j=1}^{p}(\hat{y}_{ij} - y_{ij})^2 \tag{1}$$

or cumulative training error

$$E = \frac{1}{N}\sum_{i=1}^{N} E_i, \tag{2}$$

which are generally sensitive to outliers with large model errors [Kos92]. Although it is felt that a neural network training scheme that includes both training and testing might stop training before overfitting the outliers, there is no guarantee that network training will ignore outliers and fit well on good samples. Such a scheme is even less reliable when there are outliers in the test set. Further, since neural networks can be extremely nonlinear, it is experienced that neural networks are more sensitive to outliers than linear regression methods because they can bend to reach the outliers. As a consequence, it is necessary to perform outlier detection and pretreatment before training the network.

3.1 Obvious Outliers

Some outliers are so obvious that they can be identified by using prior knowledge and physical laws. For example, a temperature variable cannot reach below absolute zero, and a flowrate variable cannot be negative. Generally, one can determine the possible maximum and minimum values of a process variable based on experience. Therefore, some outliers can be identified simply by checking them against the minimum and maximum. If a sample is below the minimum or above the maximum, it is considered an outlier. Although this method is very simple, it is useful for preliminary detection of outliers. Another reason to establish maximum and minimum values is to define the operating region of interest. If the data are outside the region of interest, they should be considered outliers.

After identifying the outliers, a number of approaches can be taken to treat them. First, an outlier can be replaced by the maximum or minimum value. Since we do not know the real value of the outlier, this approach is often not reliable. A conservative approach is to replace the outlier with the mean of the process variable. The third approach is to treat them as missing data, which will be discussed later.

3.2 Nonobvious Outliers

In contrast to the obvious outliers, many outliers lie within the boundaries and are hard to identify. They do not violate the minimum and maximum, but they can cause large model errors because they violate other physical constraints. For example, the pressure and flowrate are supposed to have a quadratic relationship by physical principles, but an outlier may violate this relation. In many cases, the process data are highly correlated. If some samples appear to be uncorrelated, they are considered outliers because they violate the correlation structure. These outliers can be detected using statistical methods such as principal component analysis [WEG87]) and partial least squares [MN89]).

Outliers can also be detected by examining the signals in the frequency domain. For example, a temperature variable has a certain frequency limit.

It cannot change too frequently because temperature change is a slow process. If the variable is identified as having impossibly high frequency components, it is considered as having outliers. A properly designed filter can be used to filter out the effect of outliers. In order to overcome phase lags introduced by using low-pass filters, Piovoso et al. [PO91] used finite impulse response (FIR) median hybrid filters to extract steady state information. The filtering approach is most effective when an entire variable is heavily corrupted with noise.

3.3 Robust Backpropagation

Despite the various outlier detection techniques applied to the data, there are usually some outliers left in the data set. Therefore, it is important to use some training methods that are insensitive to outliers. As indicated earlier, a distinct feature of outliers is that they carry large individual training errors. Since the regular backpropagation algorithm uses least squares training error, it is sensitive to large training errors and thus sensitive to outliers [Kos92]. Robust backpropagation that is insensitive to outliers can be developed by borrowing the techniques from robust regression methods. Instead of having a least squares training error that amplifies large training errors, one can use the following training error that treats large errors and small errors linearly:

$$E_i = \sum_{j=1}^{p} |\hat{y}_{ij} - y_{ij}|. \qquad (3)$$

This error function can reduce the influence of outliers that carry large individual errors. Another approach that uses suppressing functions is given in [Kos92].

3.4 Missing Values

Missing values are very common in historical data bases from distributed control systems and lab-test data. In general, it is difficult for multilayer neural networks to handle missing values during training. Therefore, it is necessary to preprocess missing values before training. There are a number of ways to handle missing values. One simple way is to delete samples that have missing values. However, if the available data are limited and the missing values are not clustered together, one can apply the techniques of missing data replacement to keep as many samples as possible.

When the missing values are not clustered together, it is reasonable to apply interpolation to replace the missing values. Dayal, et al. [DMT$^+$92] apply linear interpolation when there are fewer than three consecutive missing points. Of course, it is possible to apply other interpolation techniques such as cubic splines. However, when there are more consecutive missing values, it is not reliable to apply interpolation. More advanced techniques have to be applied.

As long as the missing values are evenly distributed over the data set, there are statistical methods such as principal component regression (PCR) and partial least squares (PLS) which can work with data sets that have missing values [GK86, WEG87]. These methods perform principal component or principal factor calculations and then build regression models based on them. They allow some "holes" in the data set. Therefore, one can apply principal component analysis on the training data with some missing values, then build a neural network model based on the principal components. One can handle the missing values in the calculation of principal components. One can also integrate the partial least squares with neural networks to handle the missing value problem. Another benefit of integrating these statistical methods with neural network training is to handle the correlated variables that can cause large variance on the prediction. Details of the integration of neural networks and partial least squares can be found in [QM92b].

If the process variables are highly correlated in a nonlinear manner, autoassociative neural networks can be used to recover the missing values [Kra92]. The autoassociative neural networks are special types of feedforward networks that have identical input and output variables. The autoassociative networks reconstruct the variables from other variables that are cross-correlated in a linear or nonlinear manner. This correlation offers fundamental redundancy for recovering missing values. It should be noted that to recover a variable that has a missing value, it must be highly correlated to other variables that do not have missing values. It is not possible to recover a missing value of a variable that has no correlation with other variables. An alternative approach to the autoassociative networks is the principal curve approach integrated with neural networks [DM94]. This approach provides a natural nonlinear extension to principal component analysis.

4 Variable Selection

An industrial database usually provides all the variables that can be recorded. However, it is not necessarily true that all recorded variables are relevant to the process variables to be predicted. Instead, it is often the case that when some of the irrelevant variables are deleted from the database, the modeling results can be improved; if the irrelevant variables are kept in the model, they play a role of noise, which can potentially deteriorate the modeling results. Therefore, it is imperative to select a subset of process variables that are truly relevant to the predicted variables. Prior knowledge can be used to screen out totally irrelevant variables, but further approaches are eventually needed to select relevant variables.

The issue of variable selection has been studied in regression analysis [Hoc76]. The more the variables are used in a model, the worse the prediction variance is and the better the prediction bias. Basically, each process variable to be used in a model contributes a value to the prediction with associated variance. When a variable's contribution is larger than its related variance, it is useful to the model. However, when a variable's contribution is smaller than its related variance, it could be harmful to the model.

Variable selection can be performed by judging a variable's contribution or relevance to the predicted variables. Two typical schemes used for variable selection are forward selection and backward elimination among all variables. Both the forward selection and backward elimination schemes can be used in neural network training. In a forward selection scheme, one starts the neural network with a small set of variables and adds more variables if they contribute to the prediction. In the backward elimination, the neural network starts with all available variables and then eliminates variables that are not relevant to the prediction. However, since neural network training is involved in each selection step, these variable selection schemes are quite laborious.

A sensitivity analysis approach is proposed in [QM92a] to select variables in a neural net PLS scheme. In this method, sensitivity analysis of model outputs with respect to model inputs is conducted over the operation region where data are collected. If the sensitivity of an output variable with respect to an input variable is very small, the input variable is deleted from the model because it has little contribution to the prediction. The sensitivity analysis provides an approach to identifying less sensitive variables. The sensitivity is a measure of a variable's significance in explaining the output variable. To determine whether a variable is truly contributing to the output variable, the cross-validation method is conducted and the prediction error on the test set is calculated before and after deleting the variable. If the prediction error is not increased after deleting the variable, it will be deleted from the network input; otherwise, it will be retained in the network input. With this method, one can achieve improved accuracy by deleting irrelevant variables and keeping relevant variables in the neural network model.

A related issue is to determine the process dynamics and time delays associated with the predicted outputs. Since the process data are collected over a history of normal operation, the data always contain information about dynamics and time delays. The dynamics and time delays can be included in the model by using a time window [QM92a]. The time window contains time delayed values of input variables and output variables. By treating these delayed variables as individual variables, the variable selection techniques can be used to determine the process dynamics and time delays.

There is usually high correlation or redundancy among the measured process variables. Some correlation is due to the fact that a group of variables affect each other. Correlation can also be due to insufficient variability during the normal process operation. In many situations, however, it is not desirable to delete variables that are correlated, since the correlation offers necessary redundancy for replacing missing values and for reducing gross errors [Kra92]. One example is that one physical process variable is measured by two sensors. These two sensor measurements are certainly correlated, and using both measurements can reduce the measurement noise. Therefore, it is desirable to include correlated variables as long as they are relevant to the predicted variables.

However, the presence of correlated variables brings up the problem of collinearity. In linear statistical regression, such as the ordinary least-squares approach, collinearity can cause the solution to be ill-conditioned and the resulting model to have large prediction variance. In this case, biased regression approaches such as principal component regression and partial least squares are often used. The next section discusses the effect of collinearity on neural network modeling.

5 Effect of Collinearity on Neural Network Training

5.1 Collinearity and Network Training

It is well known that collinearity presents an ill-conditioned problem to ordinary least squares in linear regression. Here the effect of collinearity on neural network training is demonstrated. Given an input vector x and an output variable y, a multilayer neural network with one hidden layer can be represented as follows:

$$y = \sum_{i=0}^{n} \nu_i s(\phi_i), \tag{4}$$

$$\phi_i = \sum_{j=0}^{m} w_{ij} x_j; \quad i = 1, 2, \ldots, n, \tag{5}$$

where ϕ_i ($i = 1, 2, \ldots, n$) is the ith hidden unit. The w_{ij} and ν_i are weights for the input layer and output layer, respectively; $s(\cdot)$ is a nonlinear (sigmoidal) function of a hidden unit; and $x_0 \equiv 1$ and $s(\phi_0) \equiv 1$ stand for the bias units for the input and hidden layers, respectively. Given a sequence of samples $(\mathbf{x}_1, y_1), (\mathbf{x}_2, y_2), \ldots, (\mathbf{x}_N, y_N)$, the neural network can be trained by minimizing the following error:

$$E = \frac{1}{N} \sum_{p=1}^{N} [y_p - f(\mathbf{x}_p; \mathbf{v}, \mathbf{W}))]^2, \tag{6}$$

where

$$\mathbf{W} \equiv [w_{ij}] \in \mathbb{R}^{n \times (m+1)}, \tag{7}$$
$$\mathbf{v} \equiv [\nu_0, \nu_1, \ldots, \nu_n]^T \in \mathbb{R}^{n+1} \tag{8}$$

represent the weights in (4) and (5).

If the input variables are collinear, there are many sets of weights that can minimize the error function in (6). For example, assuming that the input collinearity can be described by the equation

$$\mathbf{Ax} = \mathbf{0}, \tag{9}$$

where $\mathbf{A} \in \mathbb{R}^{r \times (m+1)}$, one has the following relation for any $\mathbf{B} \in \mathbb{R}^{n \times r}$:

$$f(\mathbf{x}; \mathbf{v}, \mathbf{W}) = f(\mathbf{x}; \mathbf{v}, \mathbf{W} + \mathbf{BA}). \tag{10}$$

In other words, if a set of weights $(\mathbf{v}, \mathbf{W})$ minimizes the error function in (6), $(\mathbf{v}, \mathbf{W} + \mathbf{BA})$ is also a solution to (6) for any $\mathbf{B}$. In practice, although the variables may not be exactly collinear as in (9), they are often highly correlated. If a backpropagation algorithm is used for training, significantly different weight distributions could result in little change in the training error. This phenomenon is an indication of collinearity.

It is known that neural network training does not have the ill-conditioned problem as does least squares; thus, one may think that collinearity is not a problem in neural network training. However, this is not true when prediction is made on new data. Since new data always include measurement noise, the model derived from backpropagation training often results in large prediction variance. In other words, neural networks trained by regular backpropagation tend to enlarge the noise variance in the presence of collinearity. This point is illustrated by analyzing the variance of prediction error in the following subsection.

5.2 Collinearity and Prediction Variance

To illustrate how collinearity affects prediction variance, the example in [MMK+91] is used here. This example considers an idealized process with one output variable and five input variables that are exactly collinear. The real process relation is assumed to be

$$y = 1.0x_2 + e. \tag{11}$$

The objective is to build a linear model of the five input variables and the output variable. Since the input variables are exactly collinear, it is obvious that an ordinary least squares approach yields an ill-conditioned problem. When the PLS method is used, the following model results:

$$(PLS): \quad y = 0.2x_1 + 0.2x_2 + 0.2x_3 + 0.2x_4 + 0.2x_5. \tag{12}$$

When a linear network model without hidden layers is built, three different models result from different initial conditions:

$$(NN1): \quad y = 0.63x_1 + 0.36x_2 + 0.09x_3 + 0.22x_4 - 0.30x_5,$$
$$(NN2): \quad y = -0.43x_1 - 0.05x_2 + 0.92x_3 - 0.02x_4 + 0.58x_5, \quad (13)$$
$$(NN3): \quad y = 0.23x_1 + 0.35x_2 - 0.26x_3 - 0.22x_4 + 0.91x_5.$$

These three models are adequate as long as their coefficients sum to 1.0.

Considering that new data for the five inputs have independent, identically distributed measurement noise with zero mean and variance, the prediction variance of the three neural network models and the PLS model can be calculated as follows:

$$\begin{aligned}
\text{Var}(y_{NN1}) &= 0.67\sigma^2, \\
\text{Var}(y_{NN2}) &= 1.37\sigma^2, \\
\text{Var}(y_{NN3}) &= 1.12\sigma^2, \\
\text{Var}(y_{PLS}) &= 0.200\sigma^2.
\end{aligned} \quad (14)$$

One can see that all the neural network models result in much larger prediction variances than the PLS model. Although the first neural net model reduces the variance, the other two models actually enlarge the variance. This demonstrates that backpropagation is sensitive to collinearity and results in a large prediction variance.

5.3 Control of Collinear Processes

Having collinearity in a process is an indication that the number of measurements is greater than the number of degrees of freedom or variability. Some collinear relations are inherent to the process, and these do not change under normal conditions. For example, a pump outlet pressure and flow rate are highly correlated. Changes in these collinear relations indicates that abnormal events occur in the process, which is useful for process monitoring and diagnosis [Wis91]. Other collinear relations are due to lack of variability in the data. This type of correlation can change when new variability occurs in the data.

A neural net model derived from collinear data is valid only when the correlation holds. Figure 2 depicts an example of two collinear inputs and one output. The correlated data cluster occurs approximately on a line in the (x_1, x_2) plane. When a neural net model is derived from these data, it only captures the functional relation in a subspace where x_1 and x_2 are correlated, since there are no data elsewhere for training. Therefore, the derived model is valid when the collinearity conforms. However, when new data do not follow the correlation, the model is no longer valid and needs to be updated.

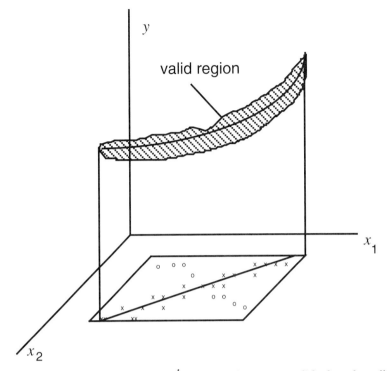

FIGURE 2. A model derived from collinear data is not valid when the collinearity changes. "x" — collinear data for modeling; "o" — new data when collinearity changes.

When a model from collinear data is used for control purposes, one has to comply with the validity of the model; that is, the variables have to be manipulated in such a way that the collinearity conforms. One cannot, for example, manipulate one variable freely and keep other variables unchanged. Another reason to change the collinearity is if a feedback loop is introduced. See [MMK+91] for further discussion. If such a model is used for inverse control, which calculates what the input values should be to achieve given output values, the inverted inputs have to follow the correlation structure to keep the model valid.

6 Integrating Neural Nets with Statistical Approaches

6.1 Modifying Backpropagation to Minimize Variance

Since there are many solutions that satisfy the training problem in the case of collinear data, we can use these extra degrees of freedom to minimize the variance. In order to minimize the output variance of the neural network given in (4) and (5), the variance of input to the hidden layer should be minimized first. Consider that each input is composed of a deterministic signal and a random noise, i.e.,

$$x_j = \bar{x}_j + \varepsilon_j, \quad (15)$$

where the measurement noise ε_j for each input variable is independent with zero mean and variance σ^2. The variance of the hidden layer inputs can be written as

$$\text{Var}(\phi_i) = \sum_{j=1}^{n} w_{ij}^2 \sigma^2; \quad i = 1, 2, \ldots, n. \quad (16)$$

To minimize the above variance while minimizing the network training error in (6), we can minimize the following error:

$$E_\lambda = E + \lambda \sum_{i=1}^{n} \sum_{j=1}^{m} w_{ij}^2 = E + \lambda \|\mathbf{W}\|_2^2, \quad (17)$$

where λ is an adjustable weighting parameter. Similarly, the output variance with respect to the output layer weights should also be minimized. Therefore, we have the following error function to minimize:

$$E_{\lambda\mu} = E + \frac{1}{2}\mu\|\mathbf{v}\|_2^2 = E + \frac{1}{2}(\mu\|\mathbf{v}\|_2^2 + \lambda\|\mathbf{W}\|_2^2), \quad (18)$$

where λ and μ are penalty factors for the magnitude of the network weights. Given many solutions that result in the same training error from (6), the above training error chooses the one with minimum norm of weights. The error in (18) can be minimized by gradient descent or conjugate gradient methods. Note that (18) is actually one application of the statistical technique known as ridge regression [Hoe70].

6.2 Example

To illustrate the effectiveness of variance reduction, the example given in the preceding section is reused here. Given that the five inputs are exactly collinear and with the constraint that the magnitude of the weights is minimized, it is easy to see that backpropagation with ridge regression results in the following relation:

$$y = 0.2x_1 + 0.2x_2 + 0.2x_3 + 0.2x_4 + 0.2x_5, \quad (19)$$

which gives a prediction variance of $0.2\sigma^2$.

Although ridge regression helps backpropagation reduce the prediction variance, it is difficult to determine the penalty factors. Practical process data may not be exactly collinear but rather highly correlated. In this case, the penalty factors induce bias in prediction while reducing the variance. The larger the penalty factors, the smaller the variance but the larger the bias. Therefore, it is important to choose the penalty factors so as to give the best model in a mean-square-error sense. Cross-validation may be used to determine the penalty factors.

6.3 Principal Components for Neural Network Modeling

Principal component analysis (PCA) has been used to remove collinearity in linear regression as principal component regression (PCR) [Jol86]. Here, the PCA is applied to remove collinearity for neural network training. To follow the notation of PCA and PLS, the input and output data are arranged into two data matrices, $\mathbf{X}$ and $\mathbf{Y}$, respectively. The basic idea of PCA is to transform the data matrix $\mathbf{X}$ into a matrix with fewer and orthogonal columns while keeping most of the variance in the data matrix $\mathbf{X}$, that is,

$$\mathbf{T} = \mathbf{XP}, \qquad (20)$$

where the columns of $\mathbf{T} \in \mathbb{R}^{m \times p}$, known as principal components, are orthogonal. Columns of $\mathbf{P}$, known as loading vectors, are orthonormal. The calculation of $\mathbf{T}$ and $\mathbf{P}$ is done by decomposing $\mathbf{X}$ into the following bilinear relation:

$$\mathbf{X} = \mathbf{TP}^T + \mathbf{R}, \qquad (21)$$

where the residual $\mathbf{R}$ is minimized during calculation.

When the residual $\mathbf{R}$ can be negligible, matrix $\mathbf{T}$ is completely representative of matrix $\mathbf{X}$. Therefore, a nonlinear relation between $\mathbf{X}$ and $\mathbf{Y}$ can be modeled in two steps: first calculating $\mathbf{T}$ and then building a neural network between $\mathbf{T}$ and $\mathbf{Y}$. The model structure can be depicted in Figure 3. The neural network training between the output variable and the principal components can be treated as a regular network training problem. When the trained model is used for prediction, it goes through a PCA calculation and a neural network calculation. The combined neural network and PCA scheme offers a viable approach to overcoming collinearity in neural network training. Since the principal components are orthogonal, they can be used for analyzing the features of the data and for monitoring process changes. A limitation of the neural net PCA approach is that PCA focuses only on the variance of inputs, which may ignore the inputs' correlation to the output. A component that is nonprincipal in PCA analysis can be significant in explaining the output. This situation can happen when some input variables carry a lot of variance but make little contribution to the output, and some other variables carry less variance but make significant contribution to the output. An integration of PLS with neural network training can overcome the limitation.

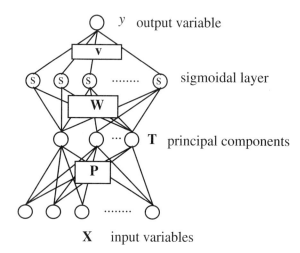

FIGURE 3. Use of principal component analysis for neural network modeling.

6.4 *A Neural Net PLS Approach*

The PLS approach decomposes both input and output data into bilinear terms as follows:

$$\mathbf{X} = \mathbf{t}_1\mathbf{p}_1^T + \mathbf{E}_1, \tag{22}$$
$$\mathbf{Y} = \mathbf{u}_1\mathbf{q}_1^T + \mathbf{F}_1, \tag{23}$$

where $\mathbf{t}_1$ and $\mathbf{u}_1$ are latent score vectors of the first PLS factor and $\mathbf{p}_1$ and $\mathbf{q}_1$ are corresponding loading vectors. These vectors are determined such that the residuals are minimized. The PLS decomposition is different from principal component analysis in that the correlation between $\mathbf{X}$ and $\mathbf{Y}$ is emphasized [GK86]. The above two equations formulate a PLS outer model. After the outer calculation, the score vectors are related by a linear inner model:

$$\mathbf{u}_1 = b_1\mathbf{t}_1 + \mathbf{r}_1, \tag{24}$$

where b_1 is a coefficient that is determined by minimizing the residual $\mathbf{r}_1$. After going through the above calculation, the residual matrices are calculated as

$$\mathbf{E}_1 = \mathbf{X} - \mathbf{t}_1\mathbf{p}_1^T \quad \text{for matrix } \mathbf{X}, \tag{25}$$
$$\mathbf{F}_1 = \mathbf{Y} - b_1\mathbf{t}_1\mathbf{q}_1^T \quad \text{for matrix } \mathbf{Y}. \tag{26}$$

Then the second factor is calculated by decomposing the residuals $\mathbf{E}_1$ and $\mathbf{F}_1$ using the same procedure as for the first factor. This procedure is repeated until the last (ath) factor is calculated, which leaves almost no information in the residual matrices $\mathbf{E}_h$ and $\mathbf{F}_h$.

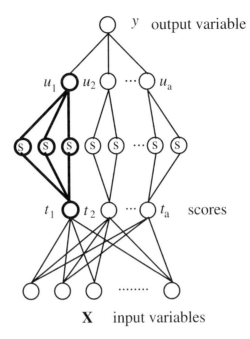

FIGURE 4. An illustrative structure of the neural net PLS approach. The input and output variables are projected onto latent space, and then the latent variables are modeled with neural networks.

When the process to be modeled has significant nonlinearity, the inner relationship between the score variables would be nonlinear. In this case, a neural net PLS (NNPLS) approach is proposed that uses neural networks as inner models while keeping the outer PLS transforms to remove collinearity [QM92b]. The framework of the integrated NNPLS method can be illustrated in Figure 4. The PLS outer transform is kept to generate score variables from the data. Then the scores ($\mathbf{u}_h$ and $\mathbf{t}_h$) are used to train the inner network models. Detailed discussion of the NNPLS algorithm can be found in [QM92b].

The NNPLS method differs from the direct network approach in that the data are not directly used to train the neural networks but are preprocessed by the PLS outer transform. The transform decomposes a multivariate regression problem into a number of univariate regressors. Each regressor is implemented by a neural network in this method. A direct benefit of doing so is that only one single-input single-output network is trained at a time. It can be shown that the score vectors $\mathbf{t}_h$, $h = 1, 2, \ldots, a$, of the NNPLS method are mutually orthogonal. As a result, the collinearity problem is removed.

7 Application to a Refinery Process

To illustrate the effectiveness of the neural net PLS method and a regular backpropagation network approach, a set of operating data from a catalytic reforming system is used. It is known that the process has five input variables and two output variables. Further, there is strong correlation among the five input variables. The data set consists of two batches; one batch has 149 samples, which are used for training; the other batch has 141 samples, which are used for testing. To include plant dynamics, the model uses several past output values and past input values, that is,

$$\mathbf{y}(t) = \mathbf{f}\big(\mathbf{y}(t-1), \mathbf{y}(t-2), \ldots, \mathbf{y}(t-n_y), \mathbf{u}(t), \mathbf{u}(t-1), \ldots, \mathbf{u}(t-n_u)\big). \quad (27)$$

The catalytic reformer data are modeled using the NNPLS approach. It is found that the time lags $n_y = 1$ and $n_u = 1$ are good choices for this application. It is also found with test-set validation that seven factors give the best prediction. The training and test root-mean-square errors (RMSEs) are given in Table 1. For comparison, the neural network approach with gradient-based training is applied to the same data set. The neural network has one sigmoidal hidden layer, and the output layer is linear. Under the same condition as in the NNPLS approach, the training error and the test error from backpropagation are listed in Table 1. It can be seen that although both approaches give similar training results, the NNPLS approach has better generalization results on the test data. The reason is that the neural net PLS approach is able to reduce variance by removing collinearity in the input data.

8 Conclusions and Recommendations

Owing to the availability of a vast amount of historical process data, neural networks have great potential in building models for process quality prediction and control. The use of neural networks as intelligent soft sensors can predict process variables that are not on-line measurable. An immediate benefit of this application is to have a timely estimation of process quality with a cost-effective approach, which allows early control of the process if its quality does not meet the desired requirements.

When the process data are usually highly correlated, a regular backpropagation training can result in large prediction variance under correlated inputs. This large prediction variance causes a large prediction mean-square

TABLE 1. Training and testing root-mean-square errors (RMSE) for a catalytic reformer using the NNPLS method and the neural network approach.

	Training RMSE	Testing RMSE
NNPLS	0.661	0.883
Neural Network	0.663	1.390

error. Several approaches to integrating neural networks with biased statistical methods, including ridge regression, principal component analysis, and partial least squares seem to be able to reduce the mean-square-error.

While there are many training algorithms available, several practical issues before and after training a neural network are the most time-consuming and deserve further study. Data preprocessing that includes outlier detection and missing value replacement is a typical problem in real applications. Advanced solutions are needed besides checking lower and upper bounds for outliers and interpolations for missing values. Since this topic has been long studied in statistics, it seems to be beneficial to combine the results in statistics with neural network training.

The issue of variable selection is very important to build a parsimonious neural network that gives minimal prediction errors. The sensitivity analysis approach discussed is adequate to calculate the relative significance of each input variable with associated time delay. A further issue is to determine the criteria for selecting or deleting a variable. The cross-validation approach seems to be effective, but it can be quite time-consuming. Alternative approaches that could quantify a variable's contribution against its contributed variance would be desirable for variable selection.

After a neural network is trained and used for prediction, it is required to check the validity of the predicted values. It is important to alert the operator if a prediction is not valid. Checking the lower and upper bounds for a variable is necessary but not sufficient to identify valid regions, particularly when the input variables are highly correlated. Any control actions based on the trained neural network should also be conducted within the valid region.

The use of neural networks as intelligent sensors seems to be a rather general application across various process industries. To build successful applications, one needs to use several techniques to handle the practical issues. Therefore, it is worthwhile to study how to package these techniques in a single software package for process engineers. Although there are many commercial neural network training packages available, most of them place emphasis on training algorithms but give little attention to preprocessing and postvalidation. An easy-to-use toolkit that has comprehensive features to address the practical issues could significantly reduce the gap between the academic research and real-world applications and thus could prevent any potential misuse of the neural network technology.

9 REFERENCES

[BMMW90] N. V. Bhat, P. A. Minderman, T. J. McAvoy, and N. S. Wang. Modeling chemical process systems via neural computation. *IEEE Control Systems Magazine*, pages 24–30, 1990.

[Cyb89] G. Cybenko. Approximation by superpositions of a sigmoidal function. *Mathematics of Controls, Signals and Systems*, 2:303–314, 1989.

[DM94] D. Dong and T. J. McAvoy. Nonlinear principal component analysis — based on principal curves and neural networks. In *Proceedings of the American Control Conference*, pages 1284–1288, June 29 - July 1, Baltimore, 1994.

[DMT+92] B. Dayal, J. F. MacGregor, P. Taylor, R. Kildaw, and S. Marcikic. Application of artificial neural networks and partial least squares regression for modeling kappa number in a continuous Kamyr digester. In *Proceedings of "Control Systems - Dreams vs. Reality"*, pages 191–196, 1992.

[GK86] P. Geladi and B. R. Kowalski. Partial least-squares regression: A tutorial. *Analytica Chimica Acta*, 185:1–17, 1986.

[HA90] E. Hernandez and Y. Arkun. Neural network modeling and extended DMC algorithm to control nonlinear systems. In *Proceedings of the American Control Conference*, pages 2454–2459, 1990.

[HH92] J. C. Hoskins and D. M. Himmelblau. Process control via artificial neural networks and reinforcement learning. *Computers & Chemical Engineering*, 16:241–251, 1992.

[Hoc76] R. R. Hocking. The analysis and selection of variables in linear regression. *Biometrics*, 32:1–49, 1976.

[Hoe70] A. E. Hoerl. Ridge regression: Biased estimation for non-orthogonal problems. *Technometrics*, 12:55–67, 1970.

[HSW89] K. Hornik, M. Stinchcombe, and H. White. Multilayer feedforward neural networks are universal approximators. *Neural Networks*, 2:359–366, 1989.

[Jol86] I. T. Jolliffe. *Principal Component Analysis*. Springer-Verlag, New York, 1986.

[Kos92] B. Kosko. *Neural Networks and Fuzzy Systems*. Prentice-Hall, Englewood Cliffs, New Jersey, 1992.

[Kra92] M. A. Kramer. Autoassociative neural networks. *Computers & Chemical Engineering*, 16:313–328, 1992.

[M+89] T. J. McAvoy, N. Wang, S. Naidu, N. Bhat, J. Gunter, and M. Simmons. Interpreting biosensor data via backpropagation. In *Proceedings of the International Joint Conference on Neural Networks, Washington, DC, June, 1989*, volume 1, pages 227–233. IEEE, 1989.

[MMK+91] J. F. MacGregor, T. E. Marlin, J. V. Kresta, and B. Skagerberg. Some comments on neural networks and other empirical modeling methods. In *Proceedings of the Chemical Process Control-IV Conference*, South Padre Island, Texas, February 18–22, 1991.

[MN89] H. Martens and T. Naes. *Multivariate Calibration*. Wiley, New York, 1989.

[NP90] K. S. Narendra and K. Parthasarathy. Identification and control of dynamic systems using neural networks. *IEEE Transactions on Neural Networks*, 1(1):4–27, 1990.

[PO91] M. Piovoso and A. J. Owens. Sensor data analysis using artificial neural networks. In *Proceedings of the Chemical Process Control-IV Conference*, pages 101–118. South Padre Island, Texas, February 18–22, 1991.

[QM92a] S. J. Qin and T. J. McAvoy. A data-based process modeling approach and its applications. In *Preprints of the 3rd IFAC Dycord+ Symposium*, pages 321–326, College Park, Maryland, April 26–29, 1992.

[QM92b] S. J. Qin and T. J. McAvoy. Nonlinear PLS modeling using neural networks. *Computers & Chemical Engineering*, 16(4):379–391, 1992.

[RHW86] D. Rumelhart, G. Hinton, and R. Williams. Learning internal representations by error propagation. In D. Rumelhart and J. L. McClelland, editors, *Parallel Distributed Processing*, pages 318–362. MIT Press, Cambridge, Massachusetts, 1986.

[UPK90] L. Ungar, B. Powell, and E. N. Kamens. Adaptive networks for fault diagnosis and process control. *Computers & Chemical Engineering*, 14:561–572, 1990.

[VVY90] V. Venkatasubramanian, R. Vaidyanathan, and Y. Yamamoto. Process fault detection and diagnosis using neural networks–I. steady state processes. *Computers & Chemical Engineering*, 14:699–712, 1990.

[WDM+91] M. J. Willis, C. DiMassimo, G. A. Montague, M. T. Tham, and J. Morris. Artificial neural networks in processing engineering. *Proceedings of the IEE, Part D*, 138(3):256–266, 1991.

[WEG87] S. Wold, K. Esbensen, and P. Geladi. Principal component analysis. *Chemometrics and Intelligent Laboratory Systems*, 2:37–52, 1987.

[Wis91] B. M. Wise. *Adapting Multivariate Analysis for Monitoring and Modeling Dynamic Systems.* Ph.D. Thesis, University of Washington, Seattle, 1991.

Chapter 9

Approximation of Time-Optimal Control for an Industrial Production Plant with General Regression Neural Network

Clemens Schäffner
Dierk Schröder

> ABSTRACT For plants that are characterized by a continuous moving web of material passing between rollers, the sections are coupled by the web. To make the web control forces smooth and provide fast response, we approximate a time-optimal control with a general regression neural network; simulation results are shown.

1 Introduction

In the paper, plastics, textile, or metal industry there are many plants that are characterized by a continuous moving web of material. The web has to pass various sections of rollers in order to enable the execution of several processing steps. The rollers are driven by electric motors. All sections of the continuous process are coupled by the web. To achieve proper transport and processing results, the web forces have to be kept within close limits. Therefore, the electrical and mechanical quantities of the drives (e.g., currents, torques, web speeds, web forces) have to be controlled by closed-loop control.

Today, most such plants are equipped with current and speed closed-loop control in cascaded structure. The web forces are controlled only indirectly in an open loop via the speed relations of neighboring rollers. This control concept suffers from known limitations of cascaded control structures.

An important improvement was the development of several linear state-space control concepts, which increased the control quality considerably [WS87]. However, it is impossible to incorporate restrictions on control variables in linear design procedures, because they represent nonlinearities. Since restrictions are unavoidable in practical applications, it is desirable to take them into account directly. The control behavior should be as good as possible with adherence to all restrictions.

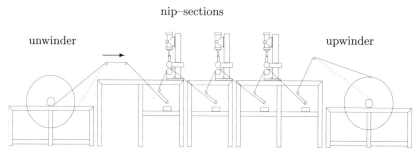

FIGURE 1. Plant.

These objectives can be met best with an approximation of time-optimal control. Classical time-optimal control laws have some severe drawbacks; the most important one is that they lead to switching controllers, which are likely to excite unmodeled high-frequency dynamics (see, e.g., [PBGM62, Pap91, AF66, BH75, Fan66, Kir70]). The resulting oscillations prevent stationary accuracy, stress the mechanical and electrical components of the plant and can — in extreme cases — destroy parts of the plant. For this reason, an approximation of the time-optimum with a smooth control surface is desirable. Radial basis networks provide smooth output functions and are therefore well suited. Because of its favorable approximation properties we use a special type of radial basis network, the *general regression neural network* (GRNN).

The problem is to create a GRNN such that the plant is controlled in an approximately time-optimal manner. This is the topic of this chapter and is described in detail below.

The chapter is organized as follows. In Sections 2 and 3 we give a brief description of the models used for the industrial production plant and the induction motor drive, respectively. The properties of GRNNs are discussed in Section 4. The most important section is Section 5, where the considerations are described in order to obtain a neural control concept approximating time-optimal control for the plant. Simulation results demonstrate the feasibility of the approach. Final remarks are given in the conclusion.

2 Description of the Plant

A scheme of the industrial production plant considered is depicted in Figure 1. Such plants with continuous moving webs of material are driven by a large number of electric motors and thus are — from a control point of view — complex, highly coupled multiple-input/multiple-output systems (MIMO systems).

The winders serve as material storage devices. The web moves with high speed from the unwinder through a large number of nip-sections (in Figure 1

9. Approximation of Time-Optimal Control

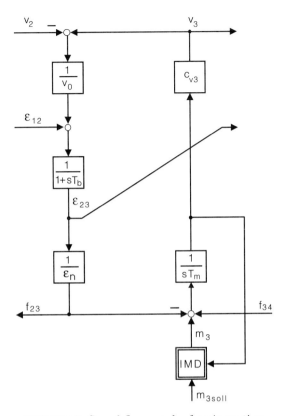

FIGURE 2. Signal flow graph of a nip-section.

there are only three nip-sections shown) to the upwinder. In every nip-section a processing step is executed; for example, the material is dried, deformed, polished, coated, printed, colored, etc. Every section is equipped with an induction motor drive.

In this chapter we assume ideal working conditions for the nip-section next to the unwinder, which means in particular that this nip-section takes over the function of the leading drive and impresses the nominal web velocity. The following nip-sections are numbered starting from one. The upwinder is controlled ideally so that the web is stored with a given constant web force, by which the upwinder is decoupled from the last section. We shall investigate the dynamic and static behavior of the nip-sections between the leading drive and the upwinder.

In order to give an idea of the physics, we want to discuss briefly the signal flow in a nip-section, for example the third section. With the above assumptions the normalized and linearized signal flow graph of this nip-section is depicted in Figure 2 [WS87, WS93]. To this section belong the reference value of the motor torque m_{3soll}, the actual motor torque m_3, the web velocity v_3 at the third roller, and the web force f_{23} between the second and the third rollers. IMD (induction motor drive) denotes a nonlinear numerical model for an induction motor drive including a voltage source inverter and torque control, see Section 3. The torque m_3 of the IMD acts on the integrator with the time constant T_m representing the combined inertia of the roller and the motor. The difference of the velocity of the second and the third roller, $v_3 - v_2$, produces a strain ε_{23} via the time constant T_b of the web. The factor v_0 is the average transport velocity of the web under steady-state conditions. The web force f_{23} is produced by ε_{23} according to Hooke's law, expressed through the factor $\frac{1}{\varepsilon_n}$. The difference $f_{34} - f_{23}$ gives the reaction of the web to the third roller and thus to the corresponding motor shaft.[1]

The variables that are to be kept within close limits and have to be controlled with closed-loop control are the web forces f_{23}, f_{34}, f_{45}, f_{56}, and f_{67} of five coupled nip-sections.[2] The control variables are the reference values of the motor torques m_{3soll}, m_{4soll}, m_{5soll}, m_{6soll}, and m_{7soll}. All quantities are deviations from steady-state conditions.

In this chapter the following normalized plant parameters are used:

$$c_{vi} = 1, \quad i = 3, \ldots, 7,$$
$$\varepsilon_n = 0.005333,$$
$$v_0 = 1,$$
$$T_i = 0.00384 \text{ s}, \quad i = 3, \ldots, 7,$$
$$T_{mi} = 0.415 \text{ s}, \quad i = 3, \ldots, 7,$$
$$T_b = 0.4875 \text{ s}.$$

3 Model of the Induction Motor Drive

In this chapter a nonlinear model for the induction motor drives is used for the numerical simulations. The drives are equipped with a voltage source inverter and are torque controlled via closed-loop control. It is not the topic of this chapter to describe this model in detail. Nevertheless, we want to give a basic idea of its behavior.

A typical time plot of step responses obtained from an experimental setup of a modern induction motor drive is depicted in Figure 3 [Hin93]. The upper half of the figure shows the reference value of the torque; the lower part shows the actual motor torque. Obviously, these transients are quite different from the often used PT_1 approximation.

The numerical model is based on [Hin93] and was developed in order to describe typical stationary and transient phenomena for torque controlled

---

[1] Please note that all signals throughout this chapter are normalized quantities, as is usual in control engineering. This means that all signals are without dimension. The basics of the normalization technique can be found in every introductory control engineering book.

[2] The signal flow graph of the coupled system will be shown in Section 5.6, Figure 19.

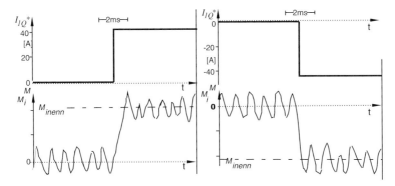

FIGURE 3. Time plots of the reference torque (top) and of the actual torque (bottom) of an experimental setup of an induction motor drive equipped with a voltage source inverter.

induction motor drives in a single-phase model:

- During steady state conditions — this means the reference value of the torque is constant — the actual torque shows ripples because of the switching behavior of the voltage source inverter.

- During transients the gradient of the torque is approximately constant. The value of the gradient varies within certain limits according to the speed and internal quantities of the drive.

Time plots of step responses obtained from simulations of the numerical model are depicted in Figure 4 and show good congruence to the experimental results of Figure 3.

4 General Regression Neural Network

Neural Networks are universal approximators for multidimensional, nonlinear static functions. A neural network can be interpreted as a box whose outputs depend on the specific inputs and on the values of the internal adjustable weights. This approximation ability is necessary for the proposed control concept. We shall show in Section 5 how to generate data that are used as input data for the neural network.

The different types of neural networks differ in their interpolation and extrapolation behavior. In this chapter the general regression neural network (GRNN) is chosen because it provides a smooth output function even with sparse data and because its interpolation behavior is predictable [Spe91, MD89]. These are desirable features especially in closed control loops. The GRNN is a one-pass learning network with a highly parallel structure. The basic idea is similar to the probabilistic neural network, which involves one-pass learning as well [Spe88].

We want to point out that our control concept does not depend on a certain type of neural network. The control concept can be applied to all

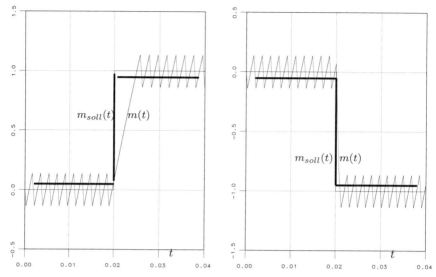

FIGURE 4. Simulation results of the numerical model of the induction motor drive equipped with a voltage source inverter: Time plots of the reference torque $m_{soll}(t)$ (thick line) and of the actual torque $m(t)$ (thin line).

neural network paradigms that implement a smooth output function like, e.g., the well-known backpropagation neural network. However, one disadvantage of backpropagation is that it needs a large number of iterations to converge to the desired solution.

The algorithmic form of the GRNN can be used for any regression problem in which an assumption of linearity is not justified. In comparison to conventional regression techniques the GRNN does not need any a priori information about the form of the regression functions. It can be implemented in parallel hardware or computed in a parallel manner, for instance on transputers. However, sequential software simulation of GRNNs requires substantial computation.

The basic equation for a GRNN with m inputs ($\boldsymbol{x} = [x_1 \ x_2 \ \ldots \ x_m]^T$) and one output ($y$ is scalar) is

$$y = y(\boldsymbol{x}) = \frac{\sum_{\nu=1}^{p} \vartheta^{\nu} \cdot \exp\left(-\frac{C_\nu}{\sigma}\right)}{\sum_{\nu=1}^{p} \exp\left(-\frac{C_\nu}{\sigma}\right)} = \boldsymbol{\vartheta}^T \cdot \boldsymbol{w}(\boldsymbol{x}), \qquad (1)$$

where

$$C_\nu = \sum_{n=1}^{m} |x_n - \chi_n^{\nu}|,$$

and σ is a smoothing parameter. The pairs $\boldsymbol{\chi}^{\nu} = [\chi_1^{\nu} \ \chi_2^{\nu} \ \ldots \ \chi_m^{\nu}]^T$ and ϑ^{ν}

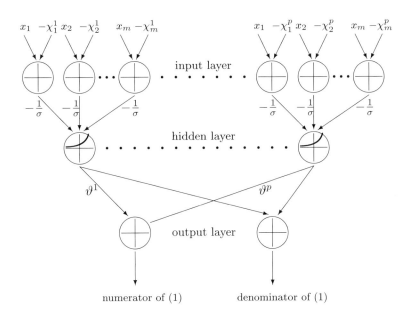

FIGURE 5. General regression neural network with neural hardware implementation.

are the data points that determine the approximation surface.

The output $y(\boldsymbol{x})$ can be considered as a weighted average of all data points $\vartheta^\nu(\boldsymbol{\chi}^\nu)$, where each data point is weighted exponentially according to the Manhattan distance[3] from $\boldsymbol{x}$. When the smoothing parameter σ is made very large, the output function $y(\boldsymbol{x})$ is forced to be smooth. As σ goes to 0, $y(\boldsymbol{x})$ assumes the value of the $\vartheta^\nu(\boldsymbol{\chi}^\nu)$ associated with the data point closest to $\boldsymbol{x}$. For intermediate values of σ the data points closer to $\boldsymbol{x}$ are given heavier weight then those further away.

A neural hardware implementation of a GRNN is depicted in Figure 5. The network consists of three layers. In the input layer the components of the Manhattan distances $|x_n - \chi_n^\nu|$ are computed. Note that the χ_n^ν are constant because they are part of the data points. The activation functions of the neurons in the input layer are all identical; they are the absolute value function. The outputs of the input layer are amplified by the factors $-\frac{1}{\sigma}$ of the synaptic weights in order to provide the corresponding $-\frac{C_\nu}{\sigma}$ as inputs for the neurons in the hidden layer. The activation function of all neurons in the hidden layer is the exponential function for the calculation of $\exp(-\frac{C_\nu}{\sigma})$. The numerator of (1) is constructed by the ϑ^ν-weighted sum of the outputs of the hidden layer. The denominator of (1) is simply the sum of all hidden-layer outputs. The final division to obtain y has to be done outside the GRNN. Extremely fast neural network hardware is available on the market to implement the GRNN, see, e.g., [FHS91]. We illustrate

---

[3]Other distance measures, such as the Euclidean distance, are suitable as well.

the approximation behavior through a simple 2-dimensional example. The data points are

$$\left\{(\boldsymbol{\chi}^\nu, \vartheta^\nu) \mid \nu = 1, 2, 3, 4\right\} = \left\{\left(\begin{bmatrix} 0 \\ 0 \end{bmatrix}, 1\right), \left(\begin{bmatrix} 0 \\ 1 \end{bmatrix}, 0\right), \left(\begin{bmatrix} 1 \\ 0 \end{bmatrix}, 0\right), \left(\begin{bmatrix} 0 \\ 0 \end{bmatrix}, 1\right)\right\},$$

which represent the XOR problem on the unit square. The result of equation (1) with $m = 2$ and $p = 4$ is shown in Figure 6. There is no over- or undershooting because the output y is bounded by the minimum and the maximum of the ϑ^ν, $\nu = 1, \ldots, p$. The output does not converge to poor solutions corresponding to local minima of an error criterion as sometimes happens with iterative techniques. Furthermore, the GRNN allows one-pass learning (however, at the cost of an increasing number of neurons); therefore, the learning procedure is very fast. These are — in this context — the main advantages relative to other nonlinear regression schemes, e.g., the well-known backpropagation neural network, whose approximation behavior is unpredictable in principle and that needs an iterative training procedure.

The only parameter that has to be adjusted according to the data set is the smoothing parameter σ. The simplest method to obtain an appropriate σ is to carry out a few trial-and-error steps, which in our case can be done easily with the help of visualization tools.

5 Control Concept

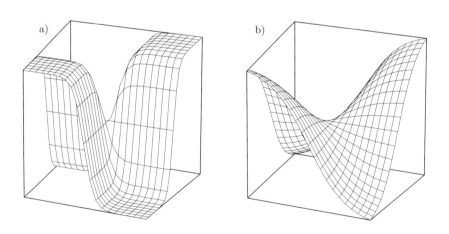

FIGURE 6. Approximation of the XOR problem on the unit-square. a) $\sigma = 0.2$, b) $\sigma = 1.0$.

5.1 Some Preliminary Remarks about Time-Optimal Control

Controller design for coupled multi-input/multi-output (MIMO) systems can be carried out easily for linear plants with well-known linear design techniques, e.g., pole placement. The resulting large matrices can be handled by modern high-end CPUs. However, from a practical point of view there are some severe drawbacks. A correspondence between system behavior and specific matrix elements often does not exist. Putting the plant into operation is a very difficult task. A change of system structure, caused for instance by a web tear, can result in completely unpredictable and indefinable system reactions.

Furthermore, it is impossible to take into account nonlinearities in the linear controller design procedure. Important nonlinearities are the restrictions on the motor torque due to voltage source inverter physics and limited power of the induction motor drive. In this chapter a nonlinear control technique is developed that takes these restrictions into account directly. The aim is to achieve time-optimal control behavior with adherence to the restrictions.

In general, time-optimal control means to drive a system from any initial state into any reference state in the shortest time possible. This task can be standardized by a linear transformation so that one has to consider only the trajectories leading to the origin of the state space.

Often, time-optimal control is designed with a single restriction only, namely the limitation of the maximum absolute value of the control variable. Under this circumstance, the time-optimal control obtained is an ideal two-point switch. The *zero-trajectory* serves as switching curve. Zero-trajectories are defined as trajectories that lead to the origin of the state space, while the controller output equals either the positive or the negative maximum value. It is well known that this type of time-optimal control results in chattering system behavior. Under stationary conditions we observe the characteristic limit cycles, because the controller output switches permanently between both maximum values.

Our control concept not only avoids these limit cycles but also takes the restriction of the maximum gradient of motor torque changes into account.

5.2 System Trajectories of an Isolated Nip-Section

In order to provide

- a transparent design procedure,
- predictable system behavior in the case of structural changes,
- the opportunity to put the plant into operation sectionwise, and
- a good approximation of time optimal control,

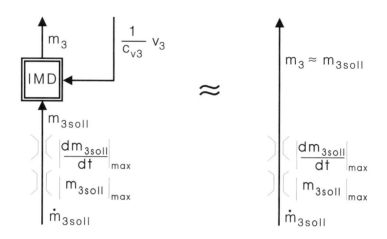

FIGURE 7. Simplification of induction motor drive.

we first consider the controller design for an isolated nip-section of low order. As a representative we consider the third section, which corresponds to the control variable $m_{3soll}(t)$, the actual torque $m_3(t)$, the web velocity $v_3(t)$, and the web force $f_{23}(t)$. After a controller for such a subsection is found we shall extend the control concept in order to handle the complete complex MIMO-system consisting of five coupled nip-sections.

An important idea of our control concept is to introduce restrictions such that the actual motor torque is just able to follow the reference value of the motor torque with reasonable accuracy. Considering Figures 3 and 4, we conclude that if we restrict the reference value of the torque as follows,

- the absolute value of $m_{3soll}(t)$ does not exceed a maximum,

$$|m_{3soll}(t)| < |m_{3soll}|_{max} \quad \forall t, \qquad |m_{3soll}|_{max} = \text{constant}, \qquad (2)$$

- the absolute value of the time derivative $\dot{m}_{3soll}(t)$ does not exceed a maximum,

$$|\dot{m}_{3soll}(t)| < |\dot{m}_{3soll}|_{max} \quad \forall t, \qquad |\dot{m}_{3soll}|_{max} = \text{constant}, \qquad (3)$$

then — provided the constant values $|m_{3soll}|_{\max}$ and $|\dot{m}_{3soll}|_{\max}$ are chosen appropriately — it is guaranteed that we take nearly full advantage of the dynamic capabilities of the induction motor drive while $m_3 \approx m_{3soll}$; see Figure 7. Note that introducing the restrictions (2) and (3) is artificial and thus can be seen as part of the controller. The resulting signal flow graph for the third isolated nip-section is shown in Figure 8. Another advantage resulting from the restrictions (2) and (3) is that now we *know* the maximum values of the actual torque and its gradient, which is important for decoupling measures; see Section 5.6. The maximum gradient would depend otherwise on the speed and on internal quantities of the IMD.

9. Approximation of Time-Optimal Control

Consequently, an important simplification can be made. If we guarantee that the controller output pays attention to (2) and (3), then it is sufficient to consider two states of the subsystem: the web velocity v_3 and the web force f_{23}. The internal states of the induction motor drive can be neglected. This enables us to use in later sections the control surface with the three dimensions v_3, f_{23}, and m_{3soll} as a visualization tool for our considerations.

The two restrictions (2) and (3) imply that there is no analytical solution possible for the time-optimal controller. Therefore, the controller has to be designed by some other means.

In principle, the time plots of time-optimal control trajectories have to look like the transient in Figure 9. Note that the transient is constructed such that the restrictions are just not violated. It is sufficient to consider two switching times, t_1 and t_2. Starting with $t = 0$, the control variable $m_{3soll}(t)$ has to be set to the maximum value as fast as possible; therefore restriction (3) is active. When the maximum value is reached, the restriction (2) becomes active. At the first switching time t_1 the control variable has to be decreased with maximum velocity (again restriction (3) is active) until the maximum negative value is reached (again restriction (2) becomes active). After the second switching time t_2 the control variable has to be reduced to zero as fast as possible, and after that the transient is completed.

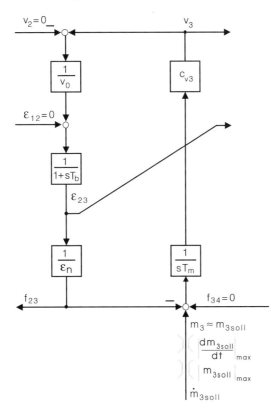

FIGURE 8. Signal flow graph of the third isolated nip-section with restrictions.

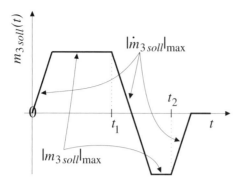

FIGURE 9. Typical time-optimal transient $m_{3soll}(t)$.

We should point out the possibility that the switching time occurs *before* the corresponding maximum value is reached; an example of this effect will be given with the simulation results.

As already mentioned before, it is the purpose of a transient like the one shown in Figure 9 to drive the system from any initial state into the origin of the state space in the shortest time possible. A specific combination of the switching times t_1 and t_2 corresponds to a certain initial state. However, this initial state is unknown. On the other hand, the final state *is* known, namely, the origin of the state space: web velocity $v_3 = 0$ and web force $f_{23} = 0$.

For this reason we integrate the system *backwards in time*. We start in the origin of the state space and use control transients of the type depicted in Figure 9 in order to obtain time-optimal system trajectories. We do this repeatedly with different switching times t_1 and t_2 in order to generate time-optimal trajectories distributed over the entire state space. A selection of six trajectories is shown in Figure 10. In the meantime we store the corresponding values of the velocity $v_3(t_\nu)$, the web force $f_{23}(t_\nu)$, and the reference torque $m_{3soll}(t_\nu)$ at time instants t_ν:

$$\chi^\nu = \begin{bmatrix} v_3(t_\nu) \\ f_{23}(t_\nu) \end{bmatrix} \quad \text{and} \quad \vartheta^\nu = m_{3soll}(t_\nu) \quad \text{for} \quad \nu = 1, \ldots, p. \quad (4)$$

It is the basic idea of our control concept that these data can be used as input data for a GRNN. The resulting control surface, implemented with a GRNN, will be shown in Section 5.5, where the simulation results are presented.

5.3 Linear-Integral Controller in a Small Region around the Stationary State

In order to provide a smooth control surface and to guarantee stationary accuracy in the presence of disturbances, the state space is divided into

9. Approximation of Time-Optimal Control

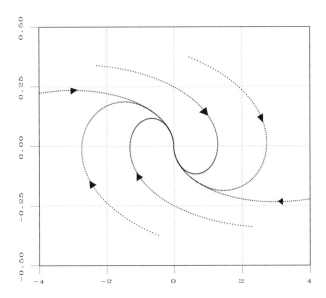

FIGURE 10. Time-optimal trajectories in the state space.

three regions (Figure 11): The first region, Ω_{in}, comprises a *small* area around the stationary state (which corresponds to the origin of the phase plane of Figure 11), where only a conventional linear state controller with an integrator is active; the second region, Ω_{trans}, is a small layer around Ω_{in}, where a smooth transition between the linear state controller and the nonlinear neural controller is provided; and the third region, Ω_{out}, comprises the rest of the state space, where only the neural controller is used.

The linear controller does not match the demand for time-optimality, but its sphere of activity is so small that it hardly deteriorates the transients. On the other hand, the linear controller has some substantial advantages: Because of the integrator, the system shows a certain degree of robustness against disturbances; the parameters of the linear controller can be optimized with respect to good disturbance rejection. Furthermore, no limit cycles around the stationary state occur.

5.4 Approximation of Time-Optimal Control for an Isolated Nip-Section

In Section 5.2 we have assumed transitions from any initial state to the origin of the state space. Now we want to allow arbitrary states as target states. Therefore, an affine transformation on the controller has to be performed. Controller inputs are then the differences m_{3d}, v_{3d}, and f_{23d} between the actual and the corresponding reference states, which are computed as follows:

$$m_{3d} = f_{23soll} - m_3, \tag{5}$$

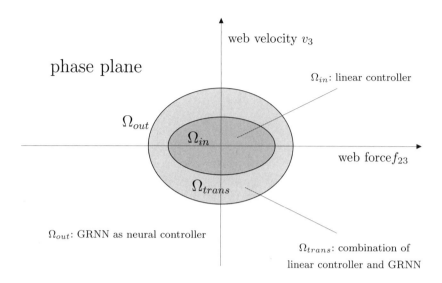

FIGURE 11. Regions in the state space.

$$v_{3d} = v_3 - f_{23soll} \cdot \varepsilon_n \cdot v_0, \quad (6)$$
$$f_{23d} = f_{23soll} - f_{23}. \quad (7)$$

To the controller output we have to add the reference value of the web force f_{23soll}. It is simple to verify this transformation through stationary considerations, as we have already mentioned that all signals are normalized quantities.

The resulting controller structure approximating time-optimal control is depicted in Figure 12. It is assumed that all system states are measurable with negligible time delay. The reference value is the web force f_{23soll}; the controlled value is the actual web force f_{23}.

5.5 Simulation Results for an Isolated Nip-Section

Numerical simulations have been done for the control structure of Figure 12. We applied a step function of the reference value of the web force:

$$f_{23soll}(t) = 0 \quad \text{for} \quad t < 0,$$
$$f_{23soll}(t) = 1 \quad \text{for} \quad t \geq 0.$$

The simulation results are shown in Figures 13–16. The controller output $m_{3soll}(t)$, Figure 13, does not violate the two restrictions (2) and (3); for example:

$$|m_{3soll}(t)| < 3,$$
$$|\dot{m}_{3soll}(t)| < 20\text{s}^{-1}.$$

The transient of the actual motor torque $m_3(t)$, Figure 14, shows very clearly the ripple on the torque because of the switching behavior of the

9. Approximation of Time-Optimal Control

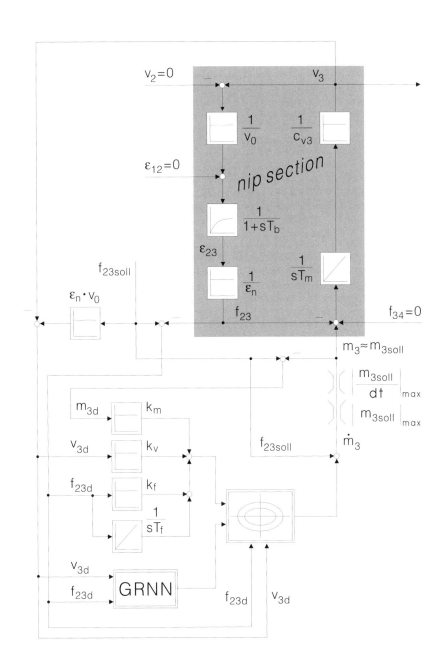

FIGURE 12. An isolated nip-section with neural control.

voltage source inverter.[4] The time plots of the web velocity $v_3(t)$, Figure 15, and of the actual web force $f_{23}(t)$, Figure 16, show very good approximation of the time-optimum. The slight overshoot of $f_{23}(t)$ is caused by the linear controller in Ω_{in}.

The resulting control surface, implemented with a GRNN, together with a typical system trajectory, is depicted in Figure 17.

5.6 Approximation of Time-Optimal Control for the Industrial Production Plant

We want to use now the results from the previous sections to control the complex plant with five highly coupled nip-sections.

In order to maintain the very good approximation of time-optimality it is necessary to introduce some decoupling measures. It is not the topic of this chapter to discuss decoupling techniques in detail. There are quite a lot of different techniques possible; in this chapter a relatively simple extension to the control concept developed so far is chosen. The resulting control concept is shown in Figures 18 and 19.

For the decoupling in the opposite moving direction of the web the signal path **a** (see Figure 18) is added, which compensates the influence of the web force f_{56} of the following nip-section (see Figure 19). For the decoupling in the moving direction of the web, the signal **b** is generated as the weighted sum of the web forces of all preceding nip-sections.

This decoupling measure has to be realized for each nip-section as it is depicted in Figure 19. Because of limited space, the control structure of each section is summarized by a nonlinear block marked with *state*/GRNN. This block contains the linear-integral state controller and the GRNN controller as well as the device for the generation of the correct final controller output according to the description in Section 5.3.

5.7 Simulation Results for the Industrial Production Plant

The control system of Figure 19 was simulated numerically. We applied a step function of the reference value of the web force of the third section:

$$f_{23soll}(t) = 0 \quad \text{for} \quad t < 0,$$
$$f_{23soll}(t) = 1 \quad \text{for} \quad t \geq 0.$$

The transients of all web forces — $f_{23}(t)$, $f_{34}(t)$, $f_{45}(t)$, $f_{56}(t)$, and $f_{67}(t)$ — are depicted in Figure 20.

We applied a step function of the reference value of the web force of the seventh section as well:

$$f_{67soll}(t) = 0 \quad \text{for} \quad t < 0,$$
$$f_{67soll}(t) = 1 \quad \text{for} \quad t \geq 0.$$

---

[4]The simulation was done with the IMD-model; see Figure 7.

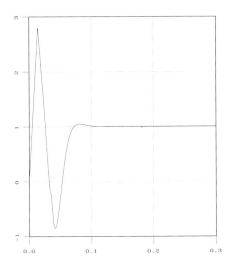

FIGURE 13. Time plot of $m_{3soll}(t)$ for a step function of $f_{23soll}(t)$; isolated nip-section.

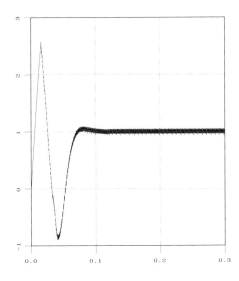

FIGURE 14. Time plot of $m_3(t)$ for a step function of $f_{23soll}(t)$; isolated nip-section.

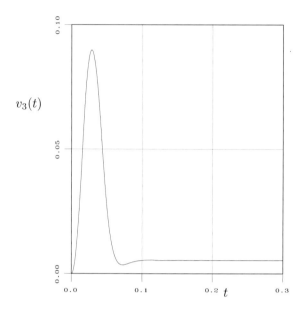

FIGURE 15. Time plot of $v_3(t)$ for a step function of $f_{23soll}(t)$; isolated nip-section.

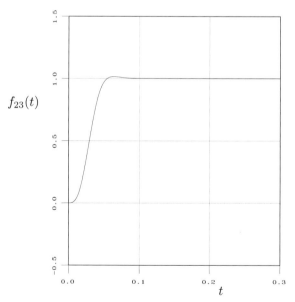

FIGURE 16. Time plot of $f_{23}(t)$ for a step function of $f_{23soll}(t)$; isolated nip-section.

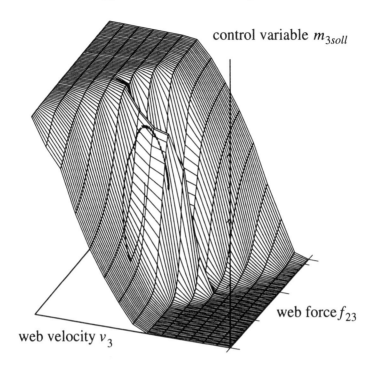

FIGURE 17. Control surface of the general regression neural network and a typical system trajectory.

The transients of all web forces are shown for this case in Figure 21.

Figure 20 as well as Figure 21 show that very good approximation of time-optimality is maintained for the MIMO case. The transients of $f_{23}(t)$ in Figure 20 and of $f_{67}(t)$ in Figure 21 are nearly the same as in Section 5.5, where the time-optimal control for an isolated nip-section was considered.

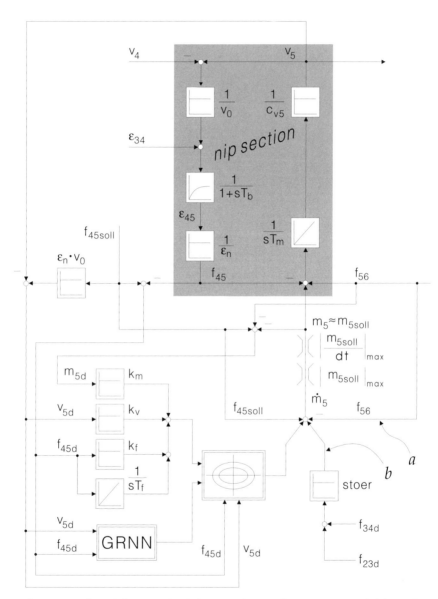

FIGURE 18. Signal flow graph of the neural control concept extended by a decoupling measure (signals a and b).

9. Approximation of Time-Optimal Control

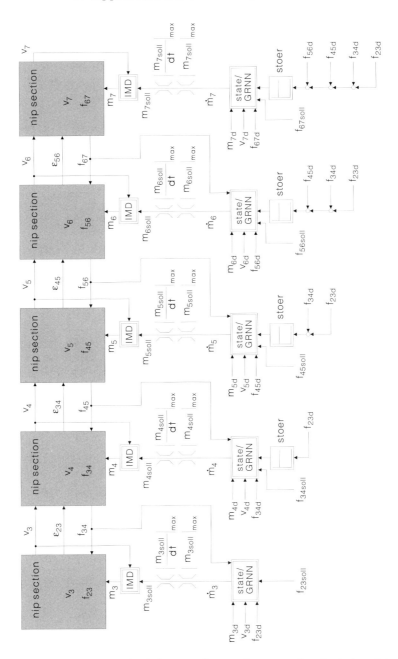

FIGURE 19. Signal flow graph of the neural control concept for the industrial production plant extended by decoupling measures.

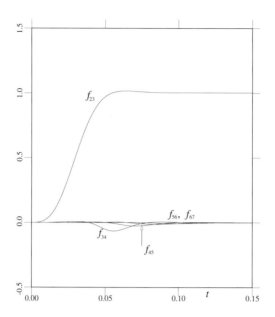

FIGURE 20. Time plot of the web forces $f_{23}(t)$, $f_{34}(t)$, $f_{45}(t)$, $f_{56}(t)$, and $f_{67}(t)$ due to a step function of $f_{23soll}(t)$.

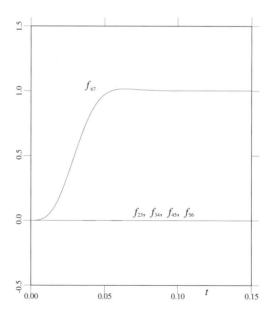

FIGURE 21. Time plot of the web forces $f_{23}(t)$, $f_{34}(t)$, $f_{45}(t)$, $f_{56}(t)$, and $f_{67}(t)$ due to a step function of $f_{67soll}(t)$.

6 Conclusion

All results and experiences with linear state control concepts for industrial plants of the type described in Section 2 show that there is a crucial objective conflict: Either the control parameters are adjusted such that the decoupling works well but at the expense of a rather bad dynamical behavior, or the other way round [WS93]. Acceptable compromises have to be found.

It is our belief that via the incorporation of *non*linear control strategies it is possible to soften this objective conflict. This means to find better compromises than are ever possible with linear techniques.

In this context we have developed a nonlinear control concept. The use of neural networks was motivated by the fact that they can represent arbitrary nonlinear static functions. In this chapter we have chosen the general regression neural network because of its favorable approximation properties.

We have demonstrated how to gain data for the neural network in order to achieve a control hypersurface that approximates time-optimal control. Simulation results have been presented for an isolated nip-section of the plant and the entire complex industrial production plant with five nip-sections. We have demonstrated that our neural control concept approximates the time-optimum very well. Note that the data acquisition described in Section 5.2 depends on the system parameters. However, in the described type of industrial production plant there are some parameters that vary considerably during operation, for example, the proportional factor e_n between the web force and the strain. Therefore, it is necessary to adapt the control surface according to these parameter changes. For this purpose error signals have to be generated, and a stable learning law has to be derived in order to incorporate adaptivity in our control concept. Further research work will show whether very promising approaches for stable online learning [SSL95, SS95, Sch96] can be used.

7 REFERENCES

[AF66] M. Athans and P. L. Falb. *Optimal Control*. McGraw–Hill, New York, 1966.

[BH75] A. E. Bryson and Y.-C. Ho. *Applied Optimal Control*. Wiley, New York, 1975.

[Fan66] L.-T. Fan. *The Continuous Maximum Principle*. Wiley, New York, 1966.

[FHS91] W. A. Fisher, R. J. Hashimoto, and R. C. Smithson. A programmable neural network processor. *IEEE Transactions on Neural Networks*, 2(2):222–229, March 1991.

[Hin93]   D. Hintze. A single–phase numerical model for induction motor drives. Internal report, Institute for Electrical Drives, Technical University of Munich, Munich, 1993.

[Kir70]   D. E. Kirk. *Optimal Control Theory, An Introduction.* Prentice-Hall, Englewood Cliffs, New Jersey, 1970.

[MD89]   J. Moody and C. J. Darken. Fast learning in networks of locally–tuned processing units. *Neural Computation*, 1(2):281–294, 1989.

[Pap91]   M. Papageorgiou. *Optimierung.* Oldenbourg Verlag, München Wien, 1991.

[PBGM62]   L. S. Pontryagin, V. Boltyanskii, R. Gamkrelidze, and E. Mischenko. *The Mathematical Theory of Optimal Processes.* Wiley(Interscience), New York, 1962.

[Sch96]   C. Schäffner. *Analyse und Synthese Neuronaler Regelungsverfahren.* Ph.D. Thesis, Herbert Utz Verlag Wissenschaft, München, 1996. ISBN 3–931327–52–3.

[Spe88]   D. Specht. Probabilistic neural networks for classification, mapping or associative memory. In *Proceedings of the International Conference on Neural Networks, June, 1988*, volume 1, pages 525–532. Lawrence Erlbaum, Hillsdale, New Jersey, 1989.

[Spe91]   D. Specht. A general regression neural network. *IEEE Transactions on Neural Networks*, 2(6):568–576, November 1991.

[SS95]   C. Schäffner and D. Schröder. Stable nonlinear observer design with neural network. In *IFAC–Workshop on Motion Control*, pages 567–574, Munich, Germany, 1995.

[SSL95]   C. Schäffner, D. Schröder, and U. Lenz. Application of neural networks to motor control. In *International Power Electronic Conference, IPEC 1995*, volume 1, pages 46–51, Yokohama, Japan, 1995.

[WS87]   W. Wolfermann and D. Schröder. Application of decoupling and state space control in processing machines with continuous moving webs. In *Preprints of the IFAC 1987 World Congress on Automatic Control*, volume 3, pages 100–105, 1987.

[WS93]   W. Wolfermann and D. Schröder. New decentralized control in processing machines with continuous moving webs. In *Proceedings of the Second International Conference on Web Handling*, pages 96–116. Web Handling Research Center, Oklahoma State University, Stillwater, Oklahoma, June, 1993.

Chapter 10
Neuro-Control Design: Optimization Aspects

H. Ted Su
Tariq Samad

ABSTRACT This chapter views neural-network-based control system design as a nonlinear optimization problem. Depending on the role of a neural network in the system, the neuro-control problems are classified into a few categories. A unifying framework for neuro-control design is presented to view neural network training as a nonlinear optimization problem. This chapter then outlines a new neuro-control concept, referred to as *parameterized neuro-nontroller* (PNC) and discusses the optimization complexities it poses. To demonstrate the unique characteristics of this new control design concept, simulation results are presented at the end of this chapter.

1 Introduction

The recent and considerable interest in neuro-control has resulted in a number of different approaches to using neural networks in control system design. For applications in the process industries, the most relevant of these are based on using neural networks as identifiers or optimizing neural network controllers using a process model. Successful practical applications of this approach are now in operation, e.g., [TSSM92, Sta93].

This chapter views neural-network-based control system design as a nonlinear optimization problem. Depending on the characteristics of the application, it also shows how different optimization algorithms may be appropriate. A fundamental distinction is made between gradient-based and non-gradient-based algorithms. In the former case, a variety of techniques are available for gradient computation. Although gradient-based algorithms can be expected to be significantly more efficient than non-gradient-based ones, there are applications where gradient computation is infeasible. For example, desired control performance criteria are not always differentiable, local minima may render strict gradient methods useless, and not all process models allow analytical derivatives. This chapter reviews a number of nongradient algorithms that have recently been used in neuro-control.

To illustrate the appropriateness for nongradient algorithms, this chapter outlines a new neuro-control concept and discusses the optimization complexities it poses. This concept is referred to as "parameterized neuro-controller" (PNC) [SF93]. PNCs designed with two types of external parameters are considered: process parameters that provide the controller with information regarding the dynamical characteristics of the process (e.g., dead time, gain) and control parameters that indicate characteristics of the desired closed-loop behavior (e.g., maximum overshoot, desired settling time). These two types of parameters make a PNC a *generic* controller. It is generic in two respects: 1) a PNC is applicable to different processes, and 2) a PNC is adjustable, or tunable, for its closed-loop control performance.

This chapter presents simulation results showing the pervasiveness of local minima for this application. The algorithm used in this study integrates the population-based search of a genetic algorithm with the random-walk aspect of chemotaxis. Experimental results on a low-order model appropriate for chemical processes are presented.

2 Neuro-Control Systems

This section briefly reviews various approaches in current neuro-control design. Although there are other ways to classify these approaches (e.g., [HSZG92]) this chapter nevertheless adopts one similar to adaptive control theory: 1) indirect neuro-control and 2) direct neuro-control.

In the indirect neuro-control scheme, a neural network does *not* send a control signal *directly* to the process. Instead, a neural network is often used as an indirect process characteristics indicator. This indicator can be a process model that mimics the process behavior or a controller auto-tuner that produces appropriate controller settings based upon the process behavior. In this category, the neuro-control approaches can be roughly distinguished as follows: 1) neural network model-based control, 2) neural network inverse model-based control, and 3) neural network auto-tuner development.

In the direct neuro-control scheme, a neural network is employed as a feedback controller, and it sends control signals *directly* to the process. Depending on the design concept, the direct neuro-control approaches can be categorized into: 1) controller modeling, 2) model-free neuro-control design, 3) model-based neuro-control design, and 4) robust model-based neuro-control design.

Regardless of these distinctions, a unifying framework for neuro-control is to view neural network training as a nonlinear optimization problem,

$$NN : \min_{w} J(w), \qquad (1)$$

in which one tries to find an optimal representation of the neural network that minimizes an objective function J over the network weight space w.

Here, NN indicates that the optimization problem formulation involves a neural network. The role a neural network plays in the objective function is then a key to distinguishing the various neuro-control design approaches. To make this chapter more interesting, the appearance/formulation of Equation 1 will take various forms in the discussion to follow. As the main thrust of this chapter, the optimization aspects of various neuro-control design approaches are discussed based on the formulation of the objective function. The various optimization problems are not precisely but rather conceptually formulated for discussion purposes.

2.1 Indirect Neuro-Control

Neural Network Model-Based Control

Full Neural Network Model

The most popular control system application of neural networks is to use a neural network as an input–output process model. This approach is a data-driven supervised learning approach, i.e., the neural network attempts to mimic an existing process from being exposed to the process data (see Figure 1). The most commonly adopted model structure for such a purpose is the nonlinear auto-regressive and moving average with exogenous inputs (known as NARMAX) model or a simpler NARX [CBG90, SMW92]. This family of NARMAX models is a discrete-time nonlinear transfer function [Lju87]. Alternatively, one can choose to identify a continuous-time model with a dynamic neural network, e.g., a recurrent network [Wer90]. Regardless of the model structure and the control strategy, the neuro-control design in this case can be conceptually stated as follows:

$$NN : \min_{w} \quad F\{y_p - y_n(w, \ldots)\}, \qquad (2)$$

where y_p stands for plant/process output, y_n for neural network output, and w for neural network weights. Here $F\{\cdot\}$ is a functional that measures

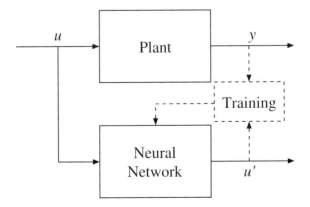

FIGURE 1. Neural network can be used as a black-box model of a process.

the performance of the optimization process. It is usually an integral or sum of the prediction errors between y_p and y_n. For example, in this model development stage, process inputs and output $\{u_p, y_p\}$ are collected over a finite period of time and used for neural network training. $F\{\cdot\}$ is usually an integral of the 2-norm of $y_p - y_n$. A typical example of Equation 2 is as follows:

$$NN : \min_w \sum_t |y_p(t) - y_n(t)|^2; \quad y_n(t) = \mathcal{N}(w, \ldots). \tag{3}$$

Once the model is developed, it can then be implemented for model-based control design.

At the implementation stage, nevertheless, the neural network model cannot be used alone. It must be incorporated with a model-based control scheme. In the chemical process industry, for example, a neural network is usually employed in a nonlinear model predictive control (MPC) scheme [SM93b, SM93a]. Figure 2 illustrates the block diagram of an MPC control system. In fact, the MPC control is also an optimization problem. The optimization problem here can often be expressed as follows:

$$\min_u F'\{y^* - y_n(u, \ldots)\}, \tag{4}$$

where y^* designates the desired closed-loop process output, u the process/model input or control signal, and y_n the predicted process output (by the neural network model). Here F' stands for an objective function that evaluates the closed-loop performance. For example, the optimization problem in the implementation stage is usually as follows:

$$\min_u \sum_t |y^*(t) - y_n(t) - d(t)|^2, \quad y_n(t) = \mathcal{N}(u, \ldots), \tag{5}$$

where $y^*(t)$ stands for desired setpoint trajectory and $d(t)$ for estimated disturbance. This optimization is performed repeatedly at each time interval during the course of feedback control. Although the constraints are

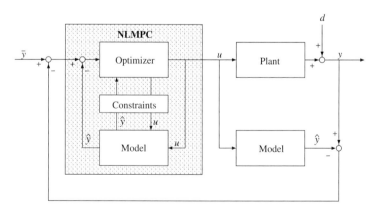

FIGURE 2. A neural network model can be incorporated into nonlinear model predictive control (MPC) scheme.

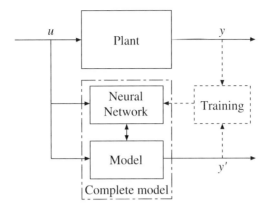

FIGURE 3. A neural network can be a parameter estimator, model structure selector, or partial elements of a physical model.

not particularly of interest in the discussion, one advantage of this indirect control design approach over the direct ones is that the constraints can be incorporated when solving the above optimization problem. For more details about MPC, refer to [SDBM91, SM93b, SM93a], for example.

As discussed later, a full neural network model can be incorporated into other neuro-control schemes, such as neural network auto-tuner design (Section 2.1), model-based controller modeling (Section 2.2) and model-based neuro-control design (Section 2.2).

Parametric or Partial Neural Network Model

In some cases, a certain degree of knowledge about the process might be available, such as model structure or particular physical phenomena that are well understood. In this case, a full black-box model might not be most desirable. For example, if the structure of the process model is available, values for the associated parameters can be determined by a neural network. Examples of these parameters can be time constants, gains, and delays or physical parameters such as diffusion rates and heat transfer coefficients. Psichogios and Ungar [PU92] employed such an approach to model a fedbatch bioreactor. Thompson and Kramer [TK94] also presented a hybrid model structure that incorporated prior knowledge into neural network models. When model structure is not known a priori, neural networks can be trained to select elements of a model structure from a predetermined set [KSF92]. These elements can then be composed into a legal structure. Lastly, in other cases where model structure is partially known, neural networks can also be integrated with such a partial model so that the process can be better modeled [SBMM92]. (See Figure 3.).

For illustration purposes, the parametric or partial neural network modeling problem can be formulated as follows:

$$NN : \min_{w} \ F\{y_p - y_m(\theta, \ldots)\}, \quad \theta = \mathcal{N}(w, \ldots), \qquad (6)$$

where y_m is the predicted output from the model and θ stands for the process parameters, model structural information, or other elements required to complete the model. Notice the only difference between Equation 6 and Equation 2 is that y_m replaces y_n. From a model-based control standpoint, this approach is essentially identical to the full black-box neural network model except that the neural network does not directly mimic the process behavior.

Neural Network Inverse Model

A neural network can be trained to develop an inverse model of the plant. The network input is the process output, and the network output is the corresponding process input (see Figure 4). In general, the optimization problem can be formulated as

$$NN : \min_{w} F\{u^*_{p-1} - u_n(w, \ldots)\}, \qquad (7)$$

where u^*_{p-1} is the process inputs. Typically, the inverse model is a steady-state/static model, which can be used for feedforward control. Given a desired process setpoint y^*, the appropriate steady-state control signal u^* for this setpoint can be immediately known:

$$u^* = \mathcal{N}(y^*, \ldots). \qquad (8)$$

Successful applications of inverse modeling are discussed in [MMW90] and [SW90]. Obviously, an inverse model exists only when the process behaves monotonically as a "forward" function at steady state. If not, this approach is inapplicable.

One can also find a few articles addressing a similar "inverse model" concept of using a nonstatic inverse neural network model for control [BM90, Yds90, UPK90, PU91]. In principle, an inverse neural network model can learn the inverse dynamics under some restrictions (e.g., minimum phase and causality are required). Then, the inverse model is arranged in a way similar to an *MPC controller*. In practice, especially for

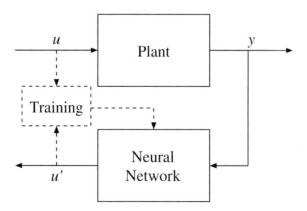

FIGURE 4. A neural network inverse model.

discrete-type dynamic models, the inverse model may not be able to learn the desired inverse dynamics. For example, Psichogios and Ungar [PU91] trained an inverse model and then performed some control case studies. The inverse model approach failed to function as expected. In many cases, a process inverse is in fact noncausal even if the process behaves monotonically as mentioned above. The noncausality of a process inverse can result from transport delay (dead time) or discretization of a continuous process in a sampled-data system. Even if an inverse model does exist, the use of a dynamic inverse model as a feedback controller will not result in a strictly proper control system. Strict properness is essential in control system design [MZ89].

In some of the open literature, "inverse-modeling" is sometimes used to refer to training a neural network "model" in a closed-loop environment, e.g., [HSZG92, PSY88]. This "inverse-modeling" approach does not lead to an "inverse model" that inverts the forward function of the process/plant of interest. For example, the process as a function maps input variables to output variables, whereas the "inverse model" does not map the same output variables or to the same input variables. Instead, it often maps the difference between the output variables and the goal (or set points) to the input variables. From this chapter's perspective, this approach belongs to the category of "neuro-control" design and will be discussed in sections to follow.

Neural Network Auto-Tuner

As in the previous case where neural networks can be used to estimate parameters of a known model, they can also be used to estimate tuning parameters of a controller whose structure is known a priori. A controller's tuning parameter estimator is often referred to as an auto-tuner. The optimization problem in this case can be formulated as follows:

$$NN : \min_{w} \quad F\{\eta^* - \eta_n(w, \ldots)\}, \tag{9}$$

where η^* denotes the controller parameters as targets and η_n stands for the predicted values by the neural network. Network input can comprise sampled process data or features extracted from it. However, these parameters η cannot be uniquely determined from the process characteristics. They also depend on the desired closed-loop control system characteristics. Usually, the controller parameters η^* are solutions to the following closed-loop control optimization:

$$\min_{\eta} \quad F'\{y^* - y_{p/m}(u, \ldots)\}; \quad u = C(\eta, \ldots) \tag{10}$$

where C is a controller with a known structure. Here, $y_{p/m}$ denotes that either a process or a model can be employed in this closed-loop control in order to find the target controller C.

Actually, the attraction of this approach is that the network can be trained in simulation. Training on actual process data is not necessary. For open-loop auto-tuning, an open-loop simulation is sufficient; otherwise, a closed-loop simulation is needed (see Figure 5). The training must be conducted over a space of process models. Ideally, this space should cover the processes that will be encountered during operation.

Most work to date in auto-tuning is directed at the PID controllers as they are still the most widely used control structure in practice. Appropriate values for the PID gains (K_c/proportional gain, K_i/reset time, K_d/derivative time) are essential if the closed-loop system is to perform in a desired manner. PID tuning is still largely a manual procedure, often relying on heuristics developed over a half century ago [ZN42]. Several auto-tuners are commercially available, but improvements are still needed. For neural-network auto-tuner design, low-order linear models with ranges for parameters are likely to suffice for most PID applications.

Developments of this concept are discussed by Swiniarski [Swi90] and Ruano et al. [RFJ92]. In Swiniarski [Swi90], the network is trained using Ziegler-Nichols heuristics for determining target PID gains. During operation, the input to the network is 128 samples of open-loop process step response. Thus the resulting auto-tuner requires open-loop operation of the process. In contrast, Ruano et al. describe a method for open-loop or closed-loop auto-tuning, accomplished by preprocessing input/output data. Optimal PID gains for training purposes are computed with a quasi-Newton optimization algorithm initialized with Ziegler-Nichols values. The optimization criterion is the minimization of the integral of time-multiplied absolute error (ITAE). The authors show how a closed-loop auto-tuner can effectively adapt PID parameters on-line in response to setpoint changes for the control loop.

2.2 Direct Neuro-Control

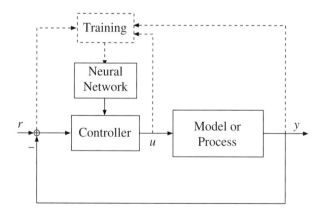

FIGURE 5. Closed-loop neural network auto-tuner.

Controller Modeling

Among the four direct neuro-control schemes, the simplest for neuro-controller development is to use a neural network to model an existing controller (see Figure 6). The input to the existing controller is the training input to the network and the controller output serves as the target. In fact, this approach is similar to the neural network modeling approach discussed in Section 2.1 except that the target here is not a process but a controller. Likewise, this neuro-control design can be formulated as follows:

$$NN : \min_{w} \; F\{u_c^* - u_n(w, \ldots)\}, \tag{11}$$

where u_{c^*} is the output of an existing controller C^*. Usually, the existing controller C^* can be a human operator or it can be obtained via

$$\min_{C} \; F'\{y^* - y_{p/m}(u, \ldots)\}; \quad u = C(\ldots). \tag{12}$$

Like a process model, a controller is generally a dynamical system and often comprises integrators or differentiators. If an algebraic feedforward network is used to model the existing controller, dynamical information must be explicitly provided as input to the network. This dynamical information can be provided either as appropriate integrals and derivatives or as tapped delay signals of process data. For example, to model a PID controller, an algebraic neural network needs not only the instantaneous error between the set-point and the process output, but also the derivative and integral of the error. Alternatively, one can train a neural network with a series of those errors and/or the controller outputs in the previous time steps. The latter approach is similar to developing an ARX (auto-regressive with exogenous inputs) type process model, except that the inputs and the outputs of the process are replaced with feedback errors and controller outputs.

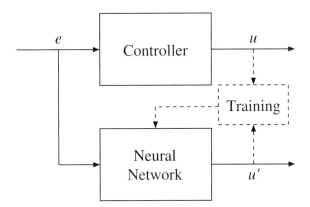

FIGURE 6. The simplest approach to neuro control design is to use a neural network to model an existing controller.

In general, this approach can result in controllers that are faster and/or cheaper than existing controllers. Using this approach, for example, Pomerleau [Pom91] presented an intriguing application, where a neural network is used to replace a human operator, i.e., an existing controller. The Carnegie Mellon University Navlab van, equipped with a video camera and an imaging laser rangefinder, is driven by a human operator at about six miles per hour for about five minutes. During this time, a neural network is trained "on-the-fly" to learn the mapping from the video and laser rangefinder inputs to steering actions. Subsequently, the network can drive the van autonomously at speeds of up to a maximum of 20 mph, which is over twice as fast as any other sensor-based autonomous system has driven the Navlab. Moreover, neural networks have been developed that are capable of driving on single-lane dirt roads, two-lane suburban neighborhood streets, and lined two-lane highways. With other sensory inputs additional capabilities have been achieved, including collision avoidance and nocturnal navigation. Pottman and Seborg [PS92] also present a neural network controller that is trained to learn an existing MPC controller. Given any setpoint change and disturbance, the MPC controller uses a neural network model and performs on-line optimization to calculate the optimal control signal. The resulting neuro-controller can then replace the "on-line optimizing" MPC and yields similar near-optimal results, except that the neuro-controller is faster since it does not need any on-line optimization.

While the benefits of this approach may be apparent when the existing controller is a human, its utility may be limited. It is applicable only when an existing controller is available, which is the case in many applications. Staib and Staib [SS92] discuss how it can be effective in a multistage training process. A neural network is trained to mimic an existing controller and then further refined in conjunction with a process model—the model-based control design concept to be discussed later.

Model-Free Neuro-Control

In the absence of an existing controller, some researchers have been inspired by the way a human operator learns to "control/operate" a process with little or no detailed knowledge of the process dynamics. Thus they have attempted to design controllers that by adaptation and learning can solve difficult control problems in the absence of process models and human design effort. In general, this model-free neuro-control design can be stated as

$$NN : \min_{w} F\{y^* - y_p(u, \ldots)\}, \quad u = \mathcal{N}(w, \ldots), \tag{13}$$

where y_p is the output from the plant. The key feature of this direct adaptive control approach is that a process model is neither known in advance nor explicitly developed during control design.

This control design problem is often referred to as "reinforcement learning." However, this chapter chooses to refer to this class of control design as "model-free neuro-control design" as it is more appropriate in the context of the discussion [SF93]. Figure 7 is a typical representation of this class of control design.

The first work in this area was the "adaptive critic" algorithm proposed by Barto et al. [BSA83]. Such an algorithm can be seen as an approximate version of dynamic programming [Wer77, BSW90]. In this work, they posed a well-known cart-pole balancing problem and demonstrated their design concept. In this class of control design, limited/poor information is often adopted as an indication of performance criteria. For example, the objective in the cart-pole balancing problem is simply to maintain the pole in a near-upright balanced position for as long as possible. The instructional feedback is limited to a "failure" signal when the controller fails to hold the pole in an upright position. The cart-pole balancing problem has become a popular test-bed for explorations of the model-free control design concept.

Despite its historical importance and intuitive appeal, model-free adaptive neuro-control is not appropriate for most real-world applications. The plant is most likely out of control during the learning process, and few industrial processes can tolerate the large number of "failures" needed to adapt the controller.

Model-Based Neuro-Control

From a practical perspective, one would prefer to let failures take place in a simulated environment (with a model) rather than in a real plant even if the failures are not disastrous or do not cause substantial losses. As opposed to the previous case, this class of neuro-control design is referred to as "model-based neuro-control design." Similar to Equation 13, as a result, the problem formulation becomes

$$NN : \min_w \ F\{y^* - y_m(u, \ldots)\}, \quad u = \mathcal{N}(w, \ldots). \tag{14}$$

Here, y_p in Equation 13 is replaced by y_m — the model's output. In this case, knowledge about the processes of interest is required. As can be seen

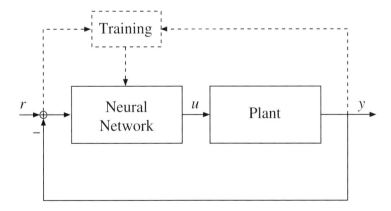

FIGURE 7. The model-free control design concept.

in Figure 8, a model replaces the plant/process in the control system (cf. Figure 7).

If a process model is not available, one can first train a second neural network to model the plant dynamics as discussed in Section 2.1. In the course of modeling the plant, the plant must be operated "normally" instead of being driven out of control. After the modeling stage, the model can then be used for control design. If a plant model is already available, a neural network controller can then be developed in a simulation in which failures cannot cause any loss but that of computer time. A neural network controller after extensive training in the simulation can then be installed in the actual control system.

In fact, these "model-based neuro-control design" approaches have not only proven effective in several studies [NW90, Tro91], but also have already produced notable economic benefits [Sta93]. These approaches can be used for both off-line control design and for on-line adaptation.

Successful demonstrations have been performed for the "truckbackerupper" problem [NW90] and a multivariable flight control problem [TGM92]. Perhaps the biggest commercial success of neuro-control to date is also based on this approach. The Intelligent Arc Furnace, developed by Neural Applications Corporation and Milltech-HOH, is a product that uses a neural network to regulate electrode position in electric arc furnaces [SS92]. A trade publication reports typical savings of over $2 million per furnace per year [Keh92]. Milltech-HOH and Neural Applications Corporation received an Outstanding Engineering Achievement Award for 1992 from the National Society of Professional Engineers.

The Intelligent Arc Furnace controller includes an interesting twist on neuro-controller development. Initially, the neural network controller is trained to mimic an existing plant controller (cf. Section 2.2). After training, the neural network then replaces the existing controller. In this latter stage, a second, pretrained, neural network is used as the process model.

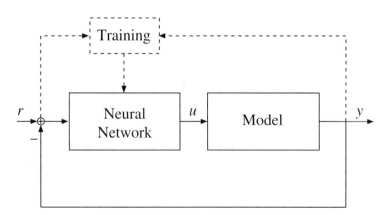

FIGURE 8. A model replaces the plant/process in the control system during the control design phase.

Both the process model network and the controller network continue to adapt on-line to compensate for plant drifts.

Nevertheless, the quality of control achieved with this approach depends crucially on the quality of the process model. If a model is not accurate enough, the trained neuro-controller is unlikely to perform satisfactorily on the real process. Without an on-line adaptive component, this neuro-controller does not allow for plant drifts or other factors that could adversely affect the performance of the control system. A controller that is highly optimized for a specific process cannot be expected to tolerate deviations from the nominal process gracefully.

Robust Model-Based Neuro-Control

The neuro-controller approaches discussed above still share a common shortcoming: A neural network must be trained for every new application. Network retraining is needed even with small changes in the control criterion, such as changes in the relative weighting of control energy and tracking response, or if the controller is to be applied to a different but similar processes. In order to circumvent such drawbacks, the concept of robustness is naturally brought into the design of a neuro-controller. In robust model-based neuro-control design, a family of process models is considered instead of just a nominal one (see Figure 9). Often such a family is specified by a range of noise models or a range of the process parameters. Robust neuro-control design can be formulated as follows:

$$NN : \min_{w} \; F\{y^* - y_{m_i}(u,\ldots)\}, \; u = \mathcal{N}(w,\ldots), \; \forall \; m_i \in \mathbf{M}, \qquad (15)$$

where m_i stands the ith member of the model family $\mathbf{M}$. Ideally, the real process to be controlled should belong to this family as well so that the controller is robust not only for the model but also for the real process.

Two aspects of robustness are commonly distinguished. Robust stability refers to a control system that is stable (qualitatively) over the entire family of processes, whereas robust performance refers to (quantitative) performance criteria being satisfied over the family [MZ89]. Not surprisingly, there is a tradeoff to achieve robustness. By optimizing a neural network controller based upon a fixed (and accurate) process model, high performance can be achieved as long as the process remains invariant, but at the likely cost of brittleness. A robust design procedure, on the other hand, is not likely to achieve the same level of nominal performance but will be less sensitive to process drifts, disturbances, and other sources of process-model mismatch.

2.3 Parameterized Neuro-Control

All the above neuro-control approaches share a common shortcoming — the need for extensive application-specific development efforts. Each application requires the optimization of the neural network controller and may

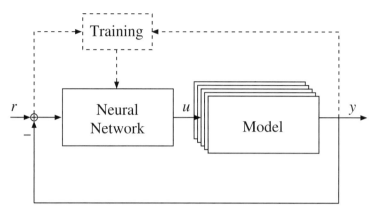

FIGURE 9. In robust control design concept, the controller is designed not only for a nominal process but also for a family of processes.

also require process model identification. The expense in time and computation is a significant barrier to widespread implementation of neuro-control systems and compares unfavorably to the cost of implementation for conventional control. Simple linear control schemes such as PID controllers, for example, enable the use of one control law in domains as diverse as building, process, and flight control.

In an attempt to avoid application-specific development, a new neuro-control design concept — parameterized neuro-control (PNC) — has evolved [SF93, SF94]. Figure 10 illustrates this PNC design strategy. The PNC controller is equipped with parameters that specify process characteristics and those that provide performance criterion information. For illustration purposes, a PNC can be conceptually formulated as follows:

$$NN : \min_{w} \; F(\xi)\{y^* - y_{m_i}(\theta, u, \ldots)\}, \; u = \mathcal{N}(w, \hat{\theta}, \xi, \ldots), \; \forall \; m_i(\theta) \in \mathbf{M}(\theta), \tag{16}$$

where ξ designates the parameter set that defines the space of performance criteria, θ stands for the process parameter set, $\hat{\theta}$ is the estimates for process parameters, and again $\mathbf{M}(\theta)$ is a family of parameterized models $m_i(\theta)$ in order to account for errors in process parameters estimates θ.

In fact, the two additional types of parameters (ξ and θ) make a PNC generic. A PNC is generic in two respects: 1) the process model parameters θ facilitate its application to different processes and 2) the performance parameters ξ allow its performance characteristics to be adjustable, or *tunable*. For example, if a PNC is designed for first-order plus delay processes, the process parameters (i.e., process gain, time constant, and dead time) will be adjustable parameters to this PNC. Once developed, this PNC requires no application-specific training or adaptation when applied to a first-order plus delay process. It only requires estimates of these process parameters. These estimates do not have to be accurate because the robustness against such inaccuracy is considered in the design phase. Notice that the parameters $\hat{\theta}$ used as input to the PNC are not identical to the parameters θ used

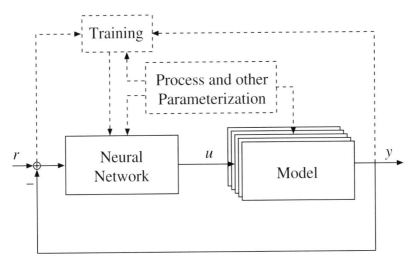

FIGURE 10. PNC control design is to design not only a robust but also a generic controller.

in the process model simulation. Parameters that specify the performance criterion can be, for example, the value of maximum allowable overshoots, desired settling times or rise times, or integral absolute errors when encountering particular setpoint changes or disturbances. The resulting controller can be featured by a tuning knob that an operator can easily understand for controlling the process. Using such tuning knobs, say a "settling time knob" (see Figure 11), an operator can set the controller so that it makes the process settle faster or slower in the presence of a disturbance. To do so, the operator does not need any sophisticated knowledge of control theory or extensive practice. Figure 11 presents a plausible easy-to-use PNC in comparison with a conventional PID controller. The performance criteria such as settling time or maximum overshoot can be directly *tunable* by an operator.

3 Optimization Aspects

The previous section has discussed various neuro-control design approaches. This section further addresses the optimization aspects in these neuro-control design problems. For convenience, all the optimization problems are listed in Table 1. In the second column of the table, "S" indicates a problem for which simple supervised learning is appropriate, whereas "C" indicates that a closed-loop system is needed in order to solve the associated optimization problem. A closed-loop system consists of a plant or a model, and a feedback controller. It is needed for evaluating the controller's performance against the criterion.

As all the neuro-control design strategies require the solution of an optimization problem, solution techniques become crucial. Depending on the

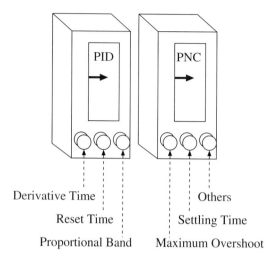

FIGURE 11. A plausible PNC can be equipped with tunable knobs, such as "Settling Time Knob" or "Maximum Overshoot Knob." With such a PNC it can be much easier for an operator to set the tuning parameters in order to achieve a desirable control performance without basic knowledge of control theory.

characteristics of the application, a particular optimization algorithm may or may not be appropriate. In particular, gradient-based optimization algorithms, which require the availability of the partial derivatives of J with respect to the decision variables, are not appropriate for applications involving nondifferentiable objective functions. To further address the optimization aspects of neuro-control design, this section hence discusses two classes of optimization algorithms: gradient and nongradient algorithms.

3.1 Gradient Algorithms

As a universal function approximator, a neural network has a remarkable feature in providing gradient information. Gradient information of its outputs with respect to its weights as well as its inputs can be easily calculated in a localized fashion. Such gradient information can be backpropagated from its outputs through its internal variables (e.g., hidden layers) and then the inputs of the network. The well-known backpropagation (BP) algorithm [RHW86] is one of the best examples to illustrate this feature. In the BP algorithm, all the gradients (Jacobian of J with respect to w) can be calculated with merely localized information.

A gradient algorithm can be applied to items labeled with "S" in Table 1. In all indirect neuro-control approaches as well as the controller modeling approach, the neural network is trained over a set of training data collected from its target. The target is an existing process or an existing controller. Using the training data, the neural network can then be trained via supervised training approaches. One key advantage of the supervised learning is that the gradient information, i.e., the Jacobian $\partial J/\partial w$, can be easily calculated. The analytical Jacobian is available as long as F or F' is in a differentiable form. Often, F takes the form of a 2-norm of its arguments (e.g. Equation 3) because of its simple differential form.

In Table 1, neuro-control problems labeled with "C" are in fact typical formulations of optimal control. These optimal control problems require a clear mathematical definition of the performance criteria. The optimization issue in these optimal control problems is in fact typical in the calculus of variations [WJ92]. The time-variant characteristics of the decision variables

TABLE 1. Objective functions of all neuro-control designs previously discussed. Note: MPC: Model Predictive Control; IM: Inverse Model; AT: Auto-Tuner; CM: Controller Modeling; MFNC: Model-Free Neuro-Control; MBNC: Model-Based Neuro-Control; RMBNC: Robust Model-Based Neuro-Control; PNC: Parameterized Neuro-Control. Notations in the second column: S: Supervised learning applied; C: Closed-loop system required.

Indirect Neuro-Control

MPC	S	$\min_{w} \; F\{y_p - y_n(w, \ldots)\}$	Eq. 2
		or $\min_{w} \; F\{y_p - y_m(\theta, \ldots)\}; \; \theta = \mathcal{N}(w, \ldots)$	Eq. 6
	C	$\min_{u} \; F'\{y^* - y_n(u, \ldots)\}$	
		or $\min_{u} \; F'\{y^* - y_m(u, \ldots)\}$	Eq. 4
IM	S	$\min_{w} \; F\{u^*_{p-1} - u_n(w, \ldots)\}$	Eq. 7
		Note: do not require optimization. $u^* = \mathcal{N}(y^*, \ldots)$	Eq. 8
AT	C	$\min_{\eta} \; F'\{y^* - y_{p/m}(u, \ldots)\}; \; u = C(\eta_c, \ldots)$	Eq. 10
	S	$\min_{w} \; F\{\eta^* - \eta_n(w, \ldots)\}$	Eq. 9

Direct Neuro-Control

CM	C	$\min_{C} \; F'\{y^* - y_{p/m}(u_c, \ldots)\}; \; u_c = C(\ldots)$	Eq. 12
	S	$\min_{w} \; F\{u_c - u_n(w, \ldots)\}$	Eq. 11
MFNC	C	$\min_{w} \; F\{y^* - y_p(u, \ldots)\}; \; u = \mathcal{N}(w, \ldots)$	Eq. 13
MBNC	C	$\min_{w} \; F\{y^* - y_m(u, \ldots)\}; \; u = \mathcal{N}(w, \ldots)$	Eq. 14
RMBNC	C	$\min_{w} \; F\{y^* - y_{m_i}(u, \ldots)\};$	
		$u = \mathcal{N}(w, \ldots), \; \forall \, m_i \in \mathbf{M}$	Eq. 15
PNC	C	$\min_{w} \; F(\xi)\{y^* - y_{m_i}(\theta, u, \ldots)\};$	
		$u = \mathcal{N}(w, \hat{\theta}, \xi, \ldots), \; \forall \, m_i(\theta) \in \mathbf{M}(\theta)$	Eq. 16

(i.e., states and control signals) make the problem extremely complex. In optimal control theory, however, the introduction of "costate" variables provides a substantial reduction of complexity — the optimization problem becomes a two-point boundary value problem. The costate variables are time-variant Lagrange multipliers as opposed to those in a time-invariant constrained optimization problem.

Among all solution techniques to the boundary value problem, the gradient descent algorithm is probably the most straightforward one. Gradient algorithms are useful as long as the Jacobian with respect to w or u is computable. A commonly adopted algorithm in neural network training for a dynamic system is "backpropagation through time" [Wer90, SMW92]. The use of an analytical Jacobian ($\partial F/\partial u$ or $\partial F/\partial w$) requires that $\partial y_p/\partial w$ or $\partial y_p/\partial u$ be available (in addition to the differentiability of F), in contrast to the "S" problems of Table 1.

Of course, the convergence rate of gradient descent can be greatly improved with higher-order gradient information, such as the Hessian (second order) With second-order derivatives, Newton methods can be used. Additionally, the Levenberg-Marquardt algorithm provides an adaptive combination of a first- and second-order search. However, the second-order gradient computation is often expensive. Without exact Hessian information, the conjugate gradient method can also provide a significant acceleration over the standard gradient descent. In fact, the "momentum" term often used with backpropagation or backpropagation through time is a simplified version of the conjugate gradient method. Recently, techniques based upon the Kalman filter have been developed and used for neuro-control design [PF94]. The recursive nature of Kalman filter methods renders these techniques well-suited for on-line adaptation. For further examples of second-order methods for neural network optimization, refer to van der Smagt [vdS94].

Gradient algorithms have played a crucial role in progress in control science, and they are also a popular choice in neuro-control design. However, these algorithms are useful only when gradients are available (e.g., F differentiable), and when cost functions are convex. For control design, the near-exclusive reliance on gradient-based optimization will result in relatively less progress on problems with any of the following characteristics — all highly relevant to neuro-control:

- Nonlinear processes or process models;

- Nonlinear control structures;

- Nonquadratic cost functions; and

- More generally, nondifferentiable cost functions.

3.2 Nongradient Algorithms

Gradient information provides a guaranteed direction of decreasing error. In its absence, an optimization algorithm inevitably takes on an exploratory, trial-and-error aspect. A parameter vector may be postulated with acceptance contingent on its cost function value.

The simplest non-gradient-based optimization algorithm may be random search, and explorations of it have a long history, e.g., [Bro58, Mat65]. One variation, the "creeping random method" is as follows [Bro58]: From some nominal starting point in parameter space, random perturbations according to some distribution are attempted until one that reduces the cost function value is found. The perturbed point is then accepted as the new nominal point and the process repeated. Several extensions of this scheme are possible:

- The mean and variance of the distribution can be adapted based on results of recent trials [Mat65]. As a special case, successful perturbations can be (recursively) reapplied — a variation that has been christened "chemotaxis" [BA91]. An application of chemotaxis to neuro-control design is described by Styer and Vemuri [SV92].

- The acceptance of a perturbed point can be a stochastic decision. In "simulated annealing," for example, perturbations that reduce the cost function are always accepted, but increased cost functions are also sometimes accepted [KGV83]. The probability of this acceptance is a function of the amount of increase and of the "temperature" of the optimization "system," which is gradually cooled from a high nondiscriminating temperature to a state in which cost function increases are deterministically rejected.

- Instead of maintaining a single nominal point, a "population" of points can be maintained [FOW66]. There are a number of variations on this "evolutionary computing" theme. Updates of the population can be based on perturbations of individual points ("mutations") and they can also rely on "multiparent" operators. The latter case is typical of genetic algorithms [Gol89], and the combinational operator is "crossover" — a splicing of coded representations of two parent individuals. An application of genetic algorithms to neuro-control design is described by Wieland [Wie91].

In some cases, non-gradient-based algorithms are more usefully viewed as extensions of methods other than simple random search. Thus, the "adaptive critic method" [BSA83] can be considered a stochastic form of dynamic programming [Wer77, BSW90]. As allusion to this line of research in the current context implies, non-gradient-based methods for model-based neuro-control design are in principle applicable to model-free neuro-control design as well. Care must be exercised, however. Real processes do not accord the same freedom as simulations for evaluating control strategies.

3.3 To General Nonlinear Control Design

An underlying theme of this chapter has been that the various approaches to the applications of neural networks in control systems can be differentiated based on the optimization problem that must be solved for each. The fact that process models or controllers are implemented as neural networks is in fact of little consequence — no assumptions have been made regarding the nature of the nonlinear structures.

An immediate consequence is that the above discussion applies to various neural network models, including radial basis function networks, multilayer perceptrons of arbitrary feedforward structure, and recurrent networks. Further, much of this chapter can be read from the perspective of general nonlinear models and controllers. Whereas the existing literature in nonlinear controls has often focused on restricted nonlinear forms that are amenable to theoretical and analytical development, this chapter has been concerned with the conceptual treatment of arbitrary nonlinear structures. This work is thus relevant not only to neuro-control, and as a second example, a brief discussion on some research in fuzzy control design within this framework is to follow.

The classical fuzzy controller model has discontinuities and is thus not everywhere differentiable (a consideration of little current relevance to neuro-control). Thus non-gradient-based optimization algorithms are of particular importance, and successful studies have been conducted with both genetic algorithms [Wig92, KG93] and with reinforcement learning methods [Lee91, BK92]. There are a variety of parameters that can be modified in the fuzzy control structure. For example, Lee [Lee91] and Wiggins [Wig92] modify peak values of membership functions whereas Karr and Gentry [KG93] search a space of trapezoidal membership functions.

In order to overcome the long convergence times that non-gradient-based algorithms can require, some researchers have used differentiable models of fuzzy controllers. One approach for realizing differentiable fuzzy controllers is to train a neural network using data from the fuzzy controller [IMT90, PP90]. Gradient-based algorithms can then be used for adapting the neuro-fuzzy controller, either as a controller modeling task (cf. Equation 11) [LL91, IFT93] or analogously to model-based neuro-control design (cf. Equation 14 or Equation 15). Werbos [Wer92] makes the general point that if differentiable fuzzy models are adopted, then a variety of existing techniques in neuro-control are readily applicable. In particular, backpropagation through time can be used for optimizing differentiable fuzzy controllers.

The availability of first-principles models has generally been assumed in off-line fuzzy control design. Empirical models can also, of course, be used, and in particular Foslien and Samad [FS94] employ a neural network process model.

4 PNC Design and Evolutionary Algorithm

To discuss the appropriateness of the nongradient algorithms as well as to demonstrate the promising advantage of the PNC design concept, a simple PNC design is presented here. As a PNC is intended to be not only a robust but also a generic controller, the complexity of PNC design is daunting. For practical purposes, the process models in this study have been single-input and single-output linear systems with delays.

4.1 PNC Problem Formulation

Particularly, this chapter chooses to demonstrate the cases where first-order linear systems plus delay

$$\frac{K_p e^{-\tau_d s}}{T_p s + 1} \tag{17}$$

are used. The class of these linear models is defined by a range of $K_{\min} < K_p < K_{\max}$, $T_{\min} < T_p < T_{\max}$, and $\tau_{\min} < \tau_d < \tau_{\max}$, respectively. In fact, $K_p = 1$ is sufficient because the process model is a linear one. By simple scaling of its output, the resultant controller can be adjusted for any first order plus delay process with any different magnitude of K_p.

In this experiment, the PNC is a feedforward neural network with dynamic inputs $e(t)$, $\int e(t)dt$, and $de(t)/dt$. In addition, the PNC also takes the two estimated model parameters $\hat{T}_p$ and $\hat{\tau}_d$ as inputs. The PNC is optimized for robustness in the continuous-time parameters space by allowing for estimation errors as follows:

$$|1 - \frac{\hat{K}_p}{K_p}| = |1 - \frac{\hat{T}_p}{T_p}| = |1 - \frac{\hat{\tau}_d}{\tau_d}| = \alpha < 1 \tag{18}$$

in order to account for robustness. The PNC also takes an additional performance parameter a_{st}. This particular performance parameter is defined as a weight factor for the importance of short settling time. The closed-loop performance cost function is defined as follows:

$$NN: \min_{w} \operatorname*{ave}_{a_{st} \in [0,1]} \{a_{st} f_1(T_{st}) + (1 - a_{st}) f_2(T_{st})\}. \tag{19}$$

Here a_{st} is the weight on settling time T_{st}, $1 - a_{st}$ is the weight on rise time T_{rt}, and $f_1(\cdot)$ and $f_2(\cdot)$ are clipped linear scaling functions as follows:

$$f_1(t) = \begin{cases} 0 & \text{if } t \leq 9, \\ (t-9)/21 & \text{if } 9 < t < 30, \\ 1 & \text{if } t \geq 30; \end{cases}$$

$$f_2(t) = \begin{cases} 0 & \text{if } t \leq 4, \\ (t-4)/2.7 & \text{if } 4 \leq\: < t < 6.7, \\ 1 & \text{if } t \geq 6.7. \end{cases} \qquad (20)$$

As can be seen in Equation 19, the larger the a_{st}, the faster the settling time the PNC should be achieving, and vice versa. The settling time T_{st} is defined as the time the process takes before its step response is within 5% of its steady-state value. The average for Equation 19 is empirically estimated over 11 values of a_{st}: 0, 0.1, 0.2, ..., 0.9, 1.0. For each run, the process setpoint is changed from 0 to 1.0. The cost function is computed from the closed-loop response to the setpoint change. For comparison, this study uses a PI controller optimized for minimum settling time over a two-dimensional space of K_c and K_i. The optimal parameters for the PI controller are $K_c = 0.243$ and $K_i = 0.177$.

4.2 Evolutionary Algorithm

Apart from the fact that response features such as settling time or overshoot are not analytically differentiable, cost function surfaces for PNC design are densely populated by local minima. Conventional neural network training methods cannot easily produce satisfactory results in this case. After initial experiments with gradient-descent techniques [RHW86], genetic algorithms [Gol89], and chemotaxis [vdS94], the algorithm finally adopted is an evolutionary optimization algorithm that incorporates chemotaxis [SS94]. The optimization algorithm is outlined below:

1. Generate an initial set of weight vectors. Evaluate the cost function for each of them.

2. Locate the weight vector w_{max} that results in the maximum cost function value J_{max}.

3. Select a weight vector w at random (uniform distribution) from the population.

4. Generate a perturbation standard deviation σ.

5. Generate a perturbation vector $\delta \in N(0, \sigma^2)$.

6. Generate a new weight vector $w_{new} = w + \delta$ by adding the perturbation vector to the selected weight vector. Evaluate the new cost function J_{new} with respect to the new weight vector.

7. If $J_{new} \leq J_{max}$, replace the weight vector w_{max} with the new weight vector w_{new}, and go to Step 6.

8. Go to Step 2 until a termination point (e.g., a time limit).

However, optimization runs are still prolonged affairs — up to a week on an HP700 series workstation — but results have been positive and promising. Although PNCs are expensive to develop, they are cheap to operate. Most of the neural networks investigated in this study seldom require more than 100 weights. An optimized neural network can be implemented as a hard-wired module in a controller that can then be installed and used on any process and for any criterion within its design space.

4.3 Results and Discussion

One advantage of a conventional PID controller is that the controller tuning parameters (the proportional band, reset time, and derivative time) are easily understandable without any sophisticated knowledge of control theory. As for the generic feature, a PID controller can be used for a wide variety of different processes as long as its parameters are properly tuned. A relatively well-trained operator can easily set these parameters such that the controller will give a satisfactory performance. However, these PID tuning parameters do not relate to performance criteria such as settling time in a directly sensible manner. On the contrary, a PNC can be designed with a tuning parameter monotonically related to a performance criterion, say settling time in this study, during the development phase. The settling time increases as the tuning parameter increases. As shown in Figure 12, where a_{st} stands for the dial of such a "settling time knob," the PNC clearly outperforms an optimal PI controller. More importantly, the settling time decreases as a_{st} increases as expected. To "tune" a "generic" PNC controller, one only has to provide a set of estimated process characteristics (such as process gain, process time constant, and delay). Once it is "tuned," an operator can simply turn a knob — a *settling time knob* — to achieve what he/she would like the controller to perform.

While a generic PNC is easy to use, it also guarantees better robustness than an optimal PI. A PNC control tuned for a nominal process can tolerate much more severe process parameter uncertainties than an optimal PI controller. In Figure 13 for example, both PNC and PI controllers are tuned for a nominal process with $T_p = 1$ and $K_p = 1$ (the nominal process parameters). As can be seen in the figure, PNC can operate on a process with process parameter uncertainties up to 300%, whereas an optimal PI can tolerate only about 60%.

5 Conclusions

This chapter briefly reviews the progress of neuro-control design in terms of an optimization problem. In the indirect neuro-control approaches, neural networks are usually used to develop a predictor. The predictor is usually a process model or a controller's parameter predictor (i.e., auto-tuner); thus it does not directly control the process. In the direct neuro-control design

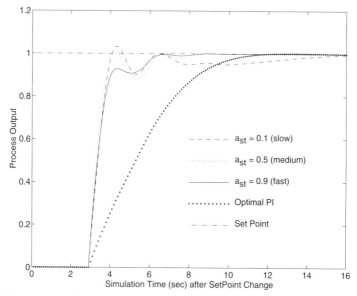

FIGURE 12. A PNC equipped with a settling time tuning factor a_{st} can be used to adjust the settling time of a process. In the design, the settling time decreases as a_{st} increases.

approaches, the concept has progressed from simple controller modeling, to model-free controller design, to nominal model-based controller design, and finally to a robust model-based controller design. This progression towards greater sophistication is accompanied by an increasing application-specific design effort requirement. Ease of use suffers.

In some ways, the concept behind the design of the "parameterized neurocontroller" is extended from robust control design. Nonetheless, the PNC design concept is intended to avoid application-specific development efforts and to provide the simplicity of implementation and ease of use. As a result, the PNC is not only a robust controller, but also a generic one. For SISO systems, PNC parameters are similar to those of a PID controller, but they are easier to understand and more directly related to desired performance criteria. The PNC design concept is also illustrated by a "settling time knob" example. In this example, results from a PNC equipped with a "settling time tuning knob" verify these claims.

As the PNC design requires an extremely complicated formulation of an optimization problem, this chapter further addresses the optimization aspects of the entire neuro-control design. It is shown that constraining assumptions that are enforced in the formulation of a control design problem can be beneficial for its solution. For appropriately formulated problems, gradient-based optimization can ensure reliable, and often rapid, convergence. An alternative control design methodology is also possible: the

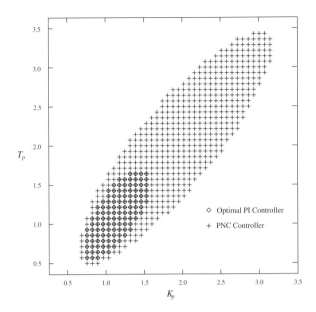

FIGURE 13. A PNC can operate on a process with much more severe parameter uncertainties than an optimal PI controller when both are tuned for a nominal process.

problem formulation can be centered around application needs. The utility of gradient-based optimization may be problematic in such cases, but non-gradient-based methods such as evolutionary optimization are now a practical recourse.

6 References

[BA91] H. J. Bremermann and R. W. Anderson. How the brain adjusts synapses—maybe. In R. S. Boyer, editor, *Automated Reasoning: Essays in Honor of Woody Bledsoe*, pages 119–147. Kluwer Academic Publishers, Dordrecht, The Netherlands, 1991.

[BK92] H. Berenji and P. Khedkar. Learning and tuning fuzzy logic controllers through reinforcements. *IEEE Transactions on Neural Networks*, 3:724–740, 1992.

[BM90] N. Bhat and T. J. McAvoy. Use of neural nets for dynamic modeling and control of chemical process systems. *Computers & Chemical Engineering*, 14(5):573–583, 1990.

[Bro58] S. H. Brooks. A discussion of random methods for seeking maxima. *Operations Research*, 6:244–251, 1958.

[BSA83]   A. G. Barto, R. S. Sutton, and C. Anderson. Neuronlike elements that can solve difficult learning control problems. *IEEE Transactions on Systems, Man, & Cybernetics*, 13:835–846, 1983.

[BSW90]   A. G. Barto, R. S. Sutton, and C. J. C. H. Watkins. Sequential decision problems and neural networks. In D. S. Touretzky, editor, *Advances in Neural Information Processing Systems*, chapter 2. Morgan Kaufmann, San Mateo, California, 1990.

[CBG90]   S. Chen, S. A. Billings, and P. M. Grant. Non-linear system identification using neural networks. *International Journal of Control*, 51(6):1191–1214, 1990.

[FOW66]   L. J. Fogel, A. J. Owens, and M. J. Walsh. *Artificial Intelligence Through Simulated Evolution*. Wiley, New York, 1966.

[FS94]   W. Foslien and T. Samad. Fuzzy controller synthesis with neural network process models. In S. Goonatilake and S. Khebbal, editors, *Intelligent Hybrid Systems*, pages 23–46. Wiley, New York, 1994.

[Gol89]   D. E. Goldberg. *Genetic Algorithms in Search, Optimization, and Machine Learning*. Addison-Wesley, Reading, Massachusetts, 1989.

[HSZG92]   K. J. Hunt, D. Sbarbaro, R. Żbikowski, and P. J. Gawthrop. Neural networks for control systems—A survey. *Automatica*, 28(6):1083–1112s, 1992.

[IFT93]   H. Ishibuchi, R. Fujioka, and H. Tanaka. Neural networks that learn from fuzzy if-then rules. *IEEE Transactions on Fuzzy Systems*, 1:85–97, 1993.

[IMT90]   T. Iwata, K. Machida, and Y. Toda. Fuzzy control using neural network techniques. In *Proceedings of the International Conference on Neural Networks, 1990*, volume 3, pages 365–370. IEEE, 1990.

[Keh92]   B. Kehoe. Eaf controller passes intelligence test. *Iron Age*, pages 28–29, March 1992.

[KG93]   C. L. Karr and J. Gentry. Fuzzy control of pH using genetic algorithms. *IEEE Transactions on Fuzzy Systems*, 1:46–53, 1993.

[KGV83]   S. Kirkpatrick, C. D. Gelatt, and M. P. Vecchi. Optimization by simulated annealing. *Science*, 220:671–680, 1983.

[KSF92] A. F. Konar, T. Samad, and W. Foslien. Hybrid neural network/algorithmic approaches to system identification. In *Proceedings of the 3rd IFAC Symposium on Dynamic and Control of Chemical Reactors, Distillation Columns, and Batch Processes*, pages 65–70. Pergamon Press, Oxford, 1992.

[Lee91] C.-C. Lee. A self-learning rule-based controller employing approximate reasoning and neural net concepts. *International Journal of Intelligent Systems*, 6:71–93, 1991.

[Lju87] L. Ljung. *System Identification: Theory for the User.* Prentice-Hall, Englewood Cliffs, New Jersey, 1987.

[LL91] C. T. Lin and C. S. G. Lee. Neural-network-based fuzzy logic control and decision system. *IEEE Transactions on Computers*, 40:1320–1336, 1991.

[Mat65] J. Matyas. Random optimization. *Automation and Remote Control* 26:246–253, 1965.

[MMW90] W. T. Miller, R. S. Sutton, and P. J. Werbos, editors. *Neural Networks for Control.* MIT Press, Cambridge, Massachusetts, 1990.

[MZ89] M. Morari and E. Zafiriou. *Robust Process Control.* Prentice-Hall, Englewood Cliffs, New Jersey, 1989.

[NW90] D. Nguyen and B. Widrow. The truck backer-upper: An example of self-learning in neural networks. In *Neural Networks for Control*, pages 287–300. W. T. Miller and P. J. Werbos, editors, MIT Press, Cambridge, Massachusetts, 1990.

[PF94] G. V. Puskorius and L. A. Feldkamp. Neurocontrol of nonlinear dynamical systems with Kalman filter-trained recurrent networks. *IEEE Transactions on Neural Networks*, volume 5, pages 279–297, 1994.

[Pom91] D. A. Pomerleau. Neural network-based vision processing for autonomous robot guidance. In S. K. Rogers, editor, *Applications of Artificial Neural Networks II*, volume 1469, pages 121–128. SPIE, Bellingham, Washington, 1991.

[PP90] A. Patrikar and J. Provence. A self-organizing controller for dynamic processes using neural networks. In *Proceedings of the International Conference on Neural Networks, 1990*, volume 3, pages 359–364. IEEE, 1990.

[PS92]   M. Pottman and D. Seborg. A nonlinear predictive control strategy based on radial basis function networks. In *Proceedings of the 3rd IFAC Symposium on Dynamic and Control of Chemical Reactor, Distillation Columns, and Batch Processes*, pages 309–314, 1992.

[PSY88]   D. Psaltis, A. Sideris, and A. A. Yamamura. A multilayered neural network controller. *IEEE Control Systems Magazine*, 8(3):17–21, April 1988.

[PU91]   D. C. Psichogios and L. H. Ungar. Direct and indirect model based control using artificial neural networks. *Industrial & Engineering Chemistry Research*, 30(12):2564–2573, 1991.

[PU92]   D. C. Psichogios and L. H. Ungar. A hybrid neural network-first principles approach to process modeling. *AIChE Journal*, 38(10):1499–1511, 1992.

[RFJ92]   A. E. B. Ruano, P. J. Fleming, and D. I. Jones. Connectionist approach to PID tuning. *IEE Proceedings, Part D*, 129:279–285, 1992.

[RHW86]   D. Rumelhart, G. Hinton, and R. Williams. Chapter 8: Error propagation and feedforward networks. In Rumelhart and McClelland, editors, *Parallel Distributed Processing, volume 1 and 2*. MIT Press, Cambridge, Massachusetts, 1986.

[SBMM92] H. T. Su, N. V. Bhat, P. A. Minderman, and T. J. McAvoy. Integrating neural networks with first principles model for dynamic modeling. In *Proceedings of the 3rd IFAC Symposium on Dynamic and Control of Chemical Reactor, Distillation Columns, and Batch Processes*, pages 77–81, 1992.

[SDBM91] J. Saint-Donat, N. Bhat, and T. J. McAvoy. Neural net based model predictive control. *International Journal of Control*, 54(6):1453–1468, 1991.

[SF93]   T. Samad and W. Foslien. Parametrized neuro-controllers. In *Proceedings of the 8th International Symposium on Intelligent Control*, pages 352–357. 1993.

[SF94]   T. Samad and W. Foslien. Neural networks as generic nonlinear controllers. In *Proceedings of the World Congress on Neural Networks, San Diego, 1994*, pages I-191–I-194. Lawrence Erlbaum, Hillsdale, New Jersey, 1994.

[SM93a]   H. T. Su and T. J. McAvoy. Applications of neural network long-range predictive models for nonlinear model predictive control. *Journal of Process Control*, 1993.

10. Neuro-Control Design 287

[SM93b]   H. T. Su and T. J. McAvoy. Neural model predictive models of nonlinear chemical processes. In *Proceedings of the 8th International Symposium on Intelligent Control*, pages 358–363, 1993.

[SMW92]   H. T. Su, T. J. McAvoy, and P. J. Werbos. Long-term predictions of chemical processes using recurrent neural networks: A parallel training approach. *Industrial & Engineering Chemistry Research*, 31:1338–1352, 1992.

[SS92]   W. E. Staib and R. B. Staib. The Intelligent Arc Furnace^{TM} controller: A neural network electrode position optimization system for the electric arc furnace. In *Proceedings of the International Conference on Neural Networks, 1992*, volume 3, pages 1–9. IEEE Press, Piscataway, New Jersey, 1992.

[SS94]   T. Samad and H. T. Su. Neural networks as process controllers—optimization aspects. In *Proceedings of the American Control Conference*, volume 3, pages 2486–2490. IEEE Press, Piscataway, New Jersey, 1994.

[Sta93]   W. E. Staib. The Intelligent Arc Furnace^{TM}: Neural networks revolutionize steelmaking. In *Proceedings of the World Congress on Neural Networks*, pages I:466–469, 1993.

[SV92]   D. L. Styer and V. Vemuri. Adaptive critic and chemotaxis in adaptive control. In *Proceedings of the Artificial Neural Networks in Engineering*, volume 1, pages 161–166. ASME Press, Fairfield, New Jersey, 1992.

[SW90]   D. A. Sofge and D. A. White. Neural network based process optimization and control. In *Proceedings of the IEEE Conference on Decision & Control*, pages 3270–3276, 1990.

[Swi90]   R. W. Swiniarski. Novel neural network based self-tuning pid controller which uses pattern recognition techniques. In *Proceedings of the American Control Conference*, pages 3023–3024, 1990.

[TGM92]   T. Troudet, S. Garg, and W. Merrill. Design and evaluation of a robust dynamic neurocontroller for a multivariable aircraft control problem. In *Proceedings of the International Conference on Neural Networks, 1992*, volume 1, pages 305–314. IEEE Press, Piscataway, New Jersey, 1992.

[TK94]   M. L. Thompson and M. A. Kramer. Modeling chemical processes using prior knowledge and neural networks. *AIChE Journal*, 40(8):1328–1340, 1994.

[Tro91]   T. Troudet. Towards practical control design using neural computation. In *Proceedings of the International Conference on Neural Networks, 1991*, volume 2, pages 675–681. IEEE Press, Piscataway, New Jersey, 1991.

[TSSM92] K. O. Temeng, P. D. Schnelle, H. T. Su, and T. J. McAvoy. Neural model predictive control of an industrial packed bed reactors, *Journal of Process Control*, 5:19–28, 1995.

[UPK90]   L. H. Ungar, B. A. Powell, and E. N. Kamens. Adaptive networks for fault diagnosis and process control. *Computers & Chemical Engineering*, 14(4/5):561–572, 1990.

[vdS94]   P. P. van der Smagt. Minimisation methods for training feedforward neural networks. *Neural Networks*, 7:1–12, 1994.

[Wer77]   P. J. Werbos. Advanced forecasting methods for global crisis warning and models of intelligence. *General Systems Yearbook*, 22:25–38, 1977.

[Wer90]   P. J. Werbos. Backpropagation through time: What it is and how to do it? *Proceedings of the IEEE*, 78:1550–1560, 1990.

[Wer92]   P. J. Werbos. Neurocontrol and fuzzy logic: Connections and designs. *International Journal of Approximate Reasoning*, 6:185–219, 1992.

[Wie91]   A. P. Wieland. Evolving neural network controllers for unstable systems. In *Proceedings of the International Conference on Neural Networks, 1991*, volume 2, pages 667–674. IEEE, Piscataway, New Jersey, 1991.

[Wig92]   R. Wiggins. Docking a truck: A genetic fuzzy approach. *AI Expert*, May 1992.

[WJ92]   D. A. White and M. I. Jordan. Optimal control: A foundation for intelligent control. In D. A. White and D. A. Sofge, editors, *Handbook of Intelligent Control: Neural, Fuzzy, and Adaptive Approaches*. Van Nostrand Reinhold, Princeton, New Jersey, 1992.

[Yds90]   B. E. Ydstie. Forecasting and control using adaptive connectionist networks. *Computers & Chemical Engineering*, 14:583–599, 1990.

[ZN42]   J. B. Ziegler and N. B. Nichols. Optimum settings for automatic controllers. *Transactions of the ASME*, 64:759–768, 1942.

Chapter 11

Reconfigurable Neural Control in Precision Space Structural Platforms

Gary G. Yen[1]

ABSTRACT The design of control algorithms for flexible space structures possessing nonlinear dynamics that are often time-varying and ill-modeled presents great challenges for all conventional methodologies. These limitations have recently led to the pursuit of autonomous neural control systems.

In this chapter, we propose the innovative use of a hybrid connectionist system as a learning controller with reconfiguration capability. The ability of connectionist systems to approximate arbitrary continuous functions provides an efficient means of vibration suppression and trajectory slewing for precision pointing of flexible space structures. Embedded with adjustable time-delays and interconnection weights, an adaptive radial basis function network offers a real-time modeling mechanism to capture most of the spatiotemporal interactions among the structure members. A fault diagnosis system is applied for health monitoring to provide the neural controller with various failure scenarios. Associative memory is incorporated into an adaptive architecture to compensate for catastrophic changes of structural parameters by providing a continuous solution space of acceptable controller configurations, which is created a priori. This chapter addresses the theoretical foundation of a feasible reconfigurable control architecture and demonstrates its applicability via specific examples.

1 Connectionist Learning System

Contemporary control design methodologies (e.g., robust, adaptive, and optimal controls) face limitations for some of the more challenging realistic systems. In particular, modern space structures are built of lightweight composites and equipped with distributive piezoelectric sensors and actuators. These flexible structures, which are likely to be highly nonlinear, time-varying, and poorly modeled, pose serious difficulties for all currently

---

[1] Adapted and revised from "Reconfigurable Learning Control in Large Space Structures" by Gary G. Yen, which appeared in *IEEE Transactions on Control Systems Technology* 2(4): pp. 362–370; December 1994.

advocated methods, as summarized in [WS92, AP92]. These control system design difficulties arise in a broad spectrum of aerospace applications such as military robots, surveillance satellites, or space vehicles. The ultimate autonomous control, intended to maintain the above acceptable performance over an extended operating range, can be especially difficult to achieve due to factors such as high dimensionality, multiple inputs and outputs, complex performance criteria, operational constraints, imperfect measurements, as well as the inevitable failures of various actuators, sensors, or other components. Indeed, an iterative and time-consuming process is required to derive a high fidelity model in order to effectively capture all of the spatiotemporal interactions among the structural members. Therefore, the controller needs to be either exceptionally robust or adaptable after deployment. Also, catastrophic changes to the structural parameters due to component failures, unpredictable uncertainties, and environmental threats require that the controller be reconfigurable.

In this chapter, we investigate a hybrid connectionist system as a means of providing a learning controller [Fu70] with reconfiguration capability. The proposed control system integrates adaptive time-delay radial basis function (ATRBF) networks, an eigenstructure bidirectional associative memory (EBAM), and a cerebellar model articulation controller (CMAC) network. A connectionist system consists of a set of interconnected processing elements and is capable of improving its performance based on past experimental information [Bar89]. An artificial neural network (herein referred to as simply "neural network") is a connectionist system that was originally proposed as a simplified model of the biological nervous system [HKP91, HN89]. Neural networks have been shown to provide an efficient means of learning concepts from past experience, abstracting features from uncorrelated data, and generalizing solutions from unforeseen inputs. Other promising advantages of neural networks are their distributed data storage and parallel information flow, which cause them to be extremely robust with respect to malfunctions of individual devices as well as being computationally efficient. Neural networks have been successfully applied to the control of various dynamic systems, including aerospace and underwater vehicles [RÖMM94, VSP92], nuclear power plants [BU92], chemical process facilities [WMA+89], and manufacturing production lines [CSH93].

There have been many architectures (i.e., schema consisting of various neuronic characteristics, interconnecting topologies, and learning rules) proposed for neural networks over the last five years (at last count over 200). Simulation experience has revealed that success is problem-dependent. Some networks are more suitable for adaptive control whereas others are more appropriate for pattern recognition, signal filtering, or associative searching. Neural networks that employ the well known backpropagation learning algorithm [RM86] are capable of approximating any continuous functions (e.g., nonlinear plant dynamics and complex control laws) with an arbitrary degree of accuracy [HSW89]. Similarly, radial basis function networks

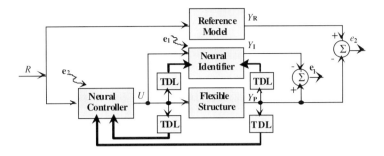

FIGURE 1. Vibration suppression neural control architecture.

[MD89] are also shown to be universal approximators [HKK90]. These model-free neural network paradigms are more effective at memory usage in solving control problems than conventional learning control approaches. An example is the BOXES algorithm, a memory intensive approach, which partitions the control law in the form of a look-up table [MC68].

Our goal is to approach structural autonomy by extending the control system's operating envelope, which has traditionally required vast memory usage. Connectionist systems, on the other hand, deliver less memory-intensive solutions to control problems and yet provide a sufficiently generalized solution space. In vibration suppression problems, we utilize the adaptive time-delay radial basis function network as a building block to allow the connectionist system to function as an indirect closed-loop controller. Prior to training the compensator, a neural identifier based on an ARMA model is utilized to identify the open-loop system. The horizon-of-one predictive controller then regulates the dynamics of the nonlinear plant to follow a prespecified reference system asymptotically as depicted in Figure 1 (i.e., the model reference adaptive control architecture) [NP90]. The reference model, which is specified by an input-output relationship $\{R, Y_R\}$, describes all desired features associated with a specific control task, e.g., a linear and highly damped system to suppress the structural vibration. As far as trajectory slewing problems are concerned, the generalized learning controller synthesized by the adaptive time-delay radial basis function network compensates the nonlinear large-space structure in a closed-loop fashion in order to follow the motion specified by the command outputs as given in Figure 2 (i.e., tapped delay lines (TDL) are incorporated to process the time-varying structural parameters as suggested in [NP90]).

The function of the neural controller is to map the states of the system into corresponding control actions in order to force the plant dynamics (Y_P) to match a certain output behavior that is specified either by the reference model (Y_R) or the command output (Y_D). However, we cannot apply the optimization procedure (e.g., gradient descent, conjugate gradient, or Newton-Raphson method) to adjust the weights of the neural controller because the desired outputs for the neural controller are not available. In [PSY88], a "specialized learning algorithm" that treats the plant as an ad-

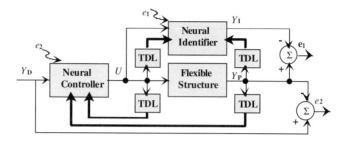

FIGURE 2. Trajectory slewing neural control architecture.

ditional unmodifiable layer of network is proposed. The output error e_2 is backpropagated through the plant to derive the neural controller output error $\hat{e}_2$. However, the authors fail to suggest a reliable way to compute $\hat{e}_2$. In [Els88], the inverse Jacobian of the plant is used to estimate $\hat{e}_2$ at each weight update, which results in a complicated and computationally expensive learning procedure. Moreover, since the plant is often not well-modeled because of modeling uncertainties, the exact partial derivatives cannot be determined. In [SS89], a "dynamic sign approximation" is utilized to determine the direction of the error surface, assuming the qualitative knowledge of the plant. This is not necessarily the case in space structure applications, which are often equipped with highly correlated parameters. To achieve the true gradient descent of the square of the error, we use "dynamic backpropagation" to accurately approximate the required partial derivatives as suggested in [NP90]. A single-layer ATRBF network is first trained to identify the open-loop system. The resulting neural identifier then serves as extended unmodifiable layers to train the compensator (i.e., another single-layer ATRBF network). If the structural dynamics are to change as a function of time, the backup neural identifier would require the learning algorithm to periodically update the network parameters accordingly [NW90].

The proposed architecture for reconfigurable neural control includes neural networks dedicated to identification and to control, structural health component assessment, and controller association retrieval and interpolation. In order to provide a clear presentation, the integration of various components in an intelligent architecture is presented first to achieve the structural reconfigurable learning control in Section 2. This is followed by discussion of each functional block in detail. For the purpose of system identification and dynamic control of flexible space structures, an adaptive time-delay radial basis function network that serves as a building block is discussed in Section 3, providing a justification to achieve real-time performance. A novel class of bidirectional associative memories synthesized by the eigenstructure decomposition algorithm is covered in Section 4 to fulfill the critical needs in real-time controller retrieval. This is followed by utilizing the cerebellar model articulation controller network for fault detection and identification of structural failures (Section 5). Specific applications to a space structural testbed are used in Section 6 to demonstrate the effectiveness of the proposed reconfigurable neural control architecture. This chapter concludes with a few pertinent observations regarding potential commercial applications in Section 7.

2 Reconfigurable Control

In the uncertain space environment, all existing methods of adaptation call for a finite time duration for exposure to the altered system as well as computational duties before a suitable controller can be determined. A controller designed for this adaptation is robust only with respect to plant variations to some degree. When the plant dynamics experience abrupt and drastic changes, the closed-loop system no longer exhibits acceptable performance and may become unstable.

Critical to autonomous system design is the development of a control scheme with globally adaptive and reconfigurable capabilities. *Reconfiguration* refers to the ability to retrieve a workable controller from the solution space (created prior to the failure). The motivation is to strive for a high degree of structural autonomy in space platforms, thereby severing the dependence of the dynamic system on *a priori* programming and perfect communications, as well as the flawless operation of the system components, while maintaining a precision pointing capability.

Existing reconfigurable control techniques often rely on computationally intensive simulations (e.g., finite-element analysis) or simple strategies such as gain scheduling [AW89] and triple modular redundancy [BCJ$^+$90]. Lately, novel design techniques — including linear quadratic control methodology [LWEB85], an adaptive control framework [MO90], knowledge-based systems [HS89], eigenstructure assignment [Jia94], and the pseudo-inverse method [GA91], to name a few — have been developed. These methods show various degrees of success in different respects. However, a common shortcoming is the computational complexity involved in reconfiguring the controller configuration while maintaining the system stability and performance level. In the present chapter, we achieve controller reconfiguration capability by integrating an eigenstructure bidirectional associative memory into a model reference adaptive control framework. In a similar spirit, bidirectional associative memory can be applied to the control of slewing flexible multibody [Yen94]. The proposed architecture is expected to maintain stability for extended periods of time without external intervention, while possibly suffering from unforeseeable perturbations. The architecture of a real-time reconfigurable control system is given in Figure 3. The adaptive control framework handles slowly varying system parameters, which commonly occur on structures exposed to an adverse space environment (e.g., increased thermal and aerodynamic load). Subsequently, as experience with the actual plant is accumulated, the learning system would be

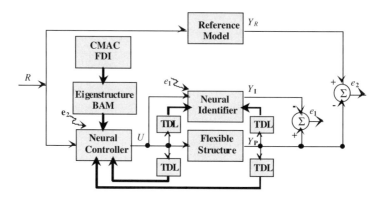

FIGURE 3. Reconfigurable neural control system architecture.

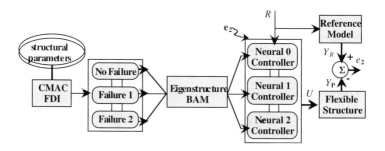

FIGURE 4. Functional diagram for controller association retrieval performed by EBAM.

used to anticipate the appropriate control or model parameters as a function of the current plant operating condition. Catastrophic changes to the system dynamics are compensated for by retrieving an acceptable controller from a *continuous* solution space, which is created beforehand and reflects a host of various system configurations. The solution space is stored within an EBAM network as opposed to a look-up table and therefore offers the capabilities of *real-time reconfiguration* and *generalization* (see Figure 4). A look-up table approach would only provide discrete controller solutions in a lengthy and sequential search. The proposed reconfiguration capability entails the design of a health monitoring procedure in detecting, isolating, and identifying adverse conditions. To achieve successful reconfiguration capabilities, we devise a reliable bidirectional associative memory (BAM) synthesized by the eigenstructure decomposition method. As pointed out in [MSY91, YM92], the eigenstructure method, which utilizes the energy function approach, guarantees the storage of a given set of desired fault scenarios/weight configurations as asymptotically stable equilibria in the state space. The assumption is made that an acceptable fault detection and identification (FDI) algorithm synthesized by the CMAC network will be used for health monitoring to provide the required information (failure index) to the eigenstructure associative memory [YK93].

3 Adaptive Time-Delay Radial Basis Function Network

Biological studies have shown that variable time-delays do occur along axons due to different conduction times and different lengths of axonal fibers. In addition, temporal properties such as temporal decays and integration occur frequently at synapses. Inspired by this observation, the time-delay backpropagation network was proposed in [WHH+88] for solving the phoneme recognition problem. In this architecture, each neuron takes into account not only the current information from all neurons of the previous layer, but also a certain amount of past information from those neurons due to delay on the interconnections. However, a fixed amount of time-delay throughout the training process has limited the usage of this architecture, possibly due to the mismatch of the temporal location in the input patterns. To overcome this limitation, Lin et al.[LDL92] developed an adaptive time-delay backpropagation network to better accommodate the varying temporal sequences and to provide more flexibility for optimization tasks. In a similar spirit, the adaptive time-delay radial basis function network is proposed in this section to take full advantages of temporal pattern matching and learning/recalling speed.

A given adaptive time-delay radial basis function network can be completely described by its interconnecting topology, neuronic characteristics, temporal delays, and learning rules. The individual processing unit performs its computations based only on local information. A generic radial basis function network is a two-layer neural network whose outputs form a linear combination of the basis functions derived from the hidden neurons. The basis function produces a localized response to input stimulus as do locally tuned receptive fields in human nervous systems. The Gaussian function network, a realization of an RBF network using Gaussian kernels, is widely used in pattern classification and function approximation. The output of a Gaussian neuron in the hidden layer is defined by

$$u_j^1 = \exp\left(-\frac{\|x - w_j^1\|^2}{2\sigma_j^2}\right), \quad j = 1, \ldots, N_1, \qquad (1)$$

where u_j^1 is the output of the jth neuron in the hidden layer (denoted by the superscript 1), x is the input vector, w_j^1 denotes the weighting vector for the jth neuron in the hidden layer (i.e., the center of the jth Gaussian kernel), σ_j^2 is the normalization parameter of the jth neuron (i.e., the width of the jth Gaussian kernel), and N_1 is the number of neurons in the hidden layer.

Equation 1 produces a radially symmetric output with a unique maximum at the center dropping off rapidly to zero for large radii. The output layer equations are described by

$$y_j = \sum_{i=1}^{N_1} w_{ji}^2 u_j^1, \quad j = 1, \ldots, N_2, \qquad (2)$$

where y_j is the output of the jth neuron in the output layer, w_{ji}^2 denotes the weight from the ith neuron in the hidden layer to the jth neuron in the output layer, u_j^1 is the output from the ith neuron in the hidden layer, and N_2 is the number of linear neurons in the output layer. Inspired by the adaptive time-delay backpropagation network, the output equation of ATRBF networks is described by

$$y_j(t_n) = \sum_{i=1}^{N_1} \sum_{l=1}^{L_{ji}} w_{ji,l}^2 u_i^1(t_n - \tau_{ji,l}^2), \quad j = 1, \ldots, N_2, \qquad (3)$$

where $w_{ji,l}^2$ denotes the weight from the ith neuron in the hidden layer to the jth neuron in the output layer with the independent time delay $\tau_{ji,l}^2$, $u_i^1(t_n - \tau_{ji,l}^2)$ is the output from the ith neuron in the hidden layer at time $t_n - \tau_{ji,l}^2$, and L_{ji} denotes the number of delay connections between the ith neuron in the hidden layer and the jth neuron in the output layer. Shared with generic radial basis function networks, adaptive time-delay Gaussian function networks have the property of undergoing *local* changes during training, unlike adaptive time-delay backpropagation networks, which experience *global* weighting adjustments due to the characteristics of sigmoidal functions. The localized influence of each Gaussian neuron allows the learning system to refine its functional approximation in a successive and efficient manner. The hybrid learning algorithm [MD89] that employs the K-means clustering for the hidden layer and the least mean square (LMS) algorithm for the output layer further ensures a faster convergence and often leads to better performance and generalization. The combination of locality of representation and linearity of learning offers tremendous computational efficiency to achieve real-time adaptive control compared to the backpropagation network, which usually takes considerable time to converge. The K-means algorithm is perhaps the most widely known clustering algorithm because of its simplicity and its ability to produce good results. The normalization parameters, σ_j^2 are obtained once the clustering algorithm is complete. They represent a measure of the spread of the data associated with each cluster. The cluster widths are then determined by the average distance between the cluster centers and the training samples

$$\sigma_j^2 = \frac{1}{M_j} \sum_{x \in \Theta_j} \|x - w_j^1\|^2, \qquad (4)$$

where Θ_j is the set of training patterns belonging to the jth cluster and M_j is the number of samples in Θ_j. This is followed by applying an LMS algorithm to adapt the time-delays and interconnecting weights in the output layer. The training set consists of input/output pairs, but now the input patterns are pre-processed by the hidden layer before being presented to the output layer. The adaptation of the output weights and time delays are derived based on error backpropagation to minimize the cost function

$$E(t_n) = \frac{1}{2} \sum_{j=1}^{N_2} (d_j(t_n) - y_j(t_n))^2, \qquad (5)$$

where $d_j(t_n)$ indicates the desired value of the jth output neuron at time t_n. The weights and time-delays are updated step by step proportionally to the opposite direction of the error gradient, respectively

$$\Delta w_{ji,l}^2 = -\eta_1 \frac{\partial E(t_n)}{\partial w_{ji,l}^2}, \qquad (6)$$

$$\Delta \tau_{ji,l}^2 = -\eta_2 \frac{\partial E(t_n)}{\partial \tau_{ji,l}^2}, \qquad (7)$$

where η_1 and η_2 are the learning rates. The mathematical derivation of this learning algorithm is straightforward. The learning rule can be summarized as follows:

$$\Delta w_{ji,l}^2 = \eta_1 (d_j(t_n) - y_j(t_n)) \sum_{l=1}^{L_{ji}} u_i^1(t_n - \tau_{ji,l}^2), \qquad (8)$$

$$\Delta \tau_{ji,l}^2 = \eta_2 (d_j(t_n) - y_j(t_n)) \sum_{l=1}^{L_{ji}} w_{ji,l}^2 (u_i^1)'(t_n - \tau_{ji,l}^2). \qquad (9)$$

4 Eigenstructure Bidirectional Associative Memory

Based on the failure scenario determined by a fault diagnosis network (to be covered in Section 5), an eigenstructure bidirectional associative memory will promptly retrieve a corresponding controller configuration from a continuous solution space. This controller configuration in the form of weighting parameters will then be loaded into the neural controller block to achieve controller reconfiguration.

Bidirectional associative memory (BAM) [Kos88] is a two-layer nonlinear feedback neural network. Unlike the Hopfield network [Hop84], bidirectional associative memory is a heteroassociative memory that provides a flexible nonlinear mapping from input data to output data. However, bidirectional associative memory does not guarantee that a network will necessarily store the desired vectors as equilibrium points. Furthermore, experience has shown that BAM networks synthesized by "correlation encoding" [Kos88] can store effectively only up to $p < \min(m, n)$ arbitrary vectors as equilibrium points, where m and n denote the number of neurons in each of the two layers. We have shown that the BAM network can be treated as a variation of a Hopfield network [Yen95]. Under appropriate assumptions, we have demonstrated that the present class of continuous BAM is a gradient system with the properties of *global stability* (i.e, for any initial condition, the trajectories of solution will tend to some equilibrium) and *structural stability* (i.e., stability persists under small weight perturbations).

The qualitative and quantitative results (equilibrium condition, asymptotic stability criteria, and estimation of trajectory bounds) that we have developed for Hopfield-type networks [MSY91, YM92, YM91, YM95] can then be extended to the BAM networks through a special arrangement of interconnection weights. Based on these results, we investigate a class of discrete-time BAM networks defined on a closed hypercube of the state space. For the present model, we establish *stability analysis*, which enables us to generalize the solutions of discrete-time systems and to characterize the set of system equilibria. In addition, we develop an efficient *synthesis procedure* utilizing the eigenstructure decomposition method for the present class of neural networks. The synthesized networks are capable of *learning* new vectors as well as *forgetting* learned vectors without the necessity of recomputing all interconnection weights and external inputs. The resulting network can easily be implemented in digital hardware. Furthermore, when simulated by a serial processor, the present system offers extremely efficient means of simulating discrete-time BAM (modeled by a system of difference equations) compared to the computational complexity required to approximate the dynamic behavior of the continuous system (modeled by a system of differential equations).

Consider now a class of neural networks described by a pair of difference equations (DN$_i$) that are defined on a closed hypercube of the state space for times $k = 0, 1, 2, \ldots$ by

$$x_i(k+1) = \text{sat}(\sum_{j=1}^{n} W_{ij} y_j(k) + I_i), \quad i = 1, \ldots, m, \quad \text{(DN}_i\text{a)}$$

$$y_j(k+1) = \text{sat}(\sum_{i=1}^{m} V_{ji} x_i(k) + J_j), \quad j = 1, \ldots, n, \quad \text{(DN}_i\text{b)}$$

where the saturation function sat(), used in modeling all the neurons, is

$$\text{sat}(\theta) = \begin{cases} 1 & \text{if } \theta \geq 1, \\ \theta & \text{if } -1 < \theta < 1, \\ -1 & \text{if } \theta \leq -1. \end{cases}$$

11. Reconfigurable Control in Space Structures

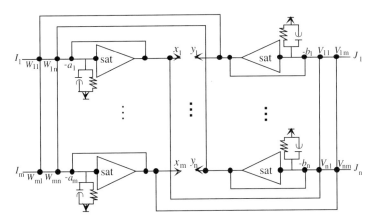

FIGURE 5. An implementation of eigenstructure bidirectional associative memory.

In contrast to the usual system defined on open subsets of $\mathbf{R}^{m+n}$, system (DN$_i$) is described on a closed hypercube

$$\mathbf{B}^{m+n} = \{x \in \mathbf{R}^m,\ y \in \mathbf{R}^n : |x_i| \leq 1, |y_j| \leq 1, i = 1, \ldots, m, j = 1, \ldots, n\}. \tag{10}$$

Figure 5 depicts an analog implementation of the eigenstructure BAM network. Within this study, vector x refers to the failure index while vector y points to the weighting parameters for the retrieving controller configuration. Equation (DN$_i$) can be put into a compact form (DN),

$$\begin{align} x(k+1) &= \operatorname{sat}(Wy(k) + I),\quad k = 0, 1, 2, \ldots, & \text{(DNa)}\\ y(k+1) &= \operatorname{sat}(Vx(k) + J),\quad k = 0, 1, 2, \ldots, & \text{(DNb)} \end{align}$$

where sat(θ) is defined componentwise, W and V are matrices denoting the interconnection weights, and I and J are vectors representing the external inputs.

The results established for system (DN) fall into one of two categories. One type of result addresses the *stability analysis* of system (DN) while the other type pertains to a *synthesis procedure* for system (DN). In [Yen95] we conduct a thorough and complete qualitative analysis of system (DN). Among other aspects, this analysis discusses the distribution of equilibrium points in the state space, the qualitative properties of the equilibrium points, global stability and structural stability properties of system (DN), and the like. For the completeness of this discussion, we briefly summarize the synthesis problem and synthesis procedure.

Synthesis Problem

Given p pairs of vectors in $\mathbf{B}^{m+n}$, say $(x^1, y^1), \ldots, (x^p, y^p)$, the problem is to design a system (DN) that satisfies the following properties:

1. $(x^1, y^1), \ldots, (x^p, y^p)$ are asymptotically stable equilibrium points of system (DN).

2. The system has no periodic solutions.

3. The total number of asymptotically stable equilibrium points of (DN) in the set $\mathbf{B}^{m+n}$ is as small as possible.

4. The domain of attraction of each (x^i, y^i), $i = 1, \ldots, p$, is as large as possible.

Based on the detailed qualitative analysis results, the above synthesis problem can be approached by the following algorithm (called the *eigenstructure decomposition method*).

Synthesis Procedure

Suppose we are given p pairs of vectors as desired library vectors to be stored as asymptotically stable equilibrium points for system (DN). We proceed as follows.

1. Form the vectors
$$\mu^i = \left[{x^i}^\top, {y^i}^\top \right]^\top, \quad i = 1, \ldots, p.$$

2. Compute the matrices $S^p = [s^1, \ldots, s^{p-1}]$, where $s^i = mu^i - \mu^p$, $i = 1, \ldots, p-1$, and the superscript p for matrix S^p denotes the number of vectors to be stored in the BAM network.

3. Perform a singular value decomposition on matrix S^p to obtain the factorization
$$S^p = U\Sigma V^\top,$$
where U and V are orthogonal matrices and Σ is a diagonal matrix with the singular values of S^p on its diagonals.[2] Let
$$L = \mathrm{Span}(s^1, \ldots, s^{p-1}),$$
$$L^a = \mathrm{Aspan}(\mu^1, \ldots, \mu^p).$$
Then L is the linear subspace spanned by the vectors $\{s^1, \ldots, s^{p-1}\}$ and $L^a = L + \mu^p$ denotes the affine subspace (i.e., the coefficients sum to 1) generated by the vectors $\{\mu^1, \ldots, \mu^p\}$.

---

[2] This can be accomplished by standard computer routines, e.g., *LSVRR* in IMSL, SingularValues in *Mathematica*, and *svd* in MATLAB or Matrix$_X$.

4. Decompose the matrix U as
$$U = [U^+ \ U^-],$$
where $U^+ = [u_1, \ldots, u_k]$, $U^- = [u_{k+1}, \ldots, u_{m+n}]$, and $k = \text{rank}(\Sigma) = \dim(L)$. From the properties of singular value decomposition, U^+ is an orthonormal basis of L and U^- is an orthonormal basis of $L^\perp$, the orthogonal complement of L.

5. Compute the matrices
$$T^+ = \sum_{i=1}^{k} u_i u_i^\top = U^+ U^{+\top},$$
$$T^- = \sum_{i=k+1}^{m+n} u_i u_i^\top = U^- U^{-\top}.$$

6. Choose parameters $\tau_1 > 1$ and $-1 < \tau_2 < 1$, and compute
$$T_\tau = \tau_1 T^+ - \tau_2 T^-,$$
$$K_\tau = \tau_1 \mu^p - T_\tau \mu^p.$$

7. Decompose matrix T_τ and vector K_τ by
$$T_\tau = \begin{bmatrix} A_1 & W_\tau \\ V_\tau & A_2 \end{bmatrix} \begin{matrix} \}m \\ \}n, \end{matrix}$$
$$K_\tau = \begin{bmatrix} I_\tau \\ J_\tau \end{bmatrix} \begin{matrix} \}m \\ \}n. \end{matrix}$$

Then all vectors in L will be stored as asymptotically stable equilibria of the synthesized system (DN$_\tau$),
$$x(k+1) = \text{sat}(W_\tau y(k) + I_\tau), k = 0, 1, 2, \ldots, \quad \text{(DN}_\tau\text{a)}$$
$$y(k+1) = \text{sat}(V_\tau x(k) + J_\tau), k = 0, 1, 2, \ldots. \quad \text{(DN}_\tau\text{b)}$$

The eigenstructure method developed above possesses several advantages since it is possible by this method to exert control over the number of spurious states, since it is possible to estimate the extent of the basin of attraction of the stable memories and since it is possible under certain circumstances to store by this method a number of desired stable vectors that by far exceeds the order of the network.

In synthesizing bidirectional associative memory, we usually assume that all desired vectors (i.e., fault scenarios) to be stored are known *a priori*. However, in the large space structure applications, this is usually not the case. Sometimes, we are also required to update the stored vectors (i.e., controller configurations) dynamically in order to accommodate new scenarios (e.g., when a novel fault condition is identified). In a similar spirit of development as [MSY91, YM92], we have successfully incorporated the *learning* and *forgetting* capabilities into the present synthesis algorithm, where learning refers to the ability of adding vectors to be stored as asymptotically stable equilibria to an existing set of stored vectors in a given network and where forgetting refers to the ability of deleting specified vectors from a given set of stored equilibria in a given network. The synthesis procedure is capable of adding an additional pattern as well as deleting an existing pattern without the necessity of *recomputing* the entire interconnection weights, i.e., W and V, and external inputs, i.e., I and J.

Making use of the updating algorithm for singular value decomposition [BN78], we can construct the required orthonormal basis set, i.e.,

$$\{u_1, \ldots, u_{m+n}\}$$

for space L, where $L = \text{Span}(s^1, \ldots, s^{p-1})$ in accordance with the new configuration. The detailed development of the learning and forgetting algorithms can be found in [Yen95]. Furthermore, the *incremental learning and forgetting* algorithm is proposed to improve the computational efficiency of the eigenstructure decomposition method by taking advantage of recursive evaluation [Yen95].

5 Fault Detection and Identification

Detection of structural failures in large-scale systems has been an interesting subject for many decades. The existing damage detection methods primarily depend on off-line destructive tests or computationally intensive finite element analysis. Quite often, these heuristic algorithms are limited to the analysis and design of a fixed structural concept or model, where the loadings, materials, and design constraints need to be specified in advance. Because of the need for time-critical response in many situations, available symptom data is either misinterpreted or unused, often leading to the incorrect removal of a system's components. Fault tolerance issues have usually been ignored or have been assumed to be handled by a simple strategy such as triple modular redundancy.

To date, relatively little *systematic* work has been pursued in connection with damage detection, isolation, and identification. Literature surveys have shown a promising potential in the application of artificial neural networks to quantify structural failures [VD92]. It has become evident that neural networks can also be trained to provide failure information based on the structural response to given payloads, so that perturbations in structural geometry and material properties can be identified by the outputs of the neural network. This information can then be fed back to the bidirectional associative network to invoke an effective neural controller before the structure breaks down. In addition, the neural-network-based fault diagnosis system developed for a certain structural component can also be used in a hierarchical manner where the same structural component is used in several places on large space structures.

We approach damage detection of flexible structures from a pattern classification perspective. In doing so, the classification of the loading to structures and the output response to such a loading are considered as an input pattern to the neural network. The output of the neural network indicates the damage index of structural members. Neural networks trained with a backpropagation learning rule have been used for various problems, including helicopter gearbox [CDL93], induction motor [MCY91], space shuttle main engine [DGM90], jet engine [DKA89] and smart sensors [RPSK91]. Simulation results show that the neural network is capable of performing fault detection and identification (FDI). Although the backpropagation algorithm proves its effectiveness in these cases, it is generally known that it takes considerable time to train the network, and the network may easily get trapped into local minima. Our proposed damage detection and identification system, which makes use of a CMAC network, is capable of incorporating new structural settings on a real-time basis (i.e., on-line learning); handling noisy and incomplete input patterns (i.e., noise-reduction); and recognizing novel structural configurations (i.e., generalization). In addition, the suggested system is not restricted to any specific problem, rather it has the potential to be adapted into a generic diagnostic tool for various complex systems.

The cerebellar model articulation controller, CMAC (also called cerebellar model arithmetic computer) network is an artificial neural network architecture based on the knowledge of the organization and functionality of the cerebellum [Alb75]. CMAC is defined by a series of mappings when nearby input patterns produce similar outputs, while distinct input patterns produce nearly independent outputs. An overlapping arrangement of the input receptive fields provides local generalization capability, as does an RBF network. The desired mapping from the input to the output can be achieved by adjusting the synaptic weights using any optimization algorithm. The output of CMAC is simply determined by summing up the weight values at each of the relevant retrieval memory locations, thus the on-line training algorithm employed by CMAC is rather simple in implementation and has a fast convergence rate.

The CMAC network is capable of learning nonlinear functions extremely fast due to the locality of representation and simplicity of mapping. However, the rectangular shape of receptive field functions produces a *staircase* functional approximation; it is a perfect justification for fault decision-making. The first novel application of CMAC networks is dedicated to the control of robotic manipulators [Alb75]. Since then CMAC has been used in real-time control of industrial robots [MGK87], pattern recognition [HMKG89], and signal filtering [GM89]. Unlike the traditional adaptive controller, CMAC assumes no prior knowledge of the controlled plants that may be subject to noise perturbation or nonlinear functionality. The capability of CMAC networks to approximate any nonlinear continuous functions has been proven using B-spline approximation theory [LHG92]. CMAC performs a nonlinear mapping $y = f(x)$ using two primary functions,

$$S : X \Rightarrow A,$$
$$P : A \Rightarrow Y,$$

where vector X denotes the sensor and actuator readings from the space structure and vector Y denotes the corresponding failure index.

Fault detection may be implemented in a hierarchical and distributed manner. At the bottom level, damage detection may be designed for parts, such as bearings, shafts, cables, sensors, or actuators. Once the appropriate failure scenarios are available, a high-level decision maker can be employed to perform a proper control action. Incorporated with the learning and forgetting capabilities of associative memory, a robust FDI system can be designed to detect, isolate, and identify evolutionary variations as well as catastrophic change of large structures on a real-time basis.

6 Simulation Studies

Example 1. Generic large space structure

To simulate the characteristics of a large space structure, the plant is chosen to possess low natural frequencies and damping as well as high modal density, and the actuators are chosen to be highly nonlinear. The plant consists of five modes with frequencies 1, 4, 5, 6, and 10 Hertz. The damping ratio for all five modes is selected to be 0.15% of critical. Two sensors, two actuators, and ten states are used in this multi-input multi-output system. The eigenvectors are arbitrarily selected under the condition that they remain linearly independent. The actuators are chosen to exhibit a combination of saturation and exponentially decaying ripple. The input/output relationship is shown in Figure 6 and is given below:

$$u(\nu) = \frac{1 - e^{-2\nu}}{1 + e^{-2\nu}} + 0.1 e^{-|\nu|} \cos(4\pi\nu).$$

A compensator is trained so that the closed-loop system containing the nonlinear actuators and lightly damped plant emulates the linear, highly damped reference model. The five natural frequencies of the reference model were set equal to those of the plant. This is realistic in a practical sense because in many cases natural frequencies of space structures can be identified with reasonable accuracy by modal testing. However, it is much more difficult to identify accurately the eigenvectors (corresponding to the mode shapes). Therefore the eigenvectors of the reference model were chosen arbitrarily, and they were different from the eigenvectors of the plant. The

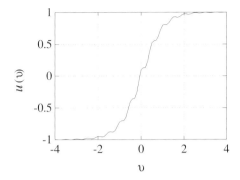

FIGURE 6. Simulated actuator input/output relationship.

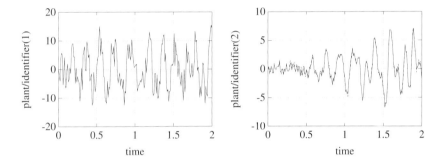

FIGURE 7. Open-loop responses (neural identifier versus nonlinear plant).

degree of damping is chosen to be 10% of critical for each of the five modes. Prior to training the compensator, an adaptive time-delay Gaussian function network consisting of 40 hidden neurons with learning rates equal to 0.001 is trained to identify the open-loop system. The resulting neural identifier assists the training of the compensator (another adaptive time-delay Gaussian function network with 40 hidden neurons) by translating the plant output error to compensator output error. These are chosen to possess four time delays from each hidden neuron to each output neurons for both neural identifier and neural controller.

Figure 7 presents the performance of the neural identifier with respect to sensors 1 and 2, respectively, in response to random inputs for two seconds after training for 100 trials. The mean square error converged to 0.01. Within the scale of the vertical axis the plant output and the neural identifier output are indistinguishable. The simulation results show that the neural identifier has successfully emulated the structural dynamics of this simulated space structure. Although the neural identifier learned to match the open-loop system very quickly, the neural compensator with learning rate 0.001 took almost an hour to converge to mean square error 0.01. The choice of a smaller learning rate ensures a monotonically decreasing mean square error in the LMS training. Figure 8 displays the closed-loop performance for two seconds with respect to sensors 1 and 2, respectively,

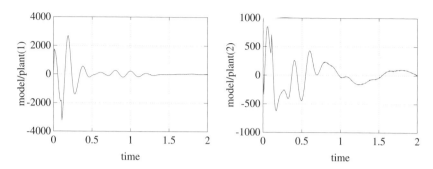

FIGURE 8. Closed-loop responses (reference model versus nonlinear plant).

in response to an impulse. Again, the reference model output and the plant output are indistinguishable. The neural controller has learned to damp out the vibration.

Example 2. ASTREX plant

The *advanced space structures technology research experiments* (ASTREX), currently located at the Phillips Laboratory, Edwards AFB, is a testbed equipped with a three-mirror space-based laser beam expander to develop, test, and validate control strategies for large space structures [ARB+92, BRSC91]; it is shown in Figure 9. The unique features of the experimental facility include a three-axis large-angle slewing maneuver capability and active tripod members with embedded piezoelectric sensors and actuators. The slewing and vibration control can be achieved with a set of reaction control thrusters, a reaction wheel, active members, control moment gyroscopes, and linear precision actuators. The test article allows three degrees of rigid body freedom, in pitch, roll, and yaw. A dedicated control and data acquisition computer is used to command and control the operations. This test article has provided a great challenge for researchers from academia and industry to implement the control strategies to maneuver and to achieve retargeting or vibration suppression.

The test article itself consists of three major sections:

1. *The Primary Structure* is a 5.5 meter diameter truss constructed of over 100 graphite epoxy tubes of 1 cm diameter with aluminum end fittings that are attached to star node connections. The primary structure includes six sets of steel plates mounted on its surface to simulate the primary mirror and two cylindrical masses mounted on its sides to simulate tracker telescopes. A pair of 30 gallon air tanks are attached inside the hub directly above the air-bearing system.

2. *The Secondary Structure* is a triangular structure that houses the reaction wheel actuators and the mass designed to simulate the secondary mirror. It is connected to the primary truss by a tripod arrangement of three 5.1 meter graphite epoxy tubes manufactured

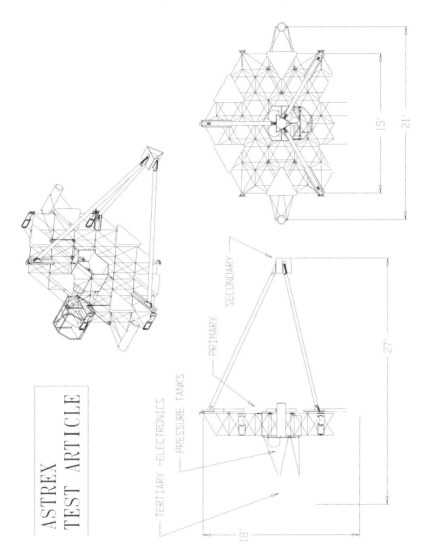

FIGURE 9. The advanced space structures technology research experiments (ASTREX) test article.

with embedded sensors and actuators.

3. *The Tertiary Structure* is a structure designed to hold the electronics and power supply for the data acquisition and control system, and other masses to balance the secondary mirror.

The finite element model (FEM) of the entire testbed consists of approximately 615 nodes and over 1000 elements. Even though the FEM has been constantly modified based on detailed modal surveys, it is not considered an accurate dynamic model. The complicated factors in this control design problem are lack of an accurate dynamic model, nonlinear thruster characteristics, and nonlinear aerodynamic effects. In the rigid-body motion model, two reference frames are employed. The base pedestal axis is an inertially fixed reference frame that points in the true vertical and true horizontal planes. The ASTREX rest position is pitch down in this coordinate system. The test article axis is the body-fixed reference frame. As shown in Figure 10, the origin for both systems is the pivot point, the location where the test article is attached to the base pedestal at the air bearing. Modeling of the physical structure is implemented by an FEM formatted as a NASTRAN data deck. The dynamic modal equation is given by

$$M\ddot{x} + E\dot{x} + Kx = f,$$

where M is the mass matrix, E denotes the viscous damping matrix, K is the stiffness matrix, vector x represents the configuration-space coordinates, and f is the force vector applied to structure. Through a mass normalization procedure on the modal matrix, the state space model of ASTREX can be obtained:

$$\dot{x} = Ax + Bu + Dw, \quad y = Cx, \quad z = Mx + Hu,$$

where A, B, C, D, M, and H are constant matrices and x, u, w, y, and z denote state, input, noise, output, and measurement vectors, respectively. The data required for the system identification are obtained from accelerometers and thrusters through finite element analysis simulations. The locations for accelerometers are carefully selected based on expectations of capturing all the relevant structural modes. For simplicity, only four accelerometers and four actuators, as described in Tables 1 and 2, are used for this preliminary study.

System identification is simulated by an adaptive time-delay Gaussian function network with 100 hidden neurons, while vibration suppression is performed by another adaptive time-delay Gaussian function network with 100 hidden neurons. The closed-loop controller regulates the dynamics of the ASTREX structure to follow a linearly and highly damped reference

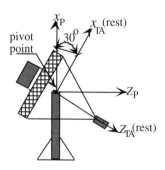

FIGURE 10. Reference frame for rigid-body motion model.

model in which the degree of damping is chosen to be 10% of critical for all modes. The five natural frequencies of the reference model were determined based upon modal test results. The eigenvectors of the reference model were arbitrarily selected under the condition that they remain linearly independent. Both the neural identifier and the neural controller with learning rate 0.01 took roughly five hours to converge to mean square error 0.01. Six time delays are used in each pair of neurons from the hidden layer to the output layer for both neural identifier and neural controller. Open-loop responses of sensors 1, 2, 3, and 4 for random inputs are given in Figure 11, while the closed-loop performance of sensors 1, 2, 3, and 4, in response to an impulse, are displayed in Figure 12.

Three possible configurations are simulated based on different fault scenarios (i.e., no fault, fault condition 1, and fault condition 2). A fault diagnosis system synthesized by a fuzzy backpropagation network is performed by mapping patterns of input sensors to damage indices of line-of-sight errors that represent fault conditions. Angular rate sensors are used at different locations for line-of-sight error measurements where failure scenarios may be evolutionarily varying or catastrophically changing. Figure 13 shows that for each fault condition, the outputs exhibit distinct thresholds crossing from the no-fault region to fault regions. The eigenstructure bidirectional associative memory, which is created prior to dynamic simulation, provides a probability for decision-making based on the information derived from the fuzzy FDI network. Figure 14 displays the closed-loop reconfiguration performance of sensor 3 when the neural controller switches from the no-fault region to fault condition 1.

7 Conclusion

The architecture proposed for reconfigurable neural control successfully demonstrates the feasibility and flexibility of connectionist learning systems for flexible space structures. The salient features associated with the proposed control strategy are discussed. In addition, a real-time autonomous control system is made possible to accommodate uncertainty through on-line interaction with nonlinear structures. In a similar spirit, the proposed architecture can be extended to the dynamic control of aeropropulsion engines, underwater vehicles, chemical processes, power plants, and manufacturing scheduling. The applicability and implementation of the present methodology to large realistic CSI structural testbeds will be pursued in our future research.

Acknowledgments: I would like to thank Dr. Joel Berg, of Orincon Corporation; Dr. Nandu Abhyankar, of Dynacs Engineering; and Dr. William Hallauer of the Air Force Academy, for discussions regarding the experimental studies in this chapter. This research was supported in part by the Air Force Office of Scientific Research.

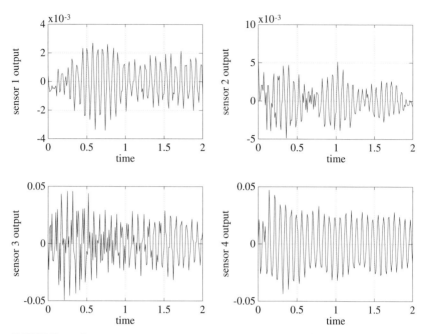

FIGURE 11. Open-loop responses of sensors 1, 2, 3, and 4 (neural identifier versus nonlinear plant).

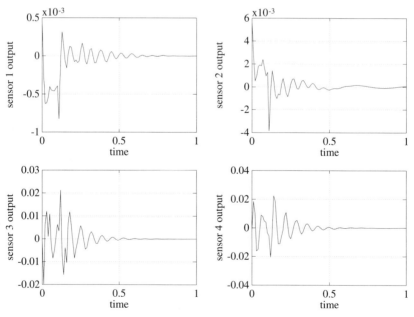

FIGURE 12. Closed-loop responses of sensors 1, 2, 3, and 4 (reference model versus nonlinear plant).

11. Reconfigurable Control in Space Structures

TABLE 1. Sensor locations in the ASTREX testbed.

Type	Location	Node	Direction
accelerometer 1	secondary section	1	(1,0,0)
accelerometer 2	secondary section	1	(0,1,0)
accelerometer 3	tripod	1525	(1,0,0)
accelerometer 4	tripod	3525	(0,1,0)

TABLE 2. Actuator locations in the ASTREX testbed.

Type	Location	Node	Direction
shaker	primary truss	62	(0.5,0,0.86)
proof mass 1	secondary section	462	(0.86,0.5,0)
proof mass 2	secondary section	461	$(-0.86,-0.5,0)$
proof mass 3	secondary section	459	(0,1,0)

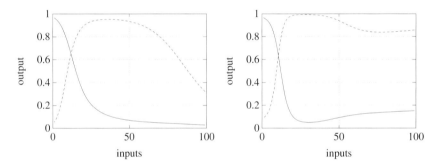

FIGURE 13. Fault conditions 1 and 2 tests (—: no fault; - - -: fault 1/fault 2).

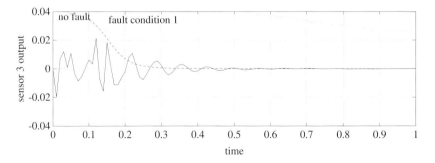

FIGURE 14. Neural reconfigurable control with fault detection and identification.

8 References

[Alb75] J. S. Albus. A new approach to manipulator control: The cerebellar model articulation controller (CMAC). *ASME Journal of Dynamic Systems, Measurement, and Control*, 97(5):220–227, September 1975.

[AP92] P. J. Antsaklis and K. M. Passino. *An Introduction to Intelligent and Autonomous Control*. Kluwer Academic, Hingham, Massachusetts, 1992.

[ARB+92] N. S. Abhyankar, J. Ramakrishnan, K. W. Byun, A. Das, F. Cossey, and J. L. Berg. Modeling, system identification and control of ASTREX. In *Proceedings of the NASA/DoD Control Structures Interaction Technology Conference*, pages 727–750, March 1992.

[Bar89] A. G. Barto. Connectionist learning for control: An overview. Technical Report 89-89, Department of Computer and Information Science, University of Massachusetts, Amherst, Massachusetts, September 1989.

[BCJ+90] R. L. Barron, R. L. Cellucci, P. R. Jordan, N. E. Beam, P. Hess, and A. R. Barron. Applications of polynomial neural networks to FDIE and reconfigurable flight control. In *Proceedings of the IEEE National Aerospace and Electronics Conference*, pages 507–519, May 1990.

[BN78] J. R. Bunch and C. P. Nielsen. Rank-one modification of the symmetric eigenproblem. *Numerische Mathematik*, 31(1):31–40, 1978.

[BRSC91] K. W. Byun, J. Ramakrishnan, R. E. Skelton, and D. F. Cossey. Covariance control of ASTREX. In *Proceedings of the AIAA Conference on Guidance, Navigation and Control Conference*, pages 1229–1235, August 1991.

[BU92] E. B. Bartlett and R. E. Uhrig. Nuclear power plant status diagnostics using an artificial neural network. *Nuclear Technology*, 97(3):272–281, March 1992.

[CDL93] H. Chin, K. Danai, and D. G. Lewicki. Pattern classification for fault diagnosis of helicopter gearboxes. *Control Engineering Practice*, 1(5):771–778, 1993.

[CSH93] M. Y. Chow, R. N. Sharpe, and J. C. Hung. On the application and detection of artificial neural networks for motor fault detection. *IEEE Transactions on Industrial Electronics*, 40(2):181–196, April 1993.

[DGM90] A. Duyar, T. H. Guo, and W. C. Merrill. A failure diagnosis system based on a neural network classifier for the space shuttle main engine. In *Proceedings of the IEEE Conference on Decision and Control*, pages 2391–2400, December 1990.

[DKA89] W. E. Dietz, E. L. Kiech, and M. Ali. Jet and rocket engine fault diagnosis in real time. *Journal of Neural Network Computing*, 1(1):5–18, Summer 1989.

[Els88] R. Elsey. A learning architecture for control based on back-propagation neural network. In *Proceedings of the International Conference on Neural Networks, July, 1988*, pages 587–594. IEEE, 1988.

[Fu70] K. S. Fu. Learning control systems — review and outlook. *IEEE Transactions on Automatic Control*, 15(2):210–221, April 1970.

[GA91] Z. Gao and P. J. Antsaklis. Reconfigurable control system design via perfect model following. *International Journal of Control*, 56(3):717–729, 1991.

[GM89] F. H. Glanz and W. T. Miller. Deconvolution and nonlinear inverse filtering using a neural network. In *Proceedings of the International Conference on Acoustics, Speech and Signal Processing*, pages 2349–2352, May 1989.

[HKK90] E. J. Hartman, J. D. Keeler, and J. M. Kowalski. Layered neural networks with gaussian hidden units as universal approximations. *Neural Computation*, 2(2):210–215, Summer 1990.

[HKP91] J. A. Hertz, A. S. Krogh, and R. G. Palmer. *Introduction to the Theory of Neural Computation*. Addison-Wesley, Redwood City, California, 1991.

[HMKG89] D. Herold, W. T. Miller, L. G. Kraft, and F. H. Glanz. Pattern recognition using a CMAC based learning system. In *Proceedings of the SPIE Conference on Automated Inspection and High Speed Vision Architectures*, volume 2, pages 100–104, November 1989.

[HN89] R. Hecht-Nielsen. *Neurocomputing*. Addison-Wesley, Reading, Massachusetts, 1989.

[Hop84] J. J. Hopfield. Neurons with graded response have collective computational properties like those of two-state neurons. *Proceedings of the National Academy of Science, U.S.A.*, 81:3088–3092, May 1984.

[HS89]   D. A. Handelman and R. F. Stengel. Combining expert system and analytical redundancy concepts for fault tolerance flight control. *Journal of Guidance, Dynamics and Control*, 12(1):39–45, January 1989.

[HSW89]   K. Hornik, M. Stinchcombe, and H. White. Multilayer feedforward networks are universal approximators. *Neural Networks*, 2(5):359–366, 1989.

[Jia94]   J. Jiang. Design of reconfigurable control systems design using eigenstructure assignments. *International Journal of Control*, 59(2):395–410, 1994.

[Kos88]   B. Kosko. Bidirectional associative memories. *IEEE Transactions on on Systems, Man and Cybernetics*, 18(1):49–60, January 1988.

[LDL92]   D. T. Lin, J. E. Dayhoff, and P. A. Ligomenides. Adaptive time-delay neural network for temporal correlation and prediction. In *Proceedings of the SPIE Conference on Biological, Neural Net, and 3-D Methods*, pages 170–181, October 1992.

[LHG92]   S. H. Lane, D. A. Handelman, and J. J. Gelfand. Theory and development of higher-order CMAC neural networks. *IEEE Control Systems Magazine*, 12(2):23–30, April 1992.

[LWEB85]   D. P. Looze, J. L. Weiss, J. S. Eterno, and N. M. Barrett. An automatic redesign approach for restructurable control systems. *IEEE Control Systems Magazine*, 5(2):16–22, April 1985.

[MC68]   D. Michie and R. A. Chambers. Boxes: An experiment in adaptive control. In E. Dale and D. Michie, editors, *Machine Intelligence*, pages 137–152. Oliver and Boyd, Redwood City, California, 1968.

[MCY91]   G. Bilbro, M. Y. Chow, and S. O. Yee. Application of learning theory to an artificial neural network that detects incipient faults in single-phase induction motors. *International Journal of Neural Systems*, 2(1):91–100, 1991.

[MD89]   J. Moody and C. J. Darken. Fast learning in networks of locally-tuned processing units. *Neural Computation*, 1(2):281–294, Summer 1989.

[MGK87]   W. T. Miller, F. H. Glanz, and L. G. Kraft. Application of a general learning algorithm to the control of robotic manipulators. *International Journal of Robotic Research*, 6(2):84–98, 1987.

[MO90]   W. D. Morse and K. A. Ossman. Model following reconfigurable flight control system for the afti/f-16. *Journal of Guidance, Dynamics and Control*, 13(4):969–976, July 1990.

[MSY91]   A. N. Michel, J. Si, and G. G. Yen. Analysis and synthesis of a class of discrete-time neural networks described on hypercubes. *IEEE Transactions on Neural Networks*, 2(1):32–46, January 1991.

[AW89]   K. J. Åström and B. Wittenmark. *Adaptive Control*. Addison-Wesley, Reading, Massachusetts, 1989.

[NP90]   K. W. Narendra and K. Parthasarathy. Identification and control of dynamical systems using neural networks. *IEEE Transactions on Neural Networks*, 1(1):4–27, March 1990.

[NW90]   D. H. Nguyen and B. Widrow. Neural networks for self learning control systems. *IEEE Control Systems Magazine*, 10(3):18–23, April 1990.

[PSY88]   D. Psaltis, A. Sideris, and A. A. Yamamura. A multilayered neural network controller. *IEEE Control Systems Magazine*, 8(3):17–21, April 1988.

[RM86]   D. E. Rumelhart and J. L. McClelland. *Parallel Distributed Processing: Explorations in the Microstructure of Cognition*, volume 1: Foundations. MIT Press, Cambridge, Massachusetts, 1986.

[RÖMM94] K. Redmill, Ü. Özgüner, J. Musgrave, and W. C. Merrill. Intelligent hierarchical thrust vector control for a space shuttle. *IEEE Control Systems Magazine*, 14(3):13–23, June 1994.

[RPSK91] T. Roppel, M. L. Padgett, S. Shaibani, and M. Kindell. Robustness of a neural network trained for sensor fault detection. In *Proceedings of the SCS Workshop on Neural Network: Academic/Industry/NASA/Defense*, pages 107–115, February 1991.

[SS89]   M. Saerens and A. Soquet. A neural controller. In *Proceedings of the IEEE International Conference on Neural Networks*, pages 211–215, October 1989.

[VD92]   M. K. Vellanki and C. H. Dagli. Automated precision assembly through neuro-vision. In *Proceedings of the SPIE Conference on Applications of Artificial Neural Networks*, volume 3, pages 493–504, April 1992.

[VSP92]   K. P. Venugopal, R. Sudhakar, and A. S. Pandya. On-line learning control of autonomous underwater vehicles using feedforward neural networks. *IEEE Journal of Oceanic Engineering*, 17(4):308–319, October 1992.

[WHH+88]   A. Waibel, T. Hanazawa, G. Hinton, K. Shikano, and K. Lang. Phoneme recognition: Neural networks versus hidden markov models. In *Proceedings of the IEEE International Conference on Acoustics, Speech and Signal Processing*, pages 107–110, April 1988.

[WMA+89]   K. Watanabe, I. Matsuura, M. Abe, M. Kubota, and D. M. Himmelblau. Incipient fault diagnosis of chemical processes via artificial neural networks. *AIChE Journal*, 35(11):1803–1812, November 1989.

[WS92]   D. A. White and D. A. Sofge. *Handbook of Intelligent Control–Neural, Fuzzy, and Adaptive Approaches*. Van Nostrand Reinhold, New York, 1992.

[Yen94]   G. G. Yen. Identification and control of large structures using neural networks. *Computers and Structures*, 52(5):859–870, September 1994.

[Yen95]   G. G. Yen. Eigenstructure bidirectional associative memory: An effective synthesis procedure. *IEEE Transactions on Neural Networks*, 6(5):1293–1297, September 1995.

[YK93]   G. G. Yen and M. K. Kwak. Neural network approach for the damage detection of structures. In *Proceedings of the AIAA/ASME/ASCE/AHS/ASC Structures, Structural Dynamics, and Material Conference*, pages 1549–1555, April 1993.

[YM91]   G. G. Yen and A. N. Michel. A learning and forgetting algorithm in associative memories: Results involving pseudo inverses. *IEEE Transactions on Circuits and Systems*, 38(10):1193–1205, October 1991.

[YM92]   G. G. Yen and A. N. Michel. A learning and forgetting algorithm in associative memories: The eigenstructure method. *IEEE Transactions on Circuits and Systems, Part II: Analog and Digital Signal Processing*, 39(4):212–225, April 1992.

[YM95]   G. G. Yen and A. N. Michel. Stability analysis and synthesis algorithm of a class of discrete-time neural networks. *Mathematical and Computer Modelling*, 21(1/2):1–29, 1995.

Chapter 12

Neural Approximations for Finite- and Infinite-Horizon Optimal Control

Riccardo Zoppoli
Thomas Parisini

ABSTRACT This chapter deals with the problem of designing a feedback control law that drives a discrete-time dynamic system (in general, nonlinear) so as to minimize a given cost function (in general, nonquadratic). The control horizon lasts a finite number N of decision stages. The model of the dynamic system is assumed to be perfectly known. Clearly, so general non-LQ optimal control problems are very difficult to solve. The proposed approximate solution is based on the following assumption: the control law is assigned a given structure in which a finite number of parameters have to be determined in order to minimize the cost function (the chosen structure is that of a multilayer feedforward neural network). Such an assumption enables us to approximate the original functional optimization problem by a nonlinear programming one. The optimal control problem is then extended from the finite to the infinite control horizon, for which a receding-horizon optimal control scheme is presented. A stabilizing regulator is derived without imposing, as is usually required by this class of control schemes, that either the origin (i.e., the equilibrium point of the controlled plant) or a suitable neighborhood of the origin be reached within a finite time. Stability is achieved by adding a proper terminal penalty function to the process cost. Also the receding-horizon regulator is approximated by means of a feedforward neural network (only one network is needed instead of a chain of N networks, as in the finite-horizon case). Simulation results show the effectiveness of the proposed approach for both finite- and infinite-horizon optimal control problems.

1 Introduction

Finite- and infinite-horizon optimal control problems are faced in two different, yet important, areas of control applications. Finite-horizon (FH) optimal control typically refers to "maneuvering problems" or "servomechanism problems," where the state vector of the dynamic system has to be driven from a given initial point to a final one in a finite number N of decision stages (discrete-time deterministic dynamic systems are considered). During such a transition, a certain number of intermediate points may have to be tracked. For infinite-horizon (IH) optimal control, we assume that there exists a given equilibrium point toward which the control device must steer the system state whenever the state has been taken away from the equilibrium point by some unpredictable action. Driving the state to such a point (i.e., the origin of the state space, without loss of generality) is usually defined as a "regulation problem."

As is well known, both the N-stage optimal control problem and the IH one can be solved analytically in only a few cases, typically under LQ assumptions (linear dynamic systems and quadratic cost functions). If such assumptions are not satisfied, a variety of numerical techniques are available for the first problem. In solving the second, greater difficulties are encountered; an attractive approach consists in approximating the IH control problem by means of the so-called "receding-horizon" (RH) optimal control problem. In this chapter we shall adopt this approximation. Even though the FH control problem is faced in a deterministic context (i.e., the model of the dynamic system is assumed to be perfectly known, no stochastic variables act on the dynamic system, and the state vector can be measured without noise), it may be important that the control law should take on a *feedback form*, i.e., that it should depend on the current state vector $\underline{x}_i$ measured at stage i. This is suggested by evident practical reasons. In the RH case, the control law is intrinsically implemented by a feedback scheme. Actually, an RH control mechanism can be described as follows. When the controlled plant is in the state $\underline{x}_t$ at time t, an N-stage optimal control problem is solved; thus the sequence of optimal control vectors $\underline{u}_t^\circ, \ldots, \underline{u}_{t+N-1}^\circ$ is derived, and the first control of this sequence becomes the control action $\underline{u}_t^{RH\circ}$ generated by the RH regulator at time t (i.e., $\underline{u}_t^{RH\circ} \triangleq \underline{u}_t^\circ$). The procedure is repeated stage after stage; then a feedback control law is obtained, as the control vector $\underline{u}_t^\circ$ depends on $\underline{x}_t$.

In the FH case, we want a little more than a feedback control law. More specifically, we request that the control law should be able to drive the system state from *any* initial state $\underline{x}_0$, belonging to a given initial set $\mathcal{A}_0$, to *any* final state $\underline{x}_N^*$, belonging to a given final set $\mathcal{A}_N$. It follows that the control law must take on the *feedback feedforward* form $\underline{u}_i = \underline{\gamma}_i(\underline{x}_i, \underline{x}_N^*)$. As is well known, to derive the optimal feedback solution of an N-stage optimal control problem, dynamic programming is the most appropriate tool, at least in principle. This procedure, however, exhibits some drawbacks. If a certain neighborhood of a given final state $\underline{x}_N^*$ must be reached, while the initial state $\underline{x}_0$ can assume any possible value on $\mathcal{A}_0$, by following the dynamic programming approach the optimal feedback law is derived in the backward phase of the procedure. This phase starts from the final stage N by defining a suitable terminal penalty function, for example, the Euclidean distance $\|\underline{x}_N^* - \underline{x}_N\|$; it terminates at the initial stage $i = 0$, when the state trajectories turn out to be optimized for all states $\underline{x}_0 \in \mathcal{A}_0$. Such a procedure, however, is not an easy numerical task, for it requires, at each decision stage, the definition of a suitable grid that is obtained by discretizing all the components of the state vector. The optimal control vectors are then determined and stored in the memory, for all the grid points; as soon as the dimension of $\underline{x}_i$ increases, this may give rise to prohibitive requirements for storage capacity. In some cases, possibly different versions of the conventional dynamic programming technique may overcome the dimensionality barrier; some of them can be found in [Sag68] and [Lar68]. As to more recent works, see, for instance, [JSS$^+$93] and the references cited therein. If $\underline{x}_N^*$ is not fixed but like $\underline{x}_0$ can take on any possible value from the final region $\mathcal{A}_N$, a set of optimal control problems can be stated, each characterized by the presence of a proper final cost parameterized by $\underline{x}_N^*$. Equivalently, we can decide to double the dimension of the state vector by introducing the augmented vector $(\underline{x}_i^T, \underline{z}_i^T)^T$, where $\underline{z}_{i+1} = \underline{z}_i = \underline{x}_N^*$.

Due to the high complexity of calculations and to the huge amount of memory generally needed to store the feedback feedforward control law (the "curse of dimensionality"), we give up dynamic programming and prefer to use an approximate approach. (However, we are careful not to state that our approach does not incur, in general, the curse of dimensionality; this will be discussed in Section 4.) The approach consists in assigning the control law a given structure in which the values of a certain number of parameters have to be determined via nonlinear programming in order to minimize the cost function. Such an approach is not new in control theory; actually, it dates back to the so-called specific optimal control problem considered by Eisenberg and Sage [ES66] for the general non-LQ case, and, under LQ assumptions, to the parametric optimal control problem faced by Kleinman and Athans [KA68] (see also the survey reported in [MT87]). Once this kind of approach has been chosen, implementing our control laws on multilayer feedforward neural networks appears quite a natural choice, since it has been shown, both experimentally (in the past few years) and theoretically (more recently [Bar93]), that these networks are very well suited to approximating nonlinear functions. This approach has been used in [ZP92, PZ94b].

Things are more complicated in the RH optimal control problem, as the asymptotic behavior of the controlled plant must be taken into account, which involves stability issues. A stabilizing regulator is proposed that seems particularly suited for applying neural approximations. Sufficient conditions for ensuring asymptotic stability, even in the case of approximate control laws, are then established.

The FH optimal control problem is addressed in Sections 2 to 5, and the RH regulator is examined in Sections 7 to 10. Simulation results are presented in Sections 6 and 10 for the FH and RH problems, respectively.

2 Statement of the Finite-Horizon Optimal Control Problem

We consider the discrete-time dynamic system (in general, nonlinear)

$$\underline{x}_{i+1} = \underline{f}_i(\underline{x}_i, \underline{u}_i), \quad i = 0, 1, \ldots, N-1, \tag{1}$$

where $\underline{x}_i \in \mathbb{R}^n$ is the state vector of the time-varying dynamic system and $\underline{u}_i \in \mathbb{R}^m$ is the control vector. The cost function (in general, nonquadratic) is given by

$$J = \sum_{i=0}^{N-1} \left[h_i(\underline{x}_i, \underline{u}_i) + \rho_{i+1}(\|\underline{x}_N^* - \underline{x}_{i+1}\|) \right], \tag{2}$$

where $\underline{x}_N^*$ is the final point to be reached, $\underline{f}_i \in \mathcal{C}^1 [\mathbb{R}^n \times \mathbb{R}^m, \mathbb{R}^n]$, $h_i \in \mathcal{C}^1 [\mathbb{R}^n \times \mathbb{R}^m, \mathbb{R}]$, and $\rho_i \in \mathcal{C}^1 [\mathbb{R}^+, \mathbb{R}]$. The transition costs $h_i(\underline{x}_i, \underline{u}_i)$, being in general nonquadratic, can take easily into account penalty or barrier functions that describe constraints (possibly time-varying) on the control and state vectors, etc. $\rho_i(z)$ are increasing functions for $z \geq 0$, with $\rho_i(0) = 0$. We assume that $\underline{x}_0$ and $\underline{x}_N^*$ can take on any value from the given compact sets $\mathcal{A}_0$ and $\mathcal{A}_N$, respectively. Then we can state the following:
Problem 1. *Find the optimal feedback feedforward control law*

$$\{\underline{u}_i^\circ = \underline{\gamma}_i^\circ(\underline{x}_i, \underline{x}_N^*), \, i = 0, 1, \ldots, N-1\}$$

that minimizes cost (2) for any pair $(\underline{x}_0, \underline{x}_N^*) \in \mathcal{A}_0 \times \mathcal{A}_N$. □

It is worth noting that the explicit introduction of the sets $\mathcal{A}_0, \mathcal{A}_N$ into the formulation of Problem 1 is due to the fact that this problem is nonlinear and nonquadratic. For its LQ version, the optimal solution is given by

$$\underline{u}_i^\circ = -L_i \underline{x}_i + F_i \underline{v}_i, \quad i = 0, 1, \ldots, N-1, \tag{3}$$

where $\underline{v}_i$ is a vector generated backward by means of the recursive equation

$$\begin{aligned} \underline{v}_i &= G_i \underline{v}_{i+1} + V_{i+1} \underline{x}_N^*, \quad i = 0, 1, \ldots, N-2, \\ \underline{v}_{N-1} &= V_N \underline{x}_N^*, \end{aligned} \tag{4}$$

and matrices L_i, F_i, G_i can be computed after solving a discrete-time Riccati equation (V_i are weighting matrices that appear in the quadratic cost function). Since these matrices are independent of $\mathcal{A}_0, \mathcal{A}_N$, the introduction of these sets into the formulation of Problem 1 is unnecessary. An extension of Problem 1 consisting in tracking a trajectory $\{\underline{x}_1^*, \ldots, \underline{x}_N^*\}$, where each vector $\underline{x}_i^*$ can take on any value from a given compact set $\mathcal{A}_i$, has been considered in [PZ94b]. We also want to remark that situations similar to the ones described above occur whenever some parameters appearing in the system model or in the cost (like $\underline{x}_N^*$ in Problem 1) are not fixed, but may take their values from given compact sets. All such situations do not differ in substance from the one considered in Problem 1, provided that it is assumed that the parameters will become known to the controller at stage $i = 0$, before the control process begins.

3 Reduction of Problem 1 to a Nonlinear Programming Problem

As stated in the Introduction, the use of dynamic programming would give rise to great difficulties in terms of computational complexity and memory requirements for storing feedback feedforward control laws. Therefore, we shall not use dynamic programming but adopt an approximate technique that consists in assigning the control law a given structure in which the values of a certain number of parameters have to be determined in order to minimize the cost function. This means that the control functions take on the form

$$\underline{u}_i = \hat{\gamma}(\underline{x}_i, \underline{x}_N^*, \underline{w}_i), \quad i = 0, 1, \ldots, N-1, \tag{5}$$

where $\hat{\gamma}$ is a known function of its arguments and $\underline{w}_0, \ldots, \underline{w}_{N-1}$ are vectors of parameters to be optimized. The function $\hat{\gamma}$ is time-invariant; the dependence on time is expressed by the time-varying vector $\underline{w}_i$. Of course, the choice of the function $\hat{\gamma}$ is quite arbitrary, and in any case, the control law (5), after the optimization of the vectors $\underline{w}_i$, constitutes an approximation for the solution of Problem 1. Among various possible structures (or approximating functions), we choose nonlinear mappings based on multilayer feedforward neural networks (it follows that the parameters appearing in (5) are the so-called synaptic weights). This choice is suggested both by practical (i.e., computational) reasons and by theoretical properties that characterize such neural approximators. This point will be discussed later on. For now, we want to remark that the method of approximating the control laws by means of a preassigned control structure was proposed, as stated previously, in the 1960s. However, it does not seem that such a method met with great success, probably because the selected structures were characterized by too small a number of free parameters to attain satisfactory approximation properties. Moreover, such structures required rather complex computational procedures to determine the optimal values of the unknown parameters.

Let us now describe in some detail the N neural networks that implement the control functions (5). We assume that each of the N neural networks is composed of L layers and that in the generic layer s, n_s neural units are active. The input/output mapping of the qth neural unit of the sth layer is given by

$$y_q^i(s) = g\left[z_q^i(s)\right], \quad s = 1, \ldots, L; \, q = 1, \ldots, n_s, \tag{6}$$

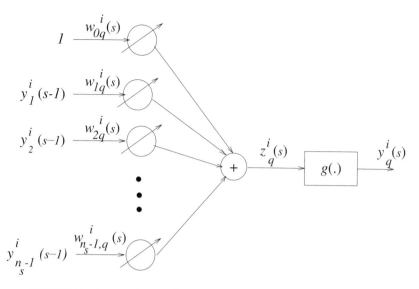

FIGURE 1. A neural unit, showing variable and weight notations.

$$z_q^i(s) = \sum_{p=1}^{n_s-1} w_{pq}^i(s) y_p^i(s-1) + w_{0q}^i(s), \quad (7)$$

where $y_q^i(s)$ is the output variable of the neural unit, $g(x) = \tanh(x)$ is a shifted sigmoidal activation function, and $w_{pq}^i(s)$ and $w_{0q}^i(s)$ are the weight and bias coefficients, respectively. All these coefficients are the components of the vector $\underline{w}_i$ appearing in the control function $\underline{u}_i = \hat{\underline{\gamma}}(\underline{x}_i, \underline{x}_N^*, \underline{w}_i)$; the variables $y_q^i(0)$ are the components of $\underline{x}_i$ and $\underline{x}_N^*$, and the variables $y_q^i(L)$ are the components of $\underline{u}_i$. For the reader's convenience, the variables and weight notations are given in Figure 1.

As shown in Figure 2, the control scheme results in a chain of N neural networks, each followed by the dynamic system. This chain is related to the control scheme proposed in [NW90]. Our structure differs from that scheme in the feedforward actions generated by the vectors $\underline{x}_N^*$ and in the fact that the neural networks are allowed to be time-dependent.

If we now substitute (5) into (1) and (2) and use the state equation repeatedly, thus eliminating the control and state vectors, the cost function

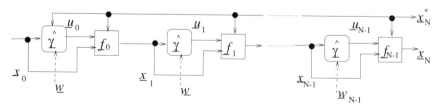

FIGURE 2. The structure of the feedback feedforward neural control law.

takes on the form $J(\underline{w}, \underline{x}_0, \underline{x}_N^*)$, where

$$\underline{w} \triangleq \operatorname{col}(\underline{w}_i, \, i = 0, 1, \ldots, N-1).$$

Since the cost function $J(\underline{w}, \underline{x}_0, \underline{x}_N^*)$ depends both on the vector $\underline{w}$ to be determined and on $\underline{x}_0, \underline{x}_N^*$, we eliminate the dependence on $\underline{x}_0, \underline{x}_N^*$ by assuming $\underline{x}_0$ and $\underline{x}_N^*$ to be mutually independent random vectors uniformly distributed on $\mathcal{A}_0$ and $\mathcal{A}_N$, respectively, and by averaging $J(\underline{w}, \underline{x}_0, \underline{x}_N^*)$ with respect to these vectors. It is worth noting that though this procedure is rather arbitrary, it is not unusual. For example, it has been applied to solve parametric LQ optimal control problems (see again, for instance, [KA68], where the gain matrix of the controller is determined after averaging the cost function with respect to the initial state, considered as a random vector). It is certainly true that another way of eliminating the dependence of $J(\underline{w}, \underline{x}_0, \underline{x}_N^*)$ on $\underline{x}_0, \underline{x}_N^*$ may consist in adopting a min–max approach, which means to maximize the cost function with respect to $\underline{x}_0 \in \mathcal{A}_0$ and $\underline{x}_N^* \in \mathcal{A}_N$. Our choice in favor of the expectation procedure is essentially motivated by the reassuring experimental results given in Section 6 and confirmed, in general, in the literature for similar optimization problems. The min–max approach, however, appears more appropriate in some practical cases, for example, when there is the danger of incurring unacceptably high costs. The possibility of using such an alternative approach should be examined carefully.

Once the expectation approach has been adopted, one has to solve the following:

Problem 2. *Find the vector $\underline{w}^\circ$ that minimizes the expected cost*

$$\mathop{\mathrm{E}}_{\underline{x}_0, \underline{x}_N^*} \left[J(\underline{w}, \underline{x}_0, \underline{x}_N^*) \right].$$

□

It follows that the functional Problem 1 has been reduced to an unconstrained nonlinear programming problem. As the solution of Problem 2 constitutes an approximation for Problem 1, in the following we shall discriminate between the corresponding solutions by calling them "optimal neural control laws" and "optimal control laws," respectively. The accuracy of this approximation will be discussed in the next section.

4 Approximating Properties of the Neural Control Law

The accuracy to which the optimal neural control law

$$\{\hat{\underline{\gamma}}(\underline{x}_i, \underline{x}_N^*, \underline{w}_i^\circ), \, i = 0, 1, \ldots, N-1\}$$

can approximate the control law $\{\underline{\gamma}_i^{\circ}(\underline{x}_i, \underline{x}_N^*), \, i = 0, 1, \ldots, N-1\}$ is clearly a crucial point of the method proposed in the chapter. In this section, we address two basic points: the first is the degree of accuracy that can be attained by using neural control laws; the second concerns the complexity that the neural networks implementing the control functions must exhibit in order to achieve a given degree of accuracy.

Suppose now that the approximating control functions $\{\hat{\underline{\gamma}}\}$ contain only one hidden layer (i.e., $L = 2$) composed of ν_i neural units, and that the output layer is composed of linear activation units. Denote such control functions by $\{\hat{\underline{\gamma}}^{(\nu_i)}(\underline{x}_i, \underline{x}_N^*, \underline{w}_i), \, i = 0, 1, \ldots, N-1\}$. As will be shown, only one hidden layer is sufficient to obtain the required approximating properties. Let us also introduce some useful notations and definitions. Given the maps

$$g_i \triangleq \underline{f}_i \left[\underline{x}_i, \underline{\gamma}_i^{\circ}(\underline{x}_i, \underline{x}_N^*)\right] : \mathbb{R}^n \times \mathbb{R}^n \to \mathbb{R}^n, \quad i = 0, 1, \ldots, N-1, \quad (8)$$

we define the following family of sets $\mathcal{B}_i \subset \mathbb{R}^n \times \mathbb{R}^n$:

$$\mathcal{B}_i \triangleq \begin{cases} \mathcal{A}_0 \times \mathcal{A}_N, & \text{for } i = 0; \\ g_{i-1}[\mathcal{B}_{i-1}] \times \mathcal{A}_N, & \text{for } i = 1, 2, \ldots, N-1. \end{cases} \quad (9)$$

Now we can state the following proposition [PZ94b]:

Proposition 1 *Assume that Problem 1 has only one solution $\underline{\gamma}_i^{\circ}(\underline{x}_i, \underline{x}_N^*) \in \mathcal{C}[\mathcal{B}_i, \mathbb{R}^m]$, $i = 0, 1, \ldots, N-1$. Then, for every $\varepsilon \in \mathbb{R}$, $\varepsilon > 0$ and every i with $0 \leq i \leq N-1$, there exist an integer ν_i and a weight vector $\underline{w}_i$ (i.e., a neural control function $\hat{\underline{\gamma}}^{(\nu_i)}(\underline{x}_i, \underline{x}_N^*, \underline{w}_i)$) such that*

$$\left\| \underline{\gamma}_i^{\circ}(\underline{x}_i, \underline{x}_N^*) - \hat{\underline{\gamma}}^{(\nu_i)}(\underline{x}_i, \underline{x}_N^*, \underline{w}_i) \right\| < \varepsilon, \quad \forall \, (\underline{x}_i, \underline{x}_N^*) \in \mathcal{B}_i. \quad (10)$$

□

Proposition 1 has been derived directly from the results reported in [HSW89, HN89, Cyb89], according to which continuous functions can be approximated to any degree of accuracy on a given compact set by feedforward neural networks based on sigmoidal functions, provided that the number ν_i of neural units is sufficiently large. It is important to note that the results presented in Proposition 1 do not necessarily involve the need for using a feedforward neural network as an approximator for the optimal control function. Actually, results like those presented in Proposition 1 are very common in approximation theory and hold true even under rather weak assumptions about the functions to be approximated. More specifically, Proposition 1 states that the functions implemented by means of feedforward neural networks are *dense* in the space of continuous functions; in a sense, this can be considered as a *necessary condition* that every approximation scheme should satisfy. Moreover, such results in themselves are not very useful, in that they do not provide any information on the rate of convergence of the approximation scheme, that is, on the rate at which the approximation error decreases as the number of parameters of the approximating structure (i.e., the number of hidden units, or, equivalently, of parameters to be determined in our neural approximators) increases.

To address this very important issue, we now apply Barron's results on neural approximation [Bar93]. To this end, let us introduce an approximating network that differs slightly from the one previously introduced to state Proposition 1. The new network is the parallel of m single-output neural networks of the type described above (i.e., containing a single hidden layer and linear output activation units). Each network generates one of the m components of the control vector $\underline{u}_i$. Denote by $\hat{\gamma}_j^{(\nu_{ij})}(\underline{x}_i, \underline{x}_N^*, \underline{w}_{ij})$ the input–output mapping of such networks, where ν_{ij} is the number of neural units in the hidden layer and $\underline{w}_{ij}$ is the weight vector. Define also as $\gamma_{ij}^\circ(\underline{x}_i, \underline{x}_N^*)$ the jth component of the vector function $\underline{\gamma}_i^\circ$. In order to characterize the ability of the functions $\hat{\gamma}_j^{(\nu_{ij})}$ to approximate the functions γ_{ij}°, we introduce the integrated square error

$$\int \left| \gamma_{ij}^\circ - \hat{\gamma}_j^{(\nu_{ij})} \right|^2 \sigma[\mathrm{d}\,(\underline{x}_i, \underline{x}_N^*)]$$

evaluated on the domain of γ_{ij}°, that is, on the compact set $\mathcal{B}_i \times \mathcal{A}_N$ (σ is a probability measure). We now need to introduce some smoothness assumptions on the optimal control functions γ_{ij}° to be approximated. Following [Bar93], we assume that each of such functions has a bound to the average of the norm of the frequency vector weighted by its Fourier transform. However, the functions γ_{ij}° have been defined on the compact sets $\mathcal{B}_i \times \mathcal{A}_N$ and not on the space $\mathbb{R}^d$, where $d \triangleq \dim[\mathrm{col}\,(\underline{x}_i, \underline{x}_N^*)] = 2n$. Then, in order to introduce the Fourier transforms, we need to "extend" the functions $\gamma_{ij}(\underline{x}_i, \underline{x}_N^*)$, defined on the compact set $\mathcal{B}_i \times \mathcal{A}_N$, from this domain to $\mathbb{R}^d$. Toward this end, we define the functions $\overline{\gamma}_{ij} : \mathbb{R}^d \to \mathbb{R}$ that coincide with $\gamma_{ij}(\underline{x}_i, \underline{x}_N^*)$ on $\mathcal{B}_i \times \mathcal{A}_N$. Finally, we define the class of functions

$$G_{c_{ij}}^i \triangleq \left\{ \overline{\gamma}_{ij} \text{ such that } \int_{\mathbb{R}^d} |\underline{\omega}|\,|\Gamma_{ij}(\underline{\omega})|\,\mathrm{d}\underline{\omega} \leq c_{ij} \right\}, \quad (11)$$

where $\Gamma_{ij}(\underline{\omega})$ is the Fourier transform of $\overline{\gamma}_{ij}$ and c_{ij} is any finite positive constant. Then, in [PZ94b], we prove the following:

Proposition 2 *Assume that Problem 1 has only one solution $\underline{\gamma}_i^\circ(\underline{x}_i, \underline{x}_N^*) \in C[\mathcal{B}_i \times \mathcal{A}_N, \mathbb{R}^m]$, $i = 0, 1, \ldots, N-1$, such that $\overline{\gamma}_{ij}^\circ \in G_{\tilde{c}_{ij}}^i$ for some finite positive scalar $\tilde{c}_{ij}$, for every j with $1 \leq j \leq m$. Then, for every i with $0 \leq i \leq N-1$, for every j with $1 \leq j \leq m$, for every probability measure σ, and for every $\nu_{ij} \geq 1$, there exist a weight vector $\underline{w}_{ij}$ (i.e., a neural* strategy $\hat{\gamma}_j^{(\nu_{ij})}(\underline{x}_i, \underline{x}_N^*, \underline{w}_{ij})$) and a positive scalar c'_{ij} such that

$$\int_{\mathcal{B}_i \times \mathcal{A}_N} \left| \gamma_{ij}^\circ(\underline{x}_i, \underline{x}_N^*) - \hat{\gamma}_j^{(\nu_{ij})}(\underline{x}_i, \underline{x}_N^*, \underline{w}_{ij}) \right|^2 \sigma\left[\mathrm{d}(\underline{x}_i, \underline{x}_N^*)\right] \leq \frac{c'_{ij}}{\nu_{ij}}, \qquad (12)$$

where $c'_{ij} = (2r_i \tilde{c}_{ij})^2$. r_i is the radius of the smallest sphere (centered in the origin) that contains $\mathcal{B}_i \times \mathcal{A}_N$.

□

It is worth noting that in a sense, Proposition 2 specifies quantitatively the content of Proposition 1. More specifically, it states that for any control function $\gamma_{ij}^\circ(\underline{x}_i, \underline{x}_N^*)$, the number of parameters required to achieve an integrated square error of order $O(1/\nu_{ij})$ is $O(\nu_{ij}d)$, which grows linearly with d, where d represents the dimension of the input vector of the neural network acting at stage i. This implies that for the functions to be approximated belonging to the class defined by (11), the risk of an exponential growth of the number of parameters (i.e., the phenomenon of the curse of dimensionality) is not incurred. This fact, however, is not completely surprising. Actually, it has been shown that a function belonging to the class defined by (11) can be written as $f(\underline{x}) = \|\underline{x}\|^{1-d} * \lambda(\underline{x})$, where $\lambda(\underline{x})$ is any function whose Fourier transform is integrable and $*$ stands for the convolution operator (the Fourier transform is assumed to be defined in the sense of generalized functions, and the convolution operator is defined accordingly) [Gir94]. Then, the "slow" growth of the number of parameters with d may be motivated by the fact that the space of functions to be approximated is more and more constrained as d increases. It is now reasonable to wonder whether the property outlined by Proposition 2 is peculiar to feedforward neural approximators or is shared by traditional linear approximation schemes (like polynomial and trigonometric expansions) as well as by other classes of nonlinear approximators.

Let us first address the case of linear approximators, that is, linear combinations of a number ν_{ij} of preassigned basis functions. In [Bar93] it is shown "that there is no choice of ν_{ij} fixed basis functions such that linear combinations of them achieve integrated square error of smaller order than $(1/\nu_{ij})^{2/d}$." This applies to functions to be approximated that belong to the previously defined class $G_{\tilde{c}_{ij}}^i$. The presence of $2/d$ instead of 1 in the exponent of $1/\nu_{ij}$ may then give rise to the curse of dimensionality. However, this fact deserves another comment. Actually, if we assume a higher degree of smoothness for the functions γ_{ij}° by requiring them to have square-integrable partial derivatives of order up to s (then γ_{ij}° belong to the Sobolev space $W_2^{(s)}$), where s is the least integer greater than $1 + \frac{d}{2}$, two results can be established: 1) there exists a scalar c_{ij}^* such that $G_{c_{ij}^*}^i \supset W_2^{(s)}$ (i.e., $W_2^{(s)}$ is a proper subset of $G_{c_{ij}^*}^i$) [Bar93], and 2) the linear schemes used to approximate functions belonging to Sobolev spaces do not suffer the curse of dimensionality [Pin86]. It follows that neural approximators should behave better than linear ones in the difference set $G^i_{c^*_{ij}} \setminus W^{(s)}_2$.

For a comparison of neural approximators with other nonlinear approximation schemes, it should be remarked that linear combinations of basis functions containing adaptable parameters may exhibit approximation properties similar to the ones that characterize the neural mappings described in this chapter. This is the case with radial basis functions (RBF) [Gir94] (for which the centers and the weighting matrices of the radial activation functions can be tuned) or with linear combinations of trigonometric basis functions [Jon92] (for which the frequencies are adaptable parameters). In general, it is important that free parameters should not appear linearly, as is the case with the coefficients of linear combinations of fixed basis functions. It is also worth noting that the approximation bound of order $O(1/\nu_{ij})$ is achieved under smoothness assumptions on the functions to be approximated that depend on the chosen nonlinear approximation schemes. The wider diffusion of feedforward neural approximators, as compared with other nonlinear approximators, is probably to be ascribed to the simplicity of the tuning algorithms (see the next section), to the robustness of such algorithms, and to other practical features.

5 Solution of Problem 2 by the Gradient Method

The unconstrained nonlinear programming Problem 2 can be solved by means of some descent algorithm. We focus our attention on methods of the gradient type, as when applied to neural networks they are simple and well suited to distributed computation. To solve Problem 2, the gradient algorithm can be written as follows:

$$\underline{w}(k+1) = \underline{w}(k) - \alpha \nabla_{\underline{w}} \operatorname*{E}_{\underline{x}_0, \underline{x}^*_N} J\left[\underline{w}(k), \underline{x}_0, \underline{x}^*_N\right], \quad k = 0, 1, \ldots, \quad (13)$$

where α is a positive, constant step size and k denotes the iteration step of the descent procedure.

However, due to the general statement of the problem, we are unable to express the average cost $\operatorname*{E}_{\underline{x}_0, \underline{x}^*_N} \left[J\left(\underline{w}, \underline{x}_0, \underline{x}^*_N\right)\right]$ in explicit form. This leads us to compute the "realization"

$$\nabla_{\underline{w}} J\left[\underline{w}(k), \underline{x}_0(k), \underline{x}^*_N(k)\right]$$

instead of the gradient appearing in (13). The sequence

$$\{[\underline{x}_0(k), \underline{x}^*_N(k)], k = 0, 1, \ldots\}$$

is generated by randomly selecting the vectors $\underline{x}_0(k), \underline{x}_N^*(k)$ from $\mathcal{A}_0$, $\mathcal{A}_N$, respectively. Then, in lieu of (13), we consider the following updating algorithm:

$$\underline{w}(k+1) = \underline{w}(k) - \alpha(k)\nabla_{\underline{w}} J\left[\underline{w}(k), \underline{x}_0(k), \underline{x}_N^*(k)\right], \qquad k = 0, 1, \ldots. \quad (14)$$

The probabilistic algorithm (14) is related to the concept of "*stochastic approximation.*" Sufficient conditions for the algorithm's convergence can be found, for instance, in [Tsy71, PT73]. Some of such conditions are related to the behavior of the time-dependent step size $\alpha(k)$, the others to the shape of the cost surface $J[\underline{w}(k), \underline{x}_0(k), \underline{x}_N^*(k)]$. To determine whether the latter conditions are fulfilled is clearly a hard task due to the high complexity of such a cost surface. As to $\alpha(k)$, we have to satisfy the following sufficient conditions for the algorithm's convergence:

$$\alpha(k) > 0, \quad \sum_{k=1}^{\infty} \alpha(k) = \infty, \quad \sum_{k=1}^{\infty} \alpha^2(k) < \infty. \quad (15)$$

In the examples given in the following, we take the step size $\alpha(k) = c_1/(c_2 + k)$, $c_1, c_2 > 0$, which satisfies conditions (15). In these examples, we also add a "momentum" $\rho\left[\underline{w}(k) - \underline{w}(k-1)\right]$ to (14), as is usually done in training neural networks (ρ is a suitable positive constant). Other acceleration techniques have been proposed in the literature, and probably they allow a faster convergence than the one achieved in the examples presented later on. However, we limit ourselves to using the simple descent algorithm described above, as the issue of convergence speed is beyond the scope of this chapter.

We now want to derive the components of $\nabla_{\underline{w}} J[\underline{w}(k), \underline{x}_0(k), \underline{x}_N^*(k)]$, i.e., the partial derivatives

$$\frac{\partial J[\underline{w}(k), \underline{x}_0(k), \underline{x}_N^*(k)]}{\partial w_{pq}^i(s)}.$$

Toward this end, we define the following two variables, which play a basic role in the development of the proposed algorithm (to simplify the notation, we drop the index k):

$$\delta_q^i(s) \triangleq \frac{\partial J(\underline{w}, \underline{x}_0, \underline{x}_N^*)}{\partial z_q^i(s)}, \quad i = 0, 1, \ldots, N-1; \ s = 1, \ldots, L; \ q = 1, \ldots, n_s; \quad (16)$$

$$\underline{\lambda}_i \triangleq \nabla_{\underline{x}_i} J(\underline{w}, \underline{x}_0, \underline{x}_N^*), \quad i = 0, 1, \ldots, N-1. \quad (17)$$

Then, by applying the well-known backpropagation updating rule (see, for instance, [RM86]), we obtain

$$\frac{\partial J(\underline{w}, \underline{x}_0, \underline{x}_N^*)}{\partial w_{pq}^i(s)} = \delta_q^i(s) y_p^i(s-1), \quad (18)$$

where $\delta_q^i(s)$ can be computed recursively by means of the equations

$$\delta_q^i(s) = g'\left[z_q^i(s)\right] \sum_{h=1}^{n_{s+1}} \delta_h^i(s+1)w_{qh}^i(s+1), \qquad s = 1,\ldots,L-1,$$

$$\delta_q^i(L) = g'\left[z_q^i(L)\right] \frac{\partial J}{\partial y_q^i(L)},$$
(19)

where g' is the derivative of the activation function. Of course, (18) implies that the partial derivatives with respect to the bias weights $w_{0q}^i(s)$ can be obtained by setting the corresponding inputs to 1.

We now have to compute the partial derivatives $\dfrac{\partial J}{\partial y_q^i(L)}$. First, we need to detail the components of $\underline{y}^i(0)$ that are the input vectors to the ith neural network. Since $\underline{y}^i(0) = \mathrm{col}(\underline{x}_N^*, \underline{x}_i)$, we let $\underline{y}_*^i(0) \stackrel{\triangle}{=} \underline{x}_N^*$ and $\underline{y}_x^i(0) \stackrel{\triangle}{=} \underline{x}_i$. Thus, the components of $\underline{x}_N^*$ correspond to the components $y_p^i(0)$, $p = 1,\ldots,n$, and the components of $\underline{x}_i$ to $y_p^i(0)$, $p = n+1,\ldots,2n$. We also define

$$\left(\frac{\partial J}{\partial \underline{y}^i(L)}\right)^T = \mathrm{col}\left[\frac{\partial J}{\partial y_q(L)}, q = 1,\ldots,m\right]$$

and, in a similar way, $\dfrac{\partial J}{\partial \underline{y}_*^i(0)} \dfrac{\partial J}{\partial \underline{y}_x^i(0)}$. Finally, we let

$$\tilde{h}_i(\underline{x}_i, \underline{u}_i, \underline{x}_N^*) \stackrel{\triangle}{=} h_i(\underline{x}_i, \underline{u}_i) + \rho_i\left(\|\underline{x}_N^* - \underline{x}_i\|\right), \quad i = 1, 2, \ldots, N-1,$$

and $\tilde{h}_0(\underline{x}_0, \underline{u}_0, \underline{x}_N^*) \stackrel{\triangle}{=} h_0(\underline{x}_0, \underline{u}_0)$. Then, we can use the following relationships, which are demonstrated in [PZ94b]:

$$\frac{\partial J}{\partial \underline{y}^i(L)} = \frac{\partial}{\partial \underline{u}_i}\tilde{h}_i(\underline{x}_i, \underline{u}_i, \underline{x}_N^*) + \underline{\lambda}_{i+1}^T \frac{\partial}{\partial \underline{u}_i}\underline{f}_i(\underline{x}_i, \underline{u}_i), \quad i = 0, 1, \ldots, N-1 \quad (20)$$

where vectors $\underline{\lambda}_i^T$ can be computed as follows:

$$\underline{\lambda}_i^T = \frac{\partial}{\partial \underline{x}_i}\tilde{h}_i(\underline{x}_i, \underline{u}_i, \underline{x}_N^*) + \underline{\lambda}_{i+1}^T \frac{\partial}{\partial \underline{x}_i}\underline{f}_i(\underline{x}_i, \underline{u}_i) + \frac{\partial J}{\partial \underline{y}_x^i(0)}, \quad i = 1,\ldots,N-1,$$
(21)

$$\underline{\lambda}_N^T = \frac{\partial}{\partial \underline{x}_N}\rho_N(\|\underline{x}_N^* - \underline{x}_N\|)$$

and

$$\frac{\partial J}{\partial \underline{y}_*^i(0)} = \mathrm{col}\left[\sum_{q=1}^{n_1} \delta_q^i(1)w_{pq}^{i'}(1), p = 1,\ldots,n\right], \quad i = 0,1,\ldots,N-1, \quad (22)$$

$$\frac{\partial J}{\partial \underline{y}_x^i(0)} = \text{col}\left[\sum_{q=1}^{n_1} \delta_q^i(1) w_{pq}^i(1), \, p = n+1, \ldots, 2n\right], \, i = 0, 1, \ldots, N-1.$$
(23)

It is worth noting that (22) is the classical adjoint equation of N-stage optimal control theory, with the addition of a term (the third) to take into account the introduction of the fixed-structure feedback control law, i.e., this term is not specific for neural networks. Instead, the presence of the feedforward neural networks is revealed by (22) and (23), which include the synaptic weights of the first layers of the networks.

It can be seen that the algorithm consists of the following two alternating "passes":

Forward pass. The state vectors $\underline{x}_0(k)$, $\underline{x}_N^*(k)$ are randomly generated from $\mathcal{A}_0$, $\mathcal{A}_N$, respectively. Then, the control sequence and the state trajectory are computed on the basis of these vectors and of $\underline{w}(k)$.

Backward pass. All the variables $\delta_q^i(s)$ and $\underline{\lambda}_i$ are computed, and the gradient $\nabla_{\underline{w}} J[\underline{w}(k), \underline{x}_0(k), \underline{x}_N^*(k)]$ is determined by using (18). Then, the new weight vector $\underline{w}(k+1)$ is generated by means of (14).

In the next section, some examples will be given to illustrate the effectiveness of the proposed method.

6 Simulation Results

We now present two examples to show the learning properties of the neural control laws. In the first example, an LQ optimal control problem is addressed to evaluate the capacity of the "optimal neural control laws" to approximate the "optimal control laws" (i.e., the solution of Problem 1, as previously defined), which in this case can be derived analytically. In the second example, a more complex non-LQ optimal control problem is dealt with, for which it is difficult to determine the optimal control law by means of conventional methods. Instead, as will be shown, the neural optimal control law can be derived quite easily. Both examples have been drawn from [PZ94b].

Example 1.
Consider the following LQ optimal control problem, where the dynamic system is given by

$$\underline{x}_{i+1} = \begin{bmatrix} 0.65 & -0.19 \\ 0 & 0.83 \end{bmatrix} \underline{x}_i + \begin{bmatrix} 7 \\ 7 \end{bmatrix} u_i,$$

where $\underline{x}_i \triangleq \text{col}(x_i, y_i)$. The cost function is

$$\sum_{i=0}^{N-1} u_i^2 + v_N \| \underline{x}_N \|^2,$$

where $v_N = 40$, $N = 10$. Note that the final set $\mathcal{A}_N$ reduces to the origin. As is well known, the optimal control is generated by the linear feedback law $u_i^\circ = -L_i \underline{x}_i$, where the matrix gain L_i is determined by solving a discrete-time Riccati equation. To evaluate the correctness of the proposed method for a problem admitting an analytical solution, we considered the control strategies $u_i = \hat{\gamma}(\underline{x}_i, \underline{w}_i)$, implemented by means of neural networks containing one hidden layer of 20 units. (In the present example, as well as in the following ones, the number of neural units was established experimentally, that is, several simulations showed that a larger number of units did not result in a significant decrease in the minimum process cost.) A momentum $\rho [\underline{w}(k) - \underline{w}(k-1)]$ was added to the right-hand side of (14), with $\rho = 0.8$. The constants of the time-dependent step size $\alpha(k)$ were $c_1 = 100$ and $c_2 = 10^5$. The parameters c_1, c_2, and η, too, were derived experimentally. More specifically, they were chosen so as to obtain a reasonable tradeoff between the convergence speed and a "regular" behavior of the learning algorithm (i.e., absence of excessive oscillations in the initial part of the learning procedure, low sensitivity to the randomly chosen initial values of the weight vector, etc.). A similar criterion was used to choose the same parameters for the following examples. The initial set was $\mathcal{A}_0 = \{(x, y) \in \mathbb{R}^2 : 2.5 \le x \le 3.5, -1 \le y \le 1\}$. Usually, the algorithm converged to the optimal solution $\underline{w}^\circ$ after 10^4 to $2 \cdot 10^4$ iterations.

The behaviors of the optimal neural state trajectories are pictorially presented in Figure 3, where four trajectories, starting from the vertices of the initial region $\mathcal{A}_0$, map $\mathcal{A}_0$ into the region $\tilde{\mathcal{A}}_1$ at stage $i = 1$, then $\tilde{\mathcal{A}}_1$ into $\tilde{\mathcal{A}}_2$ at stage $i = 2$, and so on, up to region $\tilde{\mathcal{A}}_9$ (more precisely, the set $\tilde{\mathcal{A}}_{i+1}$ is generated by the set $\tilde{\mathcal{A}}_i$ through the mapping $\underline{x}_{i+1} = \underline{f}_i [\underline{x}_i, \hat{\gamma}(\underline{x}_i, \underline{w}_i)]$).

For the sake of pictorial clarity, only the first regions are shown in the figure. Since in Figure 3 the optimal neural trajectories and the analytically derived ones practically coincide, $\mathcal{A}_0$, $\tilde{\mathcal{A}}_4$, $\tilde{\mathcal{A}}_9$ are plotted in enlarged form in Figure 4 so as to enable one to compare the neural results with the analytical ones. The continuous lines represent the optimal neural control law for different constant values of the control variable u_i ("isocontrol" lines), and the dashed lines represent the optimal control law. As can be seen, the optimal neural control law approximates the optimal one in a very satisfactory way, and this occurs not only inside the sets $\mathcal{A}_0$, $\tilde{\mathcal{A}}_4$, $\tilde{\mathcal{A}}_9$ but also outside these regions, thus pointing out the nice generalization properties of such control laws.

Example 2.

Consider the space robot presented in Figure 5, which for the sake of simplicity is assumed to move in the plane.

The robot's position with respect to the coordinate system is described by the Cartesian coordinates x, y and by the angle ϑ that its axis of symmetry (oriented in the direction of the vector $\underline{e}$ of unit length) forms with the x-axis. Two couples of thrusters, aligned with the axis of symmetry, are mounted on the robot's sides. Their thrusts, u_1 and u_2, can be modulated so as to obtain the desired intensity of the force $\underline{F}$ and the desired torque

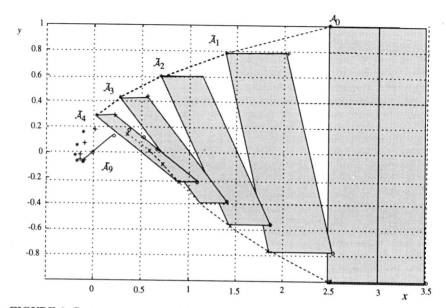

FIGURE 3. State convergence of the optimal neural trajectories from $\mathcal{A}_0$ to the origin (i.e., $\mathcal{A}_N$).

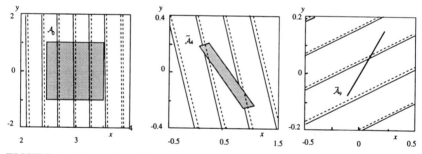

FIGURE 4. Comparison between the optimal neural control law (continuous lines) and the optimal control law (dashed lines).

T by which to control the robot's motion. We assume the mass m and the moment of inertia J to remain constant during the maneuver described in the following. Then we can write

$$\underline{F} = (u_1 + u_2)\underline{e} = m\frac{d\underline{v}}{dt}, \qquad (24)$$

$$T = (u_1 - u_2)d = J\frac{d\omega}{dt}, \qquad (25)$$

where d is the distance between the thrusters and the axis of symmetry, $\underline{v}$ is the robot's velocity, and ω is the robot's angular velocity. Let $x_1 = x$, $x_2 = \dot{x}$, $x_3 = y$, $x_4 = \dot{y}$, $x_5 = \vartheta$, $x_6 = \dot{\vartheta}$, and $\underline{x} \stackrel{\Delta}{=} \text{col}(x_i, i = 1, \ldots, 6)$. Then from (24) and (25) derive the nonlinear differential dynamic system

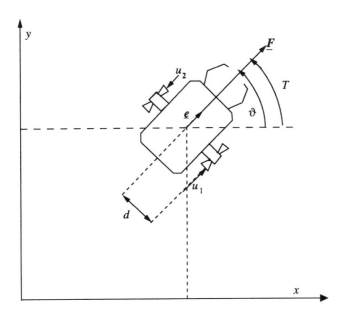

FIGURE 5. The space robot.

$$\begin{cases} \dot{x}_1 = x_2, \\ \dot{x}_2 = \dfrac{1}{m}(u_1 + u_2)\cos x_5, \\ \dot{x}_3 = x_4, \\ \dot{x}_4 = \dfrac{1}{m}(u_1 + u_2)\sin x_5, \\ \dot{x}_5 = x_6, \\ \dot{x}_6 = \dfrac{d}{J}(u_1 - u_2), \end{cases} \quad (26)$$

under the constraints
$$|u_1| \leq U, \ |u_2| \leq U, \quad (27)$$
where U is the maximum thrust value allowed.

The space robot is requested to start from any given point of the segment AB shown in Figure 6 (the parking edge of a space platform) and to reach an object moving along the segment $A'B'$ in an unpredictable way.

The dashed line shows the path of the object. When the robot is on the segment $A'B'$, it must stop with the angle $\vartheta = 0$. Then, the initial and final sets are given by

$$\mathcal{A}_0 = \left\{\underline{x} \in \mathbb{R}^6 : x_1 = 0, x_2 = 0, 1 \leq x_3 \leq 5, x_4 = 0, x_5 = 0, x_6 = 0\right\}$$

and

$$\mathcal{A}_N = \left\{\underline{x} \in \mathbb{R}^6 : x_1 = 10, x_2 = 0, 1 \leq x_3 \leq 5, x_4 = 0, x_5 = 0, x_6 = 0\right\},$$

respectively. The maneuver has to be completed at a given time t_f, and $N = 10$ control stages are allowed. The fuel consumption has to be minimized, and the robot's trajectory has to terminate "sufficiently near" the

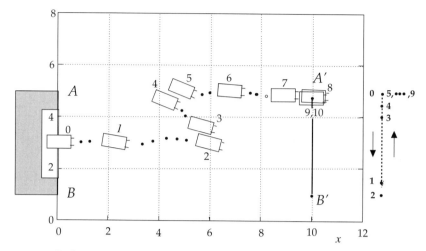

FIGURE 6. Positions of the space robot during its maneuver.

target vector $\underline{x}_N^*$. In accordance with these requirements, the cost function can be expressed as

$$J = \sum_{i=0}^{N-1} \left[c(u_{i1}) + c(u_{i2}) + \|\underline{x}_N^* - \underline{x}_i\|_V^2 \right] + \|\underline{x}_N^* - \underline{x}_N\|_{V_N}^2,$$

where $\underline{x}_i \triangleq \underline{x}(i\Delta t)$, $\underline{u}_i \triangleq \underline{u}(i\Delta t)$, and $\Delta t = t_f/N$ (for the sake of brevity, we do not write the discretized version of the differential system (26), as it is simply given by a first-order Euler's approximation for the system). Moreover, $V = \text{diag}\,[1, 0.1, 40, 0.1, 40, 0.1]$, $V_N = \text{diag}\,[40, 40, 40, 40, 40, 40]$. The cost of the fuel consumption is taken into account by the functions $c(u_{ij}) = k \left[\frac{1}{\beta} \ln(2 + e^{\beta u_{ij}} + e^{-\beta u_{ij}}) - \frac{1}{\beta} \ln(4) \right]$, $(j = 1, 2)$, which approximate (for large enough values of the parameter β) the nondifferentiable costs $k|u_{ij}|$ (it is realistic to assume the fuel consumption to be proportional to the thrust); for the present example, we took $\beta = 50$, $k = 0.01$. We also chose $c_1 = 10^{-5}$, $c_2 = 10^4$, $\eta = 0.9$. The matrices V, V_N and the constant k were chosen so as to obtain a reasonable compromise between the "attractiveness" of the vectors to be tracked and the fuel consumption. Note also that the sigmoidal functions generating the control variables u_{i1}, u_{i2} are bounded by unit values. Then, multiplying these functions by U enables us to remove constraints (27). The control functions $\hat{\underline{\gamma}}\,(\underline{x}_i, \underline{x}_N^*, \underline{w}_i)$ were implemented by means of neural networks with 12 input variables and one hidden layer of 80 units.

The positions of the space robot during its maneuver are shown in Figure 6. The effect of the feedforward action is clearly revealed by the variation occurring in the robot's trajectory when the robot perceives the "right-about turn" of the object to be reached.

7 The Infinite-Horizon Optimal Control Problem and Its Receding-Horizon Approximation

Let us consider again the discrete-time dynamic system (1) that we now assume to be time-invariant:

$$\underline{x}_{t+1} = \underline{f}(\underline{x}_t, \underline{u}_t), \quad t = 0, 1, \ldots. \tag{28}$$

We shall use indices t for the IH problems, whereas we shall go on using indices i for the FH ones. Constraints on state and control vectors are explicitly taken into account, that is, we assume $\underline{x}_t \in X \subset \mathbb{R}^n$ and $\underline{u}_t \in U \subset \mathbb{R}^m$. In general, denote by $\mathcal{Z}$ the class of compact sets $\mathcal{A} \subset \mathbb{R}^q$ containing the origin as an *internal point*. This means that $\mathcal{A} \in \mathcal{Z} \Leftrightarrow \exists \lambda \in \mathbb{R}, \lambda > 0$ such that $N(\lambda) \subset \mathcal{A}$, where $N(\lambda) \triangleq \{\underline{x} \in \mathbb{R}^q : \|\underline{x}\| \le \lambda\}$ is the closed ball with center $\underline{0}$ and radius λ. Then, assume that $X, U \in \mathcal{Z}$. The cost function is given by

$$J_{IH}(\underline{x}_t, \underline{u}_{t\infty}) = \sum_{i=t}^{+\infty} h(\underline{x}_i, \underline{u}_i), \quad t \ge 0. \tag{29}$$

In (29) and in the following, we define $\underline{u}_{t\tau} \triangleq \mathrm{col}\,(\underline{u}_t, \ldots, \underline{u}_\tau)$ for both finite and infinite values of the integer τ. Assume that $\underline{f}(\underline{0}, \underline{0}) = \underline{0}$ and $h(\underline{0}, \underline{0}) = 0$. Comparing cost (2) with cost (29), we notice that in (29) the cost terms are time-invariant functions and that the cost terms $\rho_i(\|\underline{x}_N^* - \underline{x}_i\|)$ lose their meanings and then vanish. Now we can state the following

Problem 3. *At every time instant $t \ge 0$, find the IH optimal feedback control law $\underline{u}_t^{IH\circ} = \underline{\gamma}_{IH}^\circ(\underline{x}_t) \in U$ that minimizes cost (29) for any state $\underline{x}_t \in X$.* □

As is well known, unless the dynamic system (28) is linear and cost (29) is quadratic, deriving the optimal feedback law $\underline{\gamma}_{IH}^\circ$ is a very hard, almost infeasible task. Then let us now consider an RH approximation for Problem 3. To this end, we need to define the following FH cost function

$$J_{FH}[\underline{x}_t, \underline{u}_{t,t+N-1}, N, h_F(\cdot)] = \sum_{i=t}^{t+N-1} h(\underline{x}_i, \underline{u}_i) + h_F(\underline{x}_{t+N}), \quad t \ge 0, \tag{30}$$

where $h_F(\cdot) \in \mathcal{C}^1\,[\mathbb{R}^n, \mathbb{R}^+]$, with $h_F(\underline{0}) = 0$, is a suitable terminal cost function and N is a positive integer denoting the length of the control horizon. Then we can state the following:

Problem 4. *At every time instant $t \ge 0$, find the RH optimal control law $\underline{u}_t^{RH\circ} = \underline{\gamma}_{RH}^\circ(\underline{x}_t) \in U$, where $\underline{u}_t^{RH\circ}$ is the first vector of the control sequence $\underline{u}_t^{FH\circ}, \ldots, \underline{u}_{t+N-1}^{FH\circ}$ (i.e., $\underline{u}_t^{RH\circ} \triangleq \underline{u}_t^{FH\circ}$) that minimizes cost (30) for the state $\underline{x}_t \in X$.* □

As to Problem 4, we remark that stabilizing properties of the RH regulators were established in [KP77, KP78, KBK83] under LQ assumptions. Extensions to nonlinear systems were derived by Keerthi and Gilbert [KG88] for discrete-time systems and by Mayne and Michalska [MM90, MM93] for continuous-time systems. In [MM90], the RH optimal control problem was solved under the constraint $\underline{x}_{t+N} = \underline{0}$. Such a constraint was relaxed in [MM93] by requiring that the regulator drive the system to enter a certain neighborhood W of the origin. Once the boundary of W has been reached, a linear regulator designed to stabilize the nonlinear system inside W takes over and steers the state to the origin. It is worth noting that in both approaches, the regulator computes its control actions on-line; this can be accepted only if the process is slow enough, as compared with the computation speed of the regulator itself.

As can be deduced from the statement of Problem 4, we shall derive the RH stabilizing optimal regulator without imposing either the "exact" constraint $\underline{x}_{t+N} = \underline{0}$ or the condition of reaching the neighborhood W of the origin. The stabilizing property of the RH regulator depends on proper choices of the control horizon N and of the final cost $h_F(\cdot)$ that penalizes the fact that the system state is not steered to the origin at time $t + N$.

The statement of Problem 4 does not impose any particular way of computing the control vector $\underline{u}_t^{RH°}$ as a function of $\underline{x}_t$. Actually, we have two possibilities.

1) **On-line computation.** When the state $\underline{x}_t$ is reached at time t, cost (30) must be minimized at this instant (clearly, no other state belonging to X is of interest for such minimization). Problem 2 is then an open-loop optimal control problem and may be regarded as a nonlinear programming one. This problem can be solved on-line by considering the vectors $\underline{u}_t, \ldots, \underline{u}_{t+N-1}, \underline{x}_{t+1}, \ldots, \underline{x}_{t+N}$ as independent variables. The main advantage of this approach is that many well-established nonlinear programming techniques are available to solve Problem 2. On the other hand, the approach involves a huge computational load for the regulator. If the dynamics of the controlled plant are not sufficiently slow as compared with the speed of the regulator's computing system, a practical application of the RH control mechanism turns out to be infeasible (see [YP93], where a maximum time interval T_c was assigned to the control system to generate the control vector).

2) **Off-line computation.** By following the approach of off-line computation, the regulator must be able to generate *instantaneously* $\underline{u}_t^{RH°}$ for *any* state $\underline{x}_t \in X$ that may be reached at stage t. In practice, this implies that the control law $\underline{\gamma}_{RH}°(\underline{x}_t)$ has to be computed "a priori" (i.e., off-line) and stored in the regulator's memory. Clearly, the off-line computation has advantages and disadvantages that are opposite to the ones of the on-line approach. No on-line computational effort

12. Neural Approximations for Optimal Control

is requested from the regulator, but an excessive amount of computer memory may be required to store the closed-loop control law. Moreover, an N-stage functional optimization problem has to be solved instead of a nonlinear programming one. As is well known, such a functional optimization problem can be solved analytically in only a few cases, typically under LQ assumptions. As we are looking for feedback optimal control laws, dynamic programming seems to be the most efficient tool. This implies that the control function $\gamma_{RH}^{\circ}(\underline{x}_t)$ has to be computed when the backward phase of the dynamic programming procedure, starting from the final stage $t+N-1$, has come back to the initial stage t. Unfortunately, as stated in the first sections of this chapter, dynamic programming exhibits computational drawbacks that in general are very difficult to overcome. In Section 9, we shall return to the off-line solution of Problem 2 and present a neural approximation method to solve this problem.

Here we want to remark that the works by Keerthi and Gilbert [KG88] and by Mayne and Michalska [MM93] aim to determine the RH optimal control law on-line, whereas we are more interested in an off-line computational approach. For now, we do not address these computational aspects and in the next section we present a stabilizing control law to solve Problem 4.

8 Stabilizing Properties of the Receding-Horizon Regulator

As stated in Section 7, we are looking for an RH feedback regulator that solves Problem 4, while stabilizing the origin as an equilibrium point of the closed-loop controlled plant. As previously specified, we relax the exact terminal constraint $\underline{x}_{t+N} = 0$ without imposing the condition of reaching a certain neighborhood W of the origin. Toward this end, the following assumptions are introduced.

(i) The linear system $\underline{x}_{t+1} = A\underline{x}_t + B\underline{u}_t$, obtained via the linearization of system (28) in a neighborhood of the origin, i.e.,

$$A \triangleq \left.\frac{\partial \underline{f}}{\partial \underline{x}_t}\right|_{\underline{x}_t=\underline{0},\,\underline{u}_t=\underline{0}} \quad \text{and} \quad B \triangleq \left.\frac{\partial \underline{f}}{\partial \underline{u}_t}\right|_{\underline{x}_t=\underline{0},\,\underline{u}_t=\underline{0}},$$

is stabilizable.

(ii) The transition cost function $h(\underline{x},\underline{u})$ depends on both $\underline{x}$ and $\underline{u}$, and there exists a strictly increasing function $r(\cdot) \in \mathcal{C}[\mathbf{R}^+, \mathbf{R}^+]$, with $r(0) = 0$, such that $h(\underline{x},\underline{u}) \geq r(\|(\underline{x},\underline{u})\|)$, $\forall \underline{x} \in X$, $\forall \underline{u} \in U$, where $(\underline{x},\underline{u}) \triangleq \mathrm{col}\,(\underline{x},\underline{u})$.

(iii) $h_F(\cdot) \in \mathcal{H}(a,P)$, where $\mathcal{H}(a,P) \triangleq \{h_F(\cdot): h_F(\underline{x}) = a\underline{x}^T P\underline{x}\}$, for some $a \in \mathbb{R}$, $a > 0$, and for some positive-definite symmetric matrix $P \in \mathbb{R}^{n \times n}$.

(iv) There exists a compact set $X_0 \subset X$, $X_0 \in \mathcal{Z}$ with the property that for every neighborhood $N(\lambda) \subset X_0$ of the origin of the state space there exists a control horizon $M \geq 1$ such that there exists a sequence of admissible control vectors $\{\underline{u}_i \in U, i = t, \ldots, t+M-1\}$ that yield an admissible state trajectory $\underline{x}_i \in X$, $i = t, t+1, \ldots, t+M$, ending in $N(\lambda)$ (i.e., $\underline{x}_{t+M} \in N(\lambda)$) for any initial state $\underline{x}_t \in X_0$.

(v) The optimal FH feedback control functions $\gamma^\circ_{FH}(\underline{x}_i, i)$, $i = t, \ldots, t+N-1$, which minimize cost (30), are continuous with respect to $\underline{x}_i$ for any $\underline{x}_i \in X$ and for any finite integer $N \geq 1$.

Assumption (i) is related to the possibility of stabilizing the origin as an equilibrium point of the closed-loop system by using a suitable linear regulator in a neighborhood of the origin itself. In the proof of the following Proposition 3, this assumption is exploited in order to build the region of attraction for the origin when the RH regulator $\gamma^\circ_{RH}(\underline{x}_t)$ is applied and to provide useful information on the form of the FH cost function (30) that guarantees the stability properties of the control scheme [PZ95]. Assumption (i) is the discrete-time version of the one made in [MM93].

Assumption (iii) plays a key role in the development of the stability results concerning the RH regulator and is essentially related to the relaxation of the terminal state constraint $\underline{x}_{t+N} = \underline{0}$. This is quite consistent with intuition, as in practice, the constraint $\underline{x}_{t+N} = \underline{0}$ is replaced with the final cost $h_F(\cdot)$ that penalizes the fact that the system state is not driven to the origin at time $t+N$.

Assumption (iv) substantially concerns the controllability of the nonlinear system (1). In a sense, it is very similar to the Property C defined in [KG88]. However, assumption (iv) seems to be weaker than this property, which requires the existence of an admissible control sequence that forces the system state to reach the origin after a finite number of stages, starting from any initial state belonging to $\mathbb{R}^n$.

Let us now denote by $J^\circ_{IH}(\underline{x}_t) = \sum_{i=t}^{+\infty} h(\underline{x}_i^{IH^\circ}, \underline{u}_i^{IH^\circ})$ the cost associated with the IH optimal trajectory starting from $\underline{x}_t$ (i.e., $\underline{x}_t^{IH^\circ} = \underline{x}_t$). In an analogous way, let us denote by $J^\circ_{RH}[\underline{x}_t, N, h_F(\cdot)] = \sum_{i=t}^{+\infty} h(\underline{x}_i^{RH^\circ}, \underline{u}_i^{RH^\circ})$ the cost associated with the RH trajectory starting from $\underline{x}_t$ (i.e., $\underline{x}_t^{RH^\circ} = \underline{x}_t$) and with the solution of the FH control problem characterized by a control horizon N and a terminal cost function $h_F(\cdot)$. Finally, let us denote by

$$J^\circ_{FH}[\underline{x}_t, N, h_F(\cdot)] \triangleq J_{FH}[\underline{x}_t, \underline{u}^\circ_{t,t+N-1}, N, h_F(\cdot)]$$

12. Neural Approximations for Optimal Control   339

$$= \sum_{i=t}^{t+N-1} h(\underline{x}_i^{FH°}, \underline{u}_i^{FH°}) + h_F(\underline{x}_{t+N}^{FH°})$$

the cost corresponding to the optimal N-stage trajectory starting from $\underline{x}_t$. Then we present the following proposition, which is proved in [PZ95]:

Proposition 3 *If assumptions (i) to (v) are satisfied, there exist a finite integer $\tilde{N} \geq M$, a positive scalar $\tilde{a}$, and a positive-definite symmetric matrix $P \in \mathbb{R}^{n \times n}$ such that for every terminal cost function $h_F(\cdot) \in \mathcal{H}(a, P)$ with $a \in \mathbb{R}$, $a \geq \tilde{a}$, the following properties hold:*
1) The RH control law stabilizes asymptotically the origin, which is an equilibrium point of the resulting closed-loop system.
2) There exists a positive scalar β such that for any $N \geq \tilde{N}$ the set $\mathcal{W}[N, h_F(\cdot)] \in \mathcal{Z}$, $\mathcal{W}[N, h_F(\cdot)] \overset{\triangle}{=} \{\underline{x} \in X : J_{FH}^°[\underline{x}, N, h_F(\cdot)] \leq \beta\}$, is an invariant subset of X_0 and a domain of attraction for the origin, i.e., for any $\underline{x}_t \in \mathcal{W}[N, h_F(\cdot)]$ the state trajectory generated by the RH regulator remains entirely contained in $\mathcal{W}[N, h_F(\cdot)]$ and converges to the origin.
3) For any $N \geq \tilde{N} + 1$ we have

$$J_{RH}^°[\underline{x}_t, N, h_F(\cdot)] \leq J_{FH}^°[\underline{x}_t, N, h_F(\cdot)], \quad \forall \underline{x}_t \in \mathcal{W}[N, h_F(\cdot)]. \tag{31}$$

4) $\forall \delta \in \mathbb{R}, \delta > 0$, there exists an $N \geq \tilde{N} + 1$ such that

$$J_{RH}^°[\underline{x}_t, N, h_F(\cdot)] \leq J_{IH}^°(\underline{x}_t) + \delta, \quad \forall \underline{x}_t \in \mathcal{W}[N, h_F(\cdot)]. \tag{32}$$

□

Proposition 3 asserts that there exist values of a certain number of parameters, namely, $\tilde{N}$, P, and $\tilde{a}$, that ensure us the stabilizing property of the RH control law and some nice performances of this regulator, as compared with those of the IH one (see (31) and (32)). As nothing is said as to how such parameters can be found, one is authorized to believe that they can but be derived by means of some heuristic trial-and-error procedure to test whether stability has indeed been achieved. However, some preliminary results, based on the rather constructive proof of Proposition 3, as reported in [PZ95], lead us to believe that appropriate values of $\tilde{N}$, P, and $\tilde{a}$ can be computed, at least in principle, by stating and solving some suitable constrained nonlinear programming problems. We use the words "at least in principle" because the efficiencies of the related descent algorithms have still to be verified.

In deriving (both on-line and off-line) the RH control law $\underline{u}_t^{RH°} = \underline{\gamma}_{RH}^°(\underline{x}_t)$, computational errors may affect the vector $\underline{u}_t^{RH°}$ and possibly lead to a closed-loop instability of the origin; therefore, we need to establish the robustness properties of such a control law. This is done by means of the following proposition, which characterizes the stabilizing properties of the RH regulator when suboptimal control vectors $\underline{\hat{u}}_i^{RH} \in U$, $i \geq t$ are used in the RH control mechanism, instead of the optimal ones $\underline{u}_i^{RH°}$ solving

Problem 4. Let us denote by $\hat{\underline{x}}_i^{RH}$, $i > t$, the state vector belonging to the suboptimal RH trajectory starting from $\underline{x}_t$.

Proposition 4 *If assumptions (i) to (v) are satisfied, there exist a finite integer $\tilde{N}$, a positive scalar $\tilde{a}$, and a positive-definite symmetric matrix $P \in \mathbb{R}^{n \times n}$ such that for any terminal cost function $h_F(\cdot) \in \mathcal{H}(a, P)$ and for any $N \geq \tilde{N}$ the following properties hold:*
1) There exist suitable scalars $\tilde{\delta}_i \in \mathbb{R}$, $\tilde{\delta}_i > 0$, such that if

$$\left\| \underline{u}_i^{RH^\circ} - \hat{\underline{u}}_i^{RH} \right\| \leq \tilde{\delta}_i, \, i \geq t, \text{ then}$$

$$\hat{\underline{x}}_i^{RH} \in \mathcal{W}[N, h_F(\cdot)], \, \forall i > t, \quad \forall \underline{x}_t \in \mathcal{W}[N, h_F(\cdot)]. \tag{33}$$

2) For any compact set $\mathcal{W}_d \subset \mathbb{R}^n$, $\mathcal{W}_d \in \mathcal{Z}$, there exist a finite integer $T \geq t$ and suitable scalars $\bar{\delta}_i \in \mathbb{R}$, $\bar{\delta}_i > 0$, such that if

$$\left\| \underline{u}_i^{RH^\circ} - \hat{\underline{u}}_i^{RH} \right\| \leq \bar{\delta}_i, \, i \geq t, \text{ then}$$

$$\hat{\underline{x}}_i^{RH} \in \mathcal{W}_d, \quad \forall i \geq T, \quad \forall \underline{x}_t \in \mathcal{W}[N, h_F(\cdot)]. \tag{34}$$

□

The proof of Proposition 4 is a direct consequence of the regularity assumptions on the state equation (this proof is given in [PZ95]; see also [CPRZ94] for some preliminary results). Proposition 4 has the following meaning: the RH regulator can drive the state into every desired neighborhood $\mathcal{W}_d$ of the origin in finite time, provided that the errors on the control vectors are suitably bounded. Moreover, the state will remain contained in the above neighborhood at any future time instant. Clearly, if the RH regulator (generating the above-specified suboptimal control vectors) is requested to stabilize the origin asymptotically, the hybrid control mechanism described in [MM93] may be implemented. This involves designing an LQ optimal regulator that stabilizes the nonlinear system inside a proper neighborhood W of the origin. Then, if the errors affecting the control vectors generated by the RH regulator are sufficiently small, this regulator is able to drive the system state inside W (of course, the condition $\mathcal{W}_d \subseteq W$ must be satisfied). When the boundary of W is reached, the RH regulator switches to the LQ regulator. It also follows that such a hybrid control mechanism makes $\mathcal{W}[N, h_F(\cdot)]$ not only an invariant set but also a domain of attraction for the origin.

9 Neural Approximation for the Receding-Horizon Regulator

As stated in Section 7, we are mainly interested in computing the RH control law $\underline{u}_t^{RH^\circ} = \underline{\gamma}_{RH}^\circ(\underline{x}_t)$ off-line. This requires that the regulator

12. Neural Approximations for Optimal Control 341

generate the control vector $\underline{u}_t^{RH^\circ}$ instantaneously, as soon as any state belonging to the admissible set X is reached. Then, we need to derive ("a priori") an FH closed-loop optimal control law $\underline{u}_i^{FH^\circ} = \underline{\gamma}_{FH}^\circ(\underline{x}_i, i)$, $t \geq 0$, $i = t, \ldots, t + N - 1$, that minimizes cost (30) for any $\underline{x}_t \in X$. Because of the time-invariance of the dynamic system (28) and of the cost function (30), we refer to an FH optimal control problem, starting from the state $\underline{x}_t \in X$ at a generic stage $t \geq 0$. Then, instead of $\underline{u}_i^{FH} = \underline{\gamma}_{FH}(\underline{x}_i, i)$, we consider the control functions

$$\underline{u}_i^{FH} = \underline{\gamma}_{FH}(\underline{x}_i, i - t), \quad t \geq 0, \, i = t, \ldots, t + N - 1, \tag{35}$$

and state the following:
Problem 5. *Find the FH optimal feedback control law*

$$\{\underline{u}_i^{FH^\circ} = \underline{\gamma}_{FH}^\circ(\underline{x}_i, i - t) \in U, t \geq 0, i = t, \ldots, t + N - 1\}$$

that minimizes cost (30) for any $\underline{x}_t \in X$.

□

Once the solution of Problem 5 has been found, we can write

$$\underline{u}_t^{RH^\circ} = \underline{\gamma}_{RH}^\circ(\underline{x}_t) \triangleq \underline{\gamma}_{FH}^\circ(\underline{x}_t, 0), \quad \forall \underline{x}_t \in X, t \geq 0. \tag{36}$$

Dynamic programming seems, at least in principle, the most effective computational tool for solving Problem 5. However, this algorithm exhibits the well known computational drawbacks previously pointed out for the FH optimal control problem, namely, the necessity for discretizing (at each control stage) the set X into a fine enough mesh of grid points, and consequently, the possibility of incurring the curse of dimensionality, even for a small number of state components.

Unlike the requirements related to the N-stage optimal control problem described in the first sections of this chapter, it is important to remark that we are now interested in determining only the first control function of the control law that solves Problem 5, that is, $\underline{\gamma}_{FH}^\circ(\underline{x}_t, 0)$. On the other hand, we can compute (off-line) any number of open-loop optimal control sequences $\underline{u}_t^{FH^\circ}, \ldots, \underline{u}_{t+N-1}^{FH^\circ}$ (see Problem 4) for different vectors $\underline{x}_t \in X$. Therefore, we propose to approximate the function $\underline{\gamma}_{FH}^\circ(\underline{x}_t, 0)$ by means of a function $\underline{\hat{\gamma}}_{FH}(\underline{x}_t, \underline{w})$, to which we assign a given structure. $\underline{w}$ is a vector of parameters to be optimized. More specifically, we have to find a vector $\underline{w}^\circ$ that minimizes the approximation error

$$E(\underline{w}) \triangleq \int_X \left\| \underline{\gamma}_{FH}^\circ(\underline{x}_t, 0) - \underline{\hat{\gamma}}_{FH}(\underline{x}_t, \underline{w}) \right\|^2 d\underline{x}_t. \tag{37}$$

Clearly, instead of introducing approximating functions, it would be possible to subdivide the admissible set X into a regular mesh of points, as is usually done at each stage of dynamic programming, and to associate with any point $\underline{x}_t \in X$ the control vector $\underline{u}_t^{FH°}$ corresponding to the nearest point of the grid. Under the assumption that the function $\underline{\gamma}_{FH}^°(\underline{x}_t, 0)$ is continuous in X, it is evident that the mesh should be fine enough to satisfy the conditions required in Proposition 4, i.e., $\left\|\underline{u}_i^{FH°} - \underline{\hat{u}}_i^{FH}\right\| \leq \bar{\delta}_i$, $i \geq t$, where $\underline{u}_i^{FH°}$ are the "true" stabilizing optimal controls (known only for the grid points), and $\underline{\hat{u}}_i^{FH}$ are the approximate ones. It is, however, clear that the use of such a mesh would lead us again to the unwanted phenomenon of the curse of dimensionality.

For the same reasons as explained in Sections 3 and 4, we choose again a feedforward neural network to implement the approximating function $\underline{\hat{\gamma}}_{FH}(\underline{x}_t, \underline{w})$. With respect to Problem 2, it is worth noting that now i) only one network is needed, and ii) the approximation criterion is different, in that we have to minimize the approximation error (37) instead of minimizing the expected process cost. In the following, we refer to the neural mapping (7), (7), taking into account the fact that the superscript i is useless. The weight and bias coefficients $w_{pq}(s)$ and $w_{0q}(s)$ are the components of the vector $\underline{w}$ appearing in the approximating function $\underline{\hat{\gamma}}_{FH}(\underline{x}_t, \underline{w})$; the variables $y_q(0)$ are the components of $\underline{x}_t$; and the variables $y_q(L)$ are the components of $\underline{u}_t$. To sum up, once the optimal weight vector $\underline{w}°$ has been derived (off-line), the RH neural approximate control law takes on the form

$$\underline{\hat{u}}_t^{RH°} = \underline{\hat{\gamma}}_{RH}(\underline{x}_t, \underline{w}°) \stackrel{\triangle}{=} \underline{\hat{\gamma}}_{FH}(\underline{x}_t, \underline{w}°), \quad \forall \underline{x}_t \in X, \, t \geq 0. \tag{38}$$

As to the approximating properties of the RH neural regulator, results similar to the ones established in Propositions 1 and 2 can be obtained. Proposition 1 plays an important role also for the stabilizing properties of the RH regulator. We repeat it here in a suitably modified version.

Proposition 1' *Assume that in the solution of Problem 5 the first control function $\underline{\gamma}_{RH}°(\underline{x}_t) = \underline{\gamma}_{FH}°(\underline{x}_t, 0)$ of the sequence $\{\underline{\gamma}_{FH}°(\underline{x}_i, i-t), i = t, \ldots, t+N-1\}$, is unique and that it is a $C[X, \mathbb{R}^m]$ function. Then for every $\varepsilon \in \mathbb{R}, \varepsilon > 0$ there exist an integer ν and a weight vector $\underline{w}$ (i.e., a neural RH control law $\underline{\hat{\gamma}}_{RH}^{(\nu)}(\underline{x}_t, \underline{w})$) such that*

$$\left\|\underline{\gamma}_{RH}°(\underline{x}_t) - \underline{\hat{\gamma}}_{RH}^{(\nu)}(\underline{x}_t, \underline{w})\right\| < \varepsilon, \quad \forall \underline{x}_t \in X. \tag{39}$$

□

Proposition 1' enables us to state immediately the following:

Corollary [PZ95]. *If assumptions (i) to (v) are satisfied, there exists an RH neural regulator $\underline{\hat{u}}_t^{RH} = \underline{\hat{\gamma}}_{RH}^{(\nu)}(\underline{x}_t, \underline{w})$, $t \geq 0$, for which the two properties of Proposition 4 hold true. The control vectors $\underline{\hat{u}}_t^{RH}$ are constrained to take on their values from the admissible set $\bar{U} \stackrel{\triangle}{=} \{\underline{u} : \underline{u} + \Delta \underline{u} \in U, \Delta \underline{u} \in N(\varepsilon)\}$, where ε is such that $\varepsilon \leq \bar{\delta}_i$, $i \geq t$ (see the scalars in Proposition 4) and $\bar{U} \in \mathcal{Z}$.*

□

12. Neural Approximations for Optimal Control

The corollary allows us to apply the results of Proposition 4, thus obtaining an RH regulator able to drive the system state into any desired neighborhood $\mathcal{W}_d$ of the origin in a finite time. Moreover, with reference to what has been stated at the end of Section 8, a neural regulator capable of switching to an LQ stabilizing regulator when a proper neighborhood $\mathcal{W}$ of the origin is reached makes the region $\mathcal{W}[N, h_F(\cdot)]$ a domain of attraction for the origin.

It should be noted that Proposition 1' and the corollary constitute only a first step towards the design of a stabilizing neural regulator. In fact, nothing is said as to how the sequence of scalars $\bar{\delta}_i$, $i \geq t$, (hence ε) as well as the number ν of required neural units can be derived (as we did in commenting on the computation of the parameters appearing in Proposition 3, we exclude trial-and-error procedures). The determination of the scalar ε (see the corollary) is clearly a hard constrained nonlinear optimization problem. Hopefully, some algorithm to solve it may be found. To this end, research is currently being conducted.

As to the integer ν, its derivation is an open problem of neural approximation theory, at least if one remains in the class of feedforward neural networks. If other approximators are addressed, something more can be said. Consider, for example, an approximator given by a nonlinear combination of Gaussian radial basis functions of the form $g_k(\underline{x}) = e^{-\|\underline{x}-\underline{x}^{(k)}\|^2/\sigma^2}$, where $\underline{x}^{(k)}$ are fixed centers placed in the nodes of a regular mesh. Such a mesh is obtained by subdividing the n sides of the smallest hypercube containing X into $D-1$ segments of length Δ (a suitable "extension" $\bar{\gamma}_{RH}^\circ(\underline{x}_t)$ of $\gamma_{RH}^\circ(\underline{x}_t)$ outside X must be defined). The number of nodes of the mesh is then D^n and the components of the approximating function are given by

$$\hat{\gamma}_{RHj}(\underline{x}_t, \underline{w}_j) = \sum_{k=1}^{D^n} w_j^k g_k(\underline{x}_t), \, j = 1, \ldots, m,$$

where $\underline{w}_j \triangleq \text{col}\left(w_j^k, k = 1, \ldots, D^n\right)$. If the Fourier transform $\Gamma_{RHj}^\circ(\underline{\omega})$ of the jth component of $\bar{\gamma}_{RH}^\circ(\underline{x}_t)$ is absolutely integrable on $\mathbb{R}^n$ for $j = 1, \ldots, m$, it can be shown [SS92] that

$$\|\underline{\gamma}_{RH}^\circ(\underline{x}_t) - \hat{\underline{\gamma}}_{RH}^{(\nu)}(\underline{x}_t, \underline{w})\| \leq \psi, \quad \forall \underline{x}_t \in X, \tag{40}$$

where ψ can be made arbitrarily small by suitably choosing the number D of nodes on the mesh side (or equivalently, the mesh size Δ) and the variance σ^2. The important result given in [SS92] lies in the fact that such parameters can be determined quantitatively on the basis of the smoothness characteristics of the function $\bar{\gamma}_{RH}^\circ(\underline{x}_t)$. Such characteristics are specified by the "significant" frequency ranges of the Fourier transforms Γ_{RHj}°, $j = 1, \ldots, m$ and by $\mathcal{L}_1$ bounds to these transforms. Note that as the desired value of ψ decreases or as the degree of smoothness of the function $\bar{\gamma}_{RH}^\circ(\underline{x}_t)$

decreases, the variance σ^2 and the mesh size Δ must suitably decrease (for more details, see [SS92] again). Then the above results enable one to specify the number $\nu = m\, D^n$ of parameters required to achieve a given error tolerance. This number reveals that we pay for the possibility of computing an explicit uniform bound to the approximation error with the feared danger of incurring the curse of dimensionality.

Coming back to the feedforward neural approximators, it can be expected that given a bound to the approximation error (see (39)), a computational technique will be found to determine the number ν, on the basis of the smoothness characteristics, and to approximate functions that belong to the difference set between Barron's class of functions and Sobolev spaces. (As said in Section 4, in this difference set, feedforward neural approximators should behave better than linear ones.) Waiting for such a computational technique to be derived, and reassured by the fact that a large quantity of simulation results lead us to believe that a heuristic (i.e., experimental) determination of the integer ν is, all things considered, rather easy, we shall go on with our treatment, still considering feedforward neural networks as our basic approximators. In the next section, we shall present a method for deriving the weights of this type of network and conclude by reporting some simulation results.

10 Gradient Algorithm for Deriving the RH Neural Regulator; Simulation Results

To minimize the approximation error (37), we use again a gradient algorithm (see (13)), that is,

$$\underline{w}(k+1) = \underline{w}(k) - \alpha \nabla_{\underline{w}} E\left[\underline{w}(k)\right], \quad k = 0, 1, \ldots . \tag{41}$$

Define now the function

$$D(\underline{w}, \underline{x}_t) \triangleq \left\| \gamma^\circ_{FH}(\underline{x}_t, 0) - \hat{\gamma}_{FH}(\underline{x}_t, \underline{w}) \right\|^2 = \left\| \gamma^\circ_{RH}(\underline{x}_t) - \hat{\gamma}_{RH}(\underline{x}_t, \underline{w}) \right\|^2,$$

and note that we are able to evaluate $\gamma_{FH}(\underline{x}_t, 0)$ only pointwise, that is, by solving Problem 4 for specific values of $\underline{x}_t$. It follows that we are unable to compute the gradient $\nabla_{\underline{w}} E\left[\underline{w}(k)\right]$ in explicit form. Then, we interpret $E(\underline{w})$ as the expected value of the function $D(\underline{w}, \underline{x}_t)$ by considering $\underline{x}_t$ as a random vector uniformly distributed on X. This leads us to use again a stochastic approximation approach and to compute the "realization"

$$\nabla_{\underline{w}} D[\underline{w}, \underline{x}_t(k)]$$

instead of the gradient appearing in (41).

12. Neural Approximations for Optimal Control

We generate the sequence $\{\underline{x}_t(k),\ k=0,1,\ldots\}$ randomly, taking into account the fact that $\underline{x}_t$ is considered to be uniformly distributed on X. Then the updating algorithm becomes

$$\underline{w}(k+1) = \underline{w}(k) - \alpha(k)\nabla_{\underline{w}} D\left[\underline{w}(k), \underline{x}_t(k)\right], \quad k=0,1,\ldots. \tag{42}$$

To derive the components of

$$\nabla_{\underline{w}} D\left[\underline{w}(k), \underline{x}_t(k)\right],$$

i.e., the partial derivatives

$$\frac{\partial D[\underline{w}(k), \underline{x}_t(k)]}{\partial w_{pq}(s)},$$

the backpropagation updating rule can be applied again. In the following, we report such a procedure, taking into account the fact that only one neural network has now to be trained. To simplify the notations, we drop the index k and define

$$\delta_q(s) \triangleq \frac{\partial D\left[\underline{w}, \underline{x}_t\right]}{\partial z_q(s)}, \quad s=1,\ldots,L;\ q=1,\ldots,n_s. \tag{43}$$

Then it is easy to show that

$$\frac{\partial D\left[\underline{w}, \underline{x}_t\right]}{\partial w_{pq}(s)} = \delta_q(s) y_p(s-1). \tag{44}$$

where $\delta_q(s)$ can be computed recursively by means of the equations

$$\delta_q(s) = g'[z_q(s)] \sum_{h=1}^{n_{s+1}} \delta_h(s+1) w_{qh}(s+1), \quad s=1,\ldots,L-1 \tag{45a}$$

$$\delta_q(L) = g'[z_q(L)] \frac{\partial D}{\partial y_q(L)}. \tag{45b}$$

It can be seen that the algorithm consists of the following two "passes":
Forward pass. The initial state $\underline{x}_t(k)$ is randomly generated from X. Then, the open-loop solution of the FH Problem 4 is computed, and the first control $\underline{u}_t^{FH^\circ} = \underline{\gamma}_{FH}^\circ[\underline{x}_t(k), 0]$ is stored in the memory to determine $\dfrac{\partial D}{\partial y_q(L)}$ (see (45b)). All the variables required by (44) and (45) are stored in the memory.
Backward pass. The variables $\delta_q(s)$ are computed via (45). Then the gradient $\nabla_{\underline{w}} D\left[\underline{w}(k), \underline{x}_t(k)\right]$ is determined by using (44), and the new weight vector $\underline{w}(k+1)$ is generated by using (42).

As we said in Sections 9 and 10, further research is needed to derive a computational procedure that gives us the correct values of the parameters required for the design of a stabilizing RH regulator. However, at least to judge by the following example, determining experimentally such parameters may turn out to be quite an easy task.

Example 3.

Consider the same robot as in Example 2. The space robot is now requested to start from any point of a given region and to reach the origin of the state space, while minimizing the nonquadratic IH cost

$$J_{IH} = \sum_{i=0}^{+\infty} \left[c(u_{i1}) + c(u_{i2}) + \|\underline{x}_i\|_V^2 \right].$$

For the present example, we chose $V = \text{diag}\,[1, 80, 5, 10, 1, 0.1]$, $\beta = 50$, $k = 0.01$, $c_1 = 1$, $c_2 = 10^8$, and $\rho = 0.9$. No constraint was imposed on the state vector. Then

$$\mathcal{A} = \{\underline{x} \in \mathbb{R}^6 : -2 \leq x_1 \leq 2, -0.2 \leq x_2 \leq 0.2, -2 \leq x_3 \leq 2,$$
$$-0.2 \leq x_4 \leq 0.2, -\pi \leq x_5 \leq \pi, -1 \leq x_6 \leq 1\}$$

was chosen as a training set. The FH cost function takes on the form

$$J_{FH} = \sum_{i=t}^{t+N-1} \left[c(u_{i1}) + c(u_{i2}) + \|\underline{x}_i\|_V^2 \right] + a \|\underline{x}_N\|^2,$$

where $a = 40$ and $N = 30$. The control function $\hat{\gamma}_{FH}^\circ (\underline{x}_i, \underline{w})$, $i \geq t$, was implemented by means of a neural network with six input variables and one hidden layer of 100 units. Usually, the algorithm converged to the optimal solution $\underline{w}^\circ$ after $2 \cdot 10^5$ to $3 \cdot 10^5$ iterations.

Figures 7 and 8 show the positions of the space robot along trajectories generated by the neural RH (NRH) optimal control law. Such trajectories are almost indistinguishable from the *on-line* computed ones, after solving Problem 4 on line (we denote by ORH the corresponding optimal control law). In Figure 7, the initial velocities x_2, x_4, x_6 are all set to zero, whereas in Figure 8 even the initial velocities are not set to zero (in Figure 8a, we set $x_{t2} = x_{t4} = 0$, $x_{t6} = 0.5$, and in Figure 8b, $x_{t2} = 0.5$, $x_{t4} = 0.5$, $x_{t6} = 0.5$). It is worth noting that the initial velocities were chosen such as to launch the space robot along trajectories that were "opposite" to the one that would result from initial velocities set to zero (compare the trajectories in Figure 8 with the one shown in Figure 7a). This causes the trajectories to get out of the set $\mathcal{A}$ in the first stages. However, the control actions still appear quite effective, thus showing the nice "generalization" capabilities of the neural RH regulator (i.e., the neural network was tuned even in the neighborhood of the training set $\mathcal{A}$).

12. Neural Approximations for Optimal Control 347

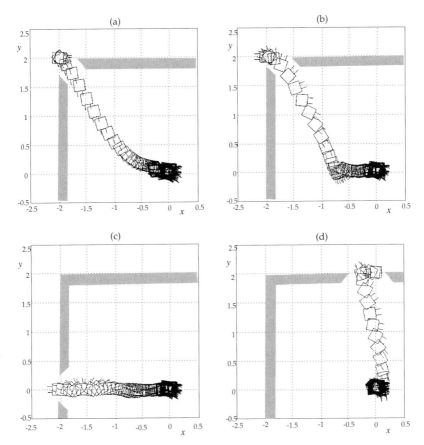

FIGURE 7. Trajectories of the space robot starting from four different initial positions at zero initial velocity.

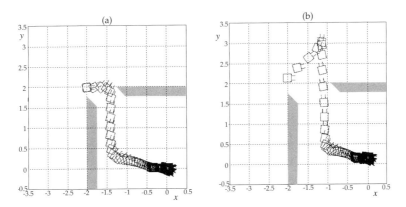

FIGURE 8. Trajectories of the space robot starting from the same positions as in the first of the previous figures but at two different sets of initial velocities.

11 Conclusions

Neural approximators have been shown to be powerful and simple approximators for solving both FH and RH optimal control problems. Bounds for the approximations have been given; this is particularly important in RH control schemes, which involve stability issues. Deterministic problems have been addressed; however, the neural approximation approach has proved effective also for the design of control devices for stochastic dynamic systems [PZ94a] and for optimal state estimators [PZ94c] in non-Gaussian contexts (i.e., outside the classical LQG framework).

As a final remark, it is worth noting that neural approximations enable us to face even so-called "nonclassical" optimal control problems, like *team control problems*, characterized by the presence of informationally decentralized organizations in which several decision makers cooperate on the accomplishment of a common goal. For this class of problems, quite typical in large-scale engineering applications, neural methods seem to constitute a very promising tool, as distributed computation, which is a peculiar property of these methods, may turn out to be a necessity and not a choice (see [PZ93] for an application in the communications area).

Acknowledgments: This work was supported by the Italian Ministry for the University and Research.

12 REFERENCES

[Bar93]   A. R. Barron. Universal approximation bounds for superpositions of a sigmoidal function. *IEEE Transactions on Information Theory*, 39:930–945, 1993.

[CPRZ94]   A. Cattaneo, T. Parisini, R. Raiteri, and R. Zoppoli. Neural approximations for receding-horizon controllers. In *Proceedings of the 1994 American Control Conference, Baltimore, Maryland*, 2144–2147, 1994.

[Cyb89]   G. Cybenko. Approximations by superpositions of a sigmoidal function. *Mathematics of Control, Signals, and Systems*, 2:303–314, 1989.

[ES66]   B. R. Eisenberg and A. P. Sage. Closed-loop optimization of fixed configuration systems. *International Journal of Control*, 3:183–194, 1966.

[Gir94]   F. Girosi. Regularization theory, radial basis functions and networks. In V. Cherkassky, J. H. Friedman, and H. Wechsler, editors, *From Statistics to Neural Networks. Theory and Pattern Recognition Applications*. Springer-Verlag, Computer and Systems Sciences, Berlin, 1994.

[HN89] R. Hecht-Nielsen. Theory of the backpropagation neural network. In *Proceedings of the IJCNN, 1989, Washington, D.C.* 593–608, 1989.

[HSW89] K. Hornik, M. Stinchcombe, and H. White. Multilayer feedforward networks are universal approximators. *Neural Networks*, 2:359–366, 1989.

[Jon92] L. K. Jones. A simple lemma on greedy approximation in Hilbert space and convergence rates for projection pursuit regression and neural network training. *Annals of Statistics*, 20:608–613, 1992.

[JSS+93] S. A. Johnson, J. D. Stedinger, C. A. Shoemaker, Y. Li, and J. A. Tejada-Guibert. Numerical solution of continuous-state dynamic programs using linear and spline interpolation. *Operations Research*, 41:484–500, 1993.

[KA68] D. L. Kleinman and M. Athans. The design of suboptimal linear time-varying systems. *IEEE Transactions on Automatic Control*, AC-13:150–159, 1968.

[KBK83] W. H. Known, A. M. Bruckstein, and T. Kailath. Stabilizing state-feedback design via the moving horizon method. *International Journal of Control*, 37:631–643, 1983.

[KG88] S. S. Keerthi and E. G. Gilbert. Optimal infinite-horizon feedback laws for a general class of constrained discrete-time systems: Stability and moving-horizon approximations. *Journal of Optimization Theory and Applications*, 57:265–293, 1988.

[KP77] W. H. Known and A. E. Pearson. A modified quadratic cost problem and feedback stabilization of a linear system. *IEEE Transactions on Automatic Control*, AC-22:838–842, 1977.

[KP78] W. H. Known and A. E. Pearson. On feedback stabilization of time-varying discrete linear systems. *IEEE Transactions on Automatic Control*, AC-23:479–481, 1978.

[Lar68] R. E. Larson. *State Increment Dynamic Programming*. American Elsevier, New York, 1968.

[MM90] D. Q. Mayne and H. Michalska. Receding horizon control of nonlinear systems. *IEEE Transactions on on Automatic Control*, 35:814–824, 1990.

[MM93] H. Michalska and D. Q. Mayne. Robust receding horizon control of constrained nonlinear systems. *IEEE Transactions on Automatic Control*, 38:1623–1633, 1993.

[MT87] P. M. Mäkilä and H. T. Toivonen. Computational methods for parametric LQ problems — A survey. *IEEE Transactions on Automatic Control*, AC-32:658–671, 1987.

[NW90] D. H. Nguyen and B. Widrow. Neural networks for self-learning control systems. *IEEE Control Systems Magazine*, 10(3):18–23, April 1990.

[Pin86] A. Pinkus. *N-Widths in Approximation Theory*. Springer-Verlag, New York, 1986.

[PT73] B. T. Polyak and Ya. Z. Tsypkin. Pseudogradient adaptation and training algorithms. *Automation and Remote Control*, 12:377–397, 1973.

[PZ93] T. Parisini and R. Zoppoli. Team theory and neural networks for dynamic routing in traffic and communication networks. *Information and Decision Technologies*, 19:1–18, 1993.

[PZ94a] T. Parisini and R. Zoppoli. Neural approximations for multistage optimal control of nonlinear stochastic systems. In *Proceedings of the 1994 American Control Conference, Baltimore* and *IEEE Transactions on Automatic Control*, 41(6):889–895, 1996.

[PZ94b] T. Parisini and R. Zoppoli. Neural networks for feedback feedforward nonlinear control systems. *IEEE Transactions on Neural Networks*, 5:436–449, 1994.

[PZ94c] T. Parisini and R. Zoppoli. Neural networks for nonlinear state estimation. *International Journal of Robust and Nonlinear Control*, 4:231–248, 1994.

[PZ95] T. Parisini and R. Zoppoli. A receding-horizon regulator for nonlinear systems and a neural approximation. *Automatica*, 31:1443–1451, 1995.

[RM86] D. E. Rumelhart and J. L. McClelland. *Parallel Distributed Processing*. MIT Press, Cambridge, Massachusetts, 1986.

[Sag68] A. P. Sage. *Optimum Systems Control*. Prentice-Hall, Englewood Cliffs, New Jersey, 1968.

[SS92] R. M. Sanner and J. J. E. Slotine. Gaussian networks for direct adaptive control. *IEEE Transactions on Neural Networks*, 3:837–863, 1992.

[Tsy71] Ya. Z. Tsypkin. *Adaptation and Learning in Automatic Systems*. Academic Press, New York, 1971.

[YP93] T. H. Yang and E. Polak. Moving horizon control of nonlinear systems with input saturation, disturbance and plant uncertainty. *International Journal of Control*, 58:875–903, 1993.

[ZP92] R. Zoppoli and T. Parisini. Learning techniques and neural networks for the solution of N-stage nonlinear nonquadratic optimal control problems. In A. Isidori and T. J. Tarn, editors, *Systems, Models and Feedback: Theory and Applications*, 193–210. Birkhäuser, Boston, 1992.

Index

A

accommodation, 117
action potential (AP), 95–97
activation
    competitive distribution of, 63
activation function, 164, 165, 171, 179, 186, 189, 197, 205, 241
Adalines, 1
adaptation, 270, 276
adaptive control, 290
adaptive critic, *see* critic, adaptive, 269, 277
adjoint equation, 330
algorithm
    EM, 47, 52
    forward-backward, 38
    Viterbi, 34, 36, 37, 41, 42, 50–52
approximation, 239
    -error, 344
        integrated square, 325, 326
        uniform, 344
    error, 341, 342
    Euler, 334
    neural, 325–348
    receding-horizon (RH), 335
    stochastic, 328, 344
approximation property, 165, 174, 191
arm
    model, 62, 63, 66–70
    movement, 62, 63
    robotic, 62
ARMA, 291
artificial neural networks, 3
ARX, 267
ASTREX, 306–308

auto-tuner, 260, 281
autoassociative neural networks, 220
autonomous control, 290
autoregulation, 117

B

backpropagation, 133, 154, 162, 171, 177, 185, 240, 242, 274, 290, 292, 295–297, 303, 328, 345
    fuzzy, 309
backpropagation through time, 276, 278
backpropagation, dynamic, 131–133, 139, 142, 155
BAM, *see* bidirectional associative memory
baroreceptor, 88–121
    Type I, 111–113
    Type II, 111–113
baroreceptor reflex, *see* baroreflex
baroreflex, 88–121
barotopical organization, 94
basis functions
    radial (RBF), 236, 289–309, 327, 343
    trigonometric, 326, 327
BDN, 120, 121
bidirectional associative memory, 290, 292–294, 297–302, 309
    eigenstructure, 297
biologically organized dynamic network, *see* BDN
blood pressure, 88–121
Boltzmann Machine, 46, 47
Brunovsky canonical form, 187, 192
building control, 272

353

354   Index

C

cardiovascular system, 88, 104, 105
cart-pole balancing, 269
central pattern generation, 55
cerebellar model articulation controller, 290, 293, 303
cerebral cortex, *see* cortex
chemical process, 260
chemotaxis, 260, 277, 280
CMAC, *see* cerebellar model articulation controller
CMM, *see* Markov model, controlled
collinearity of data, 222–225, 230
conjugate gradient method, 276
connectionist system, 289–291
continuous stirred-tank reactor, *see* CSTR
control
    adaptive, 291, 293, 296, 304
    approximate time-optimal, 236
    autonomous, 309
    closed loop, 235, 238
    feedback feedforward, 320
    habituating, 105–107
    learning, 289, 291
    linear, 243, 247
    linear state, 257
    linear state-space, 235
    min–max approach, 323
    neural, 292, 293, 297, 302, 305, 309
    neural RH (NRH), 346
    optimal, *see* optimal control
    parallel, 102–121
    parametric optimal, 319
    reconfigurable, 289, 293, 309
    smooth, 236
    specific optimal, 319
    time-optimal, 236, 243, 245, 247, 253
        approximate, 257
    tracking, 167, 173, 175, 188
control system
    MISO, 103–105
    SIMO, 103, 111, 114, 116
controller modeling, 260, 267, 278, 282
cortex, 61
    motor, 63
    motor (MI), 61, 62
    proprioceptive, 61–83
    somatosensory (SI), 62
cortical columns, 75
    clusters of, 63, 83
cortical map formation
    simulation of, 70–81
costate variable, 276
creeping random method, 277
critic, 8–25
    adaptive, 21–25
        action-dependent, 24
cross-validation, 221, 231
CSTR, 119
curse of dimensionality, 319, 326, 341, 342, 344

D

data
    missing values, 215, 216, 219–220
    outliers, 216, 218–219
    preprocessing, 215, 217
    selecting variables from, 220–222
dead time, 260, 265, 272
decoupling, 250, 257
delay, *see* dead time
delayed reward, 21
direct adaptive control, 268
direct neuro-control, 260, 267, 282
DP, *see* dynamic programming
dynamic neural network, 261
dynamic programming, 25–26, 31, 34, 55, 269, 277, 318–341
    approximate, 319, 321
dynamics, 290–292, 305, 307, 308

E

EBAM, 290, 297

eigenstructure decomposition, 292, 294, 298, 300
electric arc furnace, 270
eligibility trace, 24
error dynamics, 167, 171, 172, 188, 189, 198, 200
evaluation function, 21–22, 25
evolutionary computing, 277
evolutionary optimization, 280

F

fault detection, 294, 303
feedforward networks, 129, 130, 132, 133, 139, 143, 146, 151–153, 155, 220
    multilayer, 214
FEM (finite element model), 307
FH, see optimal control, finite-horizon
finite impulse response (FIR), 219
flight control, 270, 272
function space
    Barron's, 319, 325, 344
    Sobolev, 327, 344
fuzzy control, 278

G

gain scheduling, 293
general regression neural network, see GRNN
genetic algorithm, 260, 277, 280
Golgi tendon organs, 69, 70
gradient algorithm, 274
gradient method, 327
gradient-based algorithm, 278
gradient-based optimization, 274, 282
GRNN, 236, 239–242, 257

H

habituation, 107–110
health monitoring, 289, 294, 295
heart, 89, 90, 94, 98, 100, 104, 112
hidden layer, 170, 177–179, 189
hidden-layer neurons, 164, 197, 198, 200, 203–205

HMM, see Markov model, hidden
Hodgkin-Huxley neuron models, 92
homeostasis, 87
Hopfield network, 118
hybrid learning, 296

I

identification model, 130, 139, 142, 143, 147, 149
IH, see optimal control, infinite-horizon
incremental learning, 302
indirect neuro-control, 260, 261, 274, 281
induction motor drive, 237–239, 243
industrial production plant, 250, 257
input–output, 135, 143, 146, 149, 154
Intelligent Arc Furnace, 270
intelligent sensors, 213–231
inverse modeling, 265

J

joint angle, 67, 68, 70

K

Kalman filter, 276

L

lateral inhibition, 94
lateral inhibitory network, 43
Law of Effect, 7, 8, 14
learning, 8
    competitive, 66
    Hebbian, 62, 66
    reinforcement, 7–26
    supervised, 8
learning automata, 9–11
learning control, 290, 291
learning system, 9
Levenberg-Marquardt algorithm, 276

limit cycles, 243, 247
linear model, 266
linearization, 145
LMS, 14–17, 23

M

Manhattan distance, 241
map
    computational, 61, 62, 64, 65, 82
    feature, 62, 63
    sensory feature, 62
    topographic, 61
map formation, 62, 64
Markov
    chain, 53
    process, 50, 52
Markov control, 35
Markov decision problem (MDP), 37, 49
Markov model
    controlled (CMM), 36, 39
    variable duration hidden, 47
Markov model, hidden (HMM), 31
Markov models
    controlled (CMM), 34–40
Markov process, 33, 38, 39, 50
    controlled, 32
    partially observed, 38
Mexican Hat, 62, 64
Miltech-NOH, 270
MIMO, 243, 244
model predictive control, 262
model-based control, 262
model-based controller, 282
model-based neuro-control, 260, 268, 269, 278
model-free controller, 282
model-free neuro-control, 260, 268, 278
models
    input–output, 129–155
    state space, 130, 131, 142, 155
momentum, 276
motor neurons

postganglionic, 104, 105
    vagal, 104, 105
MPC controller, 262, 264, 268
multi-input/multi-output, *see* MIMO
multilayer Perceptron, 278
muscle, 62–83
    abductor and adductor, 63, 66, 67
    agonist and antagonist, 68
    antagonist, 61
    flexor and extensor, 63, 66, 67, 77, 80
    length and tension, 70–75
    stretch, 62
    tension, 62

N

NARMAX, 261
NARX, 261
Neural Applications Corporation, 270
neural control, 289, 291
neural network
    feed-forward, 34, 55
neural network auto-tuner, 260, 265
neural network inverse model, 264
neural network inverse model-based control, 260
neural network model-based control, 260, 261
neural networks
    multilayer feedforward, 319, 321
neuron model
    Hopfield, 118
nip-section, 237, 244, 253, 257
non-gradient algorithm, 274
non-gradient-based optimization, 259, 278, 283
nongradient algorithm, 277
nonlinear optimization, 259, 260
nonlinear programming, 317, 319, 321, 323, 327, 336, 337
NTS, 90–94, 116

Index   357

nucleus tractus solitarii, *see* NTS

O

objective function, 261
observability, 131–155
    generic, 135–155
    strong, 135
observability matrix, 135
observability, generic, 137, 149, 151
observability, strong, 136, 145, 146, 149
optimal control
    finite-horizon (FH), 317–346
    infinite-horizon (IH), 318–346
    linear-quadratic, 318, 320, 335
    receding-horizon (RH), 318–348

P

parameterized neuro-controller, 260
parameterized neuro-controller, 259, 271, 272, 282
parametric neural network, 263
parasympathetic system, 104, 105, 112
partial least squares
    neural network (NNPLS), 229–230
partial least squares (PLS), 220, 228
partitioned neural networks, 185, 186
passivity of neural network, 162, 166, 167, 183, 185
    strict, 161, 166, 169, 183, 185, 207
PCA, *see* principal component analysis
perceptron, 1, 3
peripheral resistance, 98, 104, 105, 112
phase plane, 247
phoneme, 45, 47, 55
PI controller, 280

PID controller, 266, 267, 272, 273, 281, 282
PLS, *see* partial least squares
PNC, *see* paramaterized neuro-controller
polymerization, 105, 106, 108
potassium current, 96
predictive control, 291
principal component analysis (PCA), 215, 218, 227
principal component regression (PCR), 220, 222
process control, 272
process model, 259–261, 263, 266–268, 270, 271, 276, 279, 281
process model mismatch, 271
process soft sensors, 213, 215, 230
proprioception, map formation, 62

Q

Q-learning, 24–26
quasi-Newton optimization, 266

R

radial basis function network, 278
radial basis function networks, adaptive time-delay, 290
random search, 277
RBF, *see* basis functions
RBF network, 289–309
receptor
    stretch, 68
recurrent network(s), 129–131, 133, 134, 261, 278
redundancy, 293, 302
regression matrix, 163, 179, 207
regulation problem, 318
reinforcement learning, 268
restrictions, 243, 245
reverse engineering, 93, 106, 113
RH, *see* optimal control, 335
ridge function, 3
ridge regression, 213, 215, 226, 231
robot arm, 62

robot arm, control of, 167, 169, 173, 177–179, 186, 193
robust backpropagation, 219
robust model, 271
robust model-based neuro-control, 260, 271
robust performance, 271
robust stability, 271
robustness, 271
robustness of controller, 162, 171, 183, 189

S

scheduling algorithms, 101
second-order method, 276
self-organization– cortical maps, 62
sensitivity analysis, 221, 231
servomechanism problem, 317
sigmoids, 3
simulated annealing, 277
singular value decomposition (SVD), 300–302
skew-symmetry property– robot arm, 167, 169
space robot, 331–346
space structures, 289, 292, 303–306, 309
speech recognition, 50, 53, 55, 56
spindle, see receptor, stretch
stability
    global, 298, 299
    structural, 298, 299
state space, 150
statistical methods, 213, 218, 220, 231
stretch receptor, see baroreceptors
stretch receptors, see baroreceptors
supervised learning, 261, 273, 275
switching, 245, 246
sympathetic system, 98, 104, 105, 112
system identification, 292, 308

T

TDL (tapped delay lines), 291
time-optimal, see control, time-optimal
torque, motor, 238, 243–245, 250
tracking error, 168, 169, 173, 174, 176, 177, 185, 190, 192, 195, 196
transputers, 240
transversal, 137, 138, 149, 150, 156–158
truck backer upper, 270

U

uniformly ultimately bounded (UUB), 166, 199
unmyelinated fibers, 111

V

VDHMM, see Markov model, variable duration hidden
vibration suppression, 289, 291, 306
Viterbi algorithm, see algorithm
Viterbi score, 34, 37, 41

W

web, 235, 236, 243
web force(s), 235, 238
Wiener, Norbert, 1

X

XOR problem, 242

Z

zero-trajectory, 243

ISBN 0-12-526430-5